Erik Hill

Charles Wohlforth is a lifelong Alaska resident and author of Los Angeles Times Book Prize–winning *The Whale and the Super-computer,* as well as many other books and articles. With his wife, Barbara, and their four children, Wohlforth lives in Anchorage during the winter, where he is an avid cross-country skier, and in the summer on a remote, off-the-grid Kachemak Bay shoreline reachable only by boat.

ALSO BY CHARLES WOHLFORTH

The Whale and the Supercomputer

Additional Praise for *The Fate of Nature*

"A heart-wrenching journey through the tumultuous history of [Alaska] and its fragile land and seascape, from the complex, mysterious culture of killer whales through the clash of Native worldview and Hobbesian self-interest with the arrival of Europeans, the origins of the conservation movement and its ongoing battle with development, and the devastating Valdez oil spill. Wohlforth concludes, optimistically, provocatively, but convincingly, that 'stepping off the material treadmill isn't denial, it's freedom.' "

—*Publishers Weekly* (starred review)

"Intellectual, philosophical, and packed with feeling, Wohlforth's hopeful arguments for preserving our natural world are also practical and ring true as a bell, a gentle pause in the noise that often takes the place of civilized debate on the topic."

—Deanna Larson, *BookPage*

"No one does a better job of bringing the real Alaska to life than Charles Wohlforth. In *The Fate of Nature,* he has combined compelling storytelling with a provocative contribution to our national environmental debate. I don't agree with everything Charles has to say, but his eye-opening book is an invaluable read for anyone who cares about my state and our planet and wants to leave it better than we found it."

—U.S. Senator Mark Begich, Alaska

"This is a beautifully written book. Like a well-crafted novel, you get a vivid sense of the characters involved (their motivations, passions, and foibles) and a deep sense of place. You also gain a renewed faith in humanity. Wohlforth shows us where altruism, cooperation, and compassion have blossomed throughout history and continue to thrive now, in various cultures and social settings."

—Sidney Stevens, *Mother Nature Network*

"Wohlforth has an immersive prose style that's engaging from the first page, and his obvious emotional investment in the natural beauty of Alaska, as well as his shame at the damage wrought on its environment, keeps the book anchored."
—Leonard Pierce, *Onion A.V. Club*

"The hidden truths in *The Fate of Nature* gradually come into focus through the adventures, stories, and exhilarating experiences conveyed with masterful grace and deep understanding of ancient wisdom and modern realities by Charles Wohlforth. A must-read for all who care about securing an enduring future for humankind within the natural systems that sustain us."
—Sylvia A. Earle, Explorer in Residence, National Geographic Society

"From subjects as far-reaching as political history and philosophy to psychology and animal behavior, Wohlforth offers evidence that humankind still has what it takes to put an end to the 'ceaseless repetition of the same fight' and rediscover our cooperative impulses."
—*Audubon* magazine (Editors' Choice)

The Fate of
NATURE

REDISCOVERING OUR
ABILITY TO RESCUE THE EARTH

Charles Wohlforth

Picador

———

A Thomas Dunne Book
St. Martin's Press
New York

www.picadorusa.com

Picador® is a U.S. registered trademark and is used by St. Martin's Press under license from Pan Books Limited.

For information on Picador Reading Group Guides, please contact Picador.
E-mail: readinggroupguides@picadorusa.com

Map on pages xii–xiii by Don Frazier

The Library of Congress has cataloged the St. Martin's Press edition as follows:

Wohlforth, Charles P.
  The fate of nature : rediscovering our ability to rescue the earth / Charles Wohlforth.—1st ed.
    p. cm.
  "Thomas Dunne Books."
  Includes bibliographical references and index.
  ISBN 978-0-312-37737-3
  1. Human ecology. 2. Human ecology—Alaska 3. Natural history—Alaska.
4. Conservation of natural resources—Alaska. 5. Environmental protection—Alaska.
6. Alaska—Environmental conditions. I. Title.
  GF50.W64 2010
  304.209798—dc22

                                                                    2009045779

Picador ISBN 978-0-312-57297-6

First published in the United States by Thomas Dunne Books, a division of St. Martin's Press

First Picador Edition: April 2011

10 9 8 7 6 5 4 3 2 1

To Wendy Feuer

Contents

The question at issue is the outer sea, the ocean, that expanse of water which antiquity describes as the immense, the infinite, bounded only by the heavens, parent of all things; the ocean which the ancients believed was perpetually supplied with water not only by fountains, rivers, and seas, but by the clouds, and by the very stars of heaven themselves; the ocean which, although surrounding this earth, the home of the human race, with ebb and flow of its tides, can be neither seized nor inclosed; nay, which rather possess the earth than is by it possessed.

HUGO GROTIUS, *FREEDOM OF THE SEAS*, 1608
TRANSLATED FROM LATIN BY RALPH VAN DEMAN MAGOFFIN

Northern Gulf of Alaska Coast
and Prince William Sound

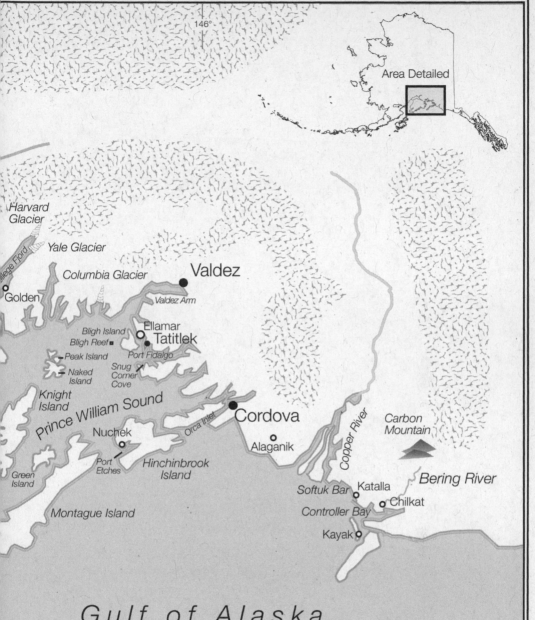

146°

Area Detailed

Harvard
Glacier

Yale Glacier

Columbia Glacier

Valdez

Valdez Arm

College Fjord

Golden

Bligh Island
Bligh Reef

Ellamar
Tatitlek

Peak Island

Port Fidalgo

Snug
Corner
Cove

Naked
Island

Knight
Island

Prince William Sound

Orca Inlet

Cordova

Nuchek

Alaganik

Copper River

Carbon
Mountain

Port
Etches

Hinchinbrook
Island

Bering River

Green
Island

Softuk Bar

Katalla

Chilkat

Controller Bay

Montague Island

Kayak

Gulf of Alaska

● Cities and towns

○ Disappeared towns and villages

FIFTY MILES

Introduction

Each of my oars produces a pair of vortices, those spiral waves that drill into the surface, and smaller swirls burble out behind the boat as I row alone on a soft summer evening along the shore of Alaska's Kachemak Bay. My small liquid spirals spin in the same shape as the deep, dark gyre of water that circles the bay and concentrates marine life there, the same as the circular storms that roll over us from Siberia every few weeks, the same as the eternal counterclockwise current that spans the North Pacific Ocean. I'm considering that similarity when a gull splashes down like a spear and pulls away with a flashing silver fish in its beak. Seal heads pop up from the water, watching me. They multiply, populating the surface like a theater audience, their eyes all on me, first five, then a dozen, and more than twenty. I don't know what they're thinking, but they're obviously thinking something or they wouldn't all be paddling in place staring at me. And I wonder. What else is down there?

The ocean is so vast, it's everything—the source and sustenance of life, the birthplace of the rain and cleanser of the air, the planet's essential medium, upon which all the land is but an island. Yet on and within the ocean every wave is different and every birth is new. To ponder its totality is like trying to think of the entire universe at once. On the other hand, the ocean can be drained of life, and in fact we're on the way to making it uninhabitable, except perhaps as a kind of farm. The causes shouldn't surprise anyone—overfishing, pollution, climate change, and habitat destruction—issues so familiar their very names create a sense of mental fatigue. But the asymmetry is surprising, of magnificent vastness threatened by the petty and mundane. On one side a scale weighs the ocean, so immense and thick with meaning that the word itself is a synonym for greatness beyond imagining, and yet the other side swings down the scale, human values overbalancing toward cheap fish-and-chips and motor oil left to dribble into storm drains. It's worth asking how we've made this choice for things that are small and fleeting over an eternal source of life.

To take up that question is my project here, and to learn if collectively we can choose a less destructive course. Can we share protection of a great good we all can touch but never individually hold? Such an investigation goes well beyond the technicalities of environmentalism, since our human species already has the technology to mitigate our downward pressure on the ocean's vitality. Instead, the answer lies within ourselves, in our nature as animals taking a part in a planetary ecosystem. What capacity for good lies in the hidden depths of people? We must look into our spirituality, economics, politics, and personal relationships. The exploration leads into the roots of caring and selfishness, well beyond the ocean's edge, eventually including our entire physical space. But the ocean can represent all that—the vastness, complexity, and fragility of the living world outside ourselves, the swirling exterior context that gives meaning to a look inside.

On our Kachemak Bay beach my older daughter, Julia, at age ten, collected dozens of hermit crabs and put them together in a mobile tide pool she had built in a red wagon. They brawled madly, one-on-one and in a furious ball of fighting crabs. Hermit crabs wear discarded snail shells as armor against other hermit crabs, whom they attack in hopes of getting a better shell. The pursuit is indiscriminate, apparently with the default assumption that any shell might be better. A crab that Julia put alone in a sand pail with two similar shells circled compulsively, like a broken computer program, switching first into one, and then into the other, and then back into the first one, and so on endlessly. The rigor of the hermit crab's competition for shells assures that none is ever satisfied and the fighting never ends. If there is a better way of living, it hasn't survived in any tide pools we've ever looked into. A peaceful hermit crab produces no heirs.

If humans were like hermit crabs, the world would be as dangerous as a tide pool, a scene of escalating struggle for stronger defenses and more control of resources by the powerful. Therein would lie our choice about how to use the oceans. Mercilessly driven by material needs that could never be fulfilled, we would be fated to overwhelm all other species and completely occupy the planetary habitat. In tragedy, fate is an arrow flying irretrievably from the preordained nature of a protagonist to his or her undoing. A world-dominant being whose nature is to be unsatisfied and seek wealth beyond surfeit is doomed to loneliness as the solitary survivor of a biosphere, accompanied only by the plants and animals raised for food or amusement. Is that our fate?

The idea that humans are such competitive beings blossomed in the seventeenth and eighteenth centuries, the Enlightenment in Europe, with the invention of the social and political mechanisms of the market economy. This hermit crab doctrine decreed freedom and opportunity for the individual, and slavery and cultural annihilation for world peoples without strong shells or the traditions of struggling for them. The success of the competitive model of human nature became its own proof. To oppose it now would seem as naïve as to run for public office opposing economic growth. Even mainstream environmentalists accept this forlorn verity about the human future, hiding their true love for wild nature to argue for ecosystems as instruments of economic power, either for service to humankind or as desirable chattel.

Science seemed to agree that people are essentially shell fighters, as evolution interpreted the ascent of our species as a sign of our competitive success. Writers cherishing ethics and generosity often retreated to the intellectual ghetto of religion, which supposedly is inaccessible to scientific reasoning. But placing our better selves in such a refuge from hard truth is like saving the whales by putting them in aquariums. If we're to imprint our goodwill on the world, those wishes have to vie in the same arena as our selfishness. And, increasingly, science does support the strength of human altruism and our will to conserve. In reality, as we always should have known, individualistic competition is not universal human nature. It's only one way of living, and a peculiar one that doesn't address the deeper needs of most people.

Evidence comes from many university departments and, most convincingly, from real life. Stronger than our greed and materialism, most of us feel a connection to other people, to animals and wild places, and when we're faced with a choice between those sources of meaning and our own material gain, we tend to prefer fairness and the bonds of the heart over getting ahead. Discovery of these positive impulses has made scientific news. In our competitive paradigm, the existence of goodness is news: we're that embedded in a system that elevates wealth above purpose and severs the connecting tissues of our community relationships. The hermit crab model of human nature may be factually wrong, but it is usually treated as self-evident truth.

The Fate of Nature seeks to unmask false assumptions about human nature. I've used the particulars of people's lives, their places and times, as well as the ideas and experiments of a wide variety of scientists. Since I began this work a historic economic wreck has helped expose the flaws of a purely competitive worldview. Presenting the alternative is more difficult, however, partly because of the cynicism we've absorbed: we're trained to discount the hope of cooperation. But saving the oceans will require cooperation. People must believe in their ability to connect as communities, and believe in the right of communities to manage their own environments for the benefit of their values. The seemingly quixotic project of diverting the earth's arrow of fate relies on the potency of these beliefs. The shared sense of right and wrong about how we use the world—our social norm—is the ground upon which institutions stand. Norms are made by individuals and personal relationships. By influencing norms, seemingly ineffectual acts can power irresistible social movements.

Therein lies my audacious purpose in writing: to prod this unseen organism of collective belief. I've started the book by asking what we are—what makes a person. The answer also tells why the lives and freedom of other animals should matter to us. After that exploration, *The Fate of Nature* follows the varied ways cultures and ideologies relate to nature. It happens that I live in a place of sublime beauty and biological wealth where conflicting worldviews have repeatedly collided and reshaped the ecosystem. The vital ferocity of the Gulf of Alaska coast seems untamable, but even these waters have been poisoned with oil and chemicals and the food web has been torn and weakened. With such damaging marks, invaders wrote of their supremacy, questionable certainties penned on the land and under the water, and engraved as generational grief upon people living along the shore.

The drama of the ocean's fate plays out here in Alaska as a world in microcosm. At this date its shape resembles tragedy, but the curtain hasn't fallen yet. There's plenty of life and wildness left in these swirling waters. Perhaps the heroic and often hopeless deeds of generous individuals can shift the very stage; such people are here and trying. I will tell their stories. Lonely summers spent picking up plastic bottles from thousands of miles of fog-swept wilderness beaches could be meaningless, or it could be the edge of a social transformation.

Not much of the book is about me. Probably, what I have to say is true for some and not for others. To feel, irresistibly, that a place or animal means something fundamentally valuable beyond one's own life, a feeling of love akin to that for a spouse or child—I wish everyone felt that, but I can't say how many people do. I would readily concede the feeling has more to do with the one perceiving than the thing perceived. None of that ultimately matters. The will to give is real, found in most of us. We do yearn to act selflessly, to leave space for other living beings. Our desire for a society similarly generous of life is all the explanation we need to begin changing the world.

Part I

THE CONTEXT: NATURAL HISTORY AND HUMAN NATURE

The Gulf of Alaska Cauldron

What seems solid and permanent swirls and disintegrates, spun by invisible forces. Sand and mist blast along the seven-mile beach of Softuk Bar, entrained in an east wind tearing the tops off roaring gray breakers and hurling them into the twilight. A lumbering brown bear, hunched against the flying grit, digs for a beach plant that the wind has buried in wet sand up to the tips of its leaves. The leading arm of a vortex hundreds of miles wide is sweeping the exposed shore of the Gulf of Alaska, east of Prince William Sound, half a day's walk from the next nearest human being. As the storms of September and October pulse unremittingly and the dark of winter thickens to impenetrability, the last person will desert these hundreds of miles of shoreline and even the blubbery, salmon-fattened bear will disappear. Sand lance, silver slivers of fish, will bury themselves in the beaches like hidden mineral veins. Humpback whales will retreat to Hawaii and stop eating, trading food for warm water. Herring will gather in the sound's fjords where cold, hidden waters may, in stillness, allow a store of fat to last through the winter. While animals wait, life recharges in the mountains above. The storm is depositing energy there, in deep fields of snow that will power the spring.

This is the starting point, within a three-mile-high semicircle of mountains along the gulf's northern rim. Remote from human experience or even detailed scientific understanding, mighty forces and abundant organisms combine in a system of life that ultimately sustains animals and distant, unknowing people and restores the air and stirs the ocean. The furious chaos vibrates with creativity and destruction, as virile and unknowable as an ancient, all-powerful god. From ingredients of water, air, light, and rock it produces the aged giants of the deep and the weightless feathers that lift terns balancing on gusts. From a swirling cauldron come delicate patterns of meaning.

Here, at this beginning, we can begin to decipher what we are, although not in any simple way. To know our own human nature we must examine our context, the ecosystem that precedes us, especially that wild part that defies human habitation or control. For example, consider the sea's alchemy that spawns tiny, crystalline herring eggs on fronds of seaweed, delivered to the lips of a woman on a warm spring afternoon next to calm water, sustaining her in body and spirit. Food, the essence of what one is, and who, and even why. She will know answers just by crushing the salty gems between her teeth—about the sea's generosity and its demand for humility. Somewhere the stormy night hides an origin for that understanding.

First the storm reaches shore, falls into the mountains' snare, screams in alpine wind tunnels, and bleeds out. Moisture rising against rock condenses into water and snow. This snowfall accretes to become the continent's largest glaciers, snow becoming dense, glassy ice and oozing downslope, grinding and shattering solid rock on the way back to the sea. Downslope from the permanent ice in the perpetually damp temperate rainforest grow enormous trees and bottomless spongy wetlands pocked with small, clear pools. In winter, all is buried under yards of heavy, wet snow. In the spring, the rivers will flood with melted snow, ice, and sediment, eventually fertilizing the ocean's plankton with iron and extending quarter-mile-wide ocean beaches.

The fresh water itself moves the ocean. Freed of their banks, the rivers turn right, pumping the coastal current toward the west. No, the water doesn't turn right. It flows straight. It's the earth that's turning. Like a marble rolling across a spinning disk, water seems to turn. The runoff from the gulf's coastal mountains follows the shore to the west, sucking behind it warmth from the American West Coast, whose energy in turn melts more snow and ice and powers stronger storms to pile more snow and ice in the mountains. Spirals within spirals: the storms bend, too. Air rushes toward their low-pressure centers, but the spinning earth diverts the wind to the right, on the long, circuitous route of a cyclonic vortex, like water going down a drain. The storm's leading edge of east wind collides with the Gulf of Alaska's coast, hurling sand west along endless, bear-tracked beaches and pushing water west, along the shore. In the dark of autumn, the waves grow into hills.

There are mountains and canyons under the sea also, along the ragged-edged continental shelf, the fringe between land and the abyss. At the center of the gulf's arc, vertical rock confuses the waves and wind, with contradictions offered by fjords, islands, channels, and spires, and within the unfathomably complex inland sea of Prince William Sound, which encloses a world of its own, water-floored corridors walled by brooding spruces leading to secret, fecund gardens of mud and flashing fish, prey for eagles. Winds funnel and focus through these mazes. Currents twist in baroque patterns, changing with each turn of the tide or season. Intricate forces entangle ecological stories into as many digressions and surprise endings as there are eddies and tide pools. But the tempo of every tale comes from the beat of the storms and the timing of the moment in the spring when the sun emerges warmly on stilled waters.

The prodigious biological productivity of the Gulf of Alaska owes everything to that moment when the surface's crop of phytoplankton is perfectly prepared for growth. The winter storms have stirred up organic nutrients from the seafloor, mainly nitrogen; few other waters in the world are as rich. The rush of fresh water from the mountains, more than the Mississippi River's annual flow by half, and all in a few months, disgorges atop the heavier salt water. Iron and other mineral nutrients arrive with the fresh water to mix with the nitrates. As the storms die and the fresh water spreads, a surface layer develops to hold blooming plankton near the sun (when the sea is mixed, the plantlike organisms fall into darkness). Now, in May, sunshine is high, gaining every day until it lasts almost all night, brightness reflecting off still-snowy shores. Water is calm and rich in fertilizer. Everything is perfect for an explosion of photosynthesis, and the phytoplankton blooms.

The energy that plankton capture from the sun over a few weeks will feed zooplankton by the billion—tiny creatures like krill and copepods, which look like shrimp, and larval forms of many other animals, such as crabs, barnacles, and other shellfish. The water clouds with them, especially where tidal currents meet, fronts between waters of different temperature or salinity that concentrate matter like invisible walls in the ocean. Forage fish such as sand lance and herring gather to feed on zooplankton in crowded schools. Gulls find the schools from the air and dive on the water, wheeling and dropping straight down, as violently as spears, then hurriedly climbing up the air again to protect a catch. Humpback whales lunge through the schools, bursting diagonally from the surface, occasionally catching a bird, too, before rolling over and sinking back again with a giant slosh. Salmon, lightning fast and bright, blaze through the schools of forage, fattening for a single spawning journey upriver. Rivers along the gulf coast reaching hundreds of miles over the mountains will receive salmon eggs and carcasses. Salmon flesh will feed bears, birds, and scavengers, whose waste will fertilize the trees, moss, and grass. Long before that can happen, during the spring, the phytoplankton bloom subsides, having consumed the winter's mixture of nutrients, but that energy flows on through the system, from mouth to mouth, up the trophic levels of the food web, and up to the floppy tops of towering hemlock trees fertilized by bear scat.

Described that way, the Gulf of Alaska ecosystem sounds like a huge organism. But it's something entirely different. Living organisms repeat processes in cycles, over days and seasons, and over life cycles that recapitulate with each generation. But unlike an organism, the Gulf of Alaska sometimes transforms itself, unpredictably, and as drastically as an otter waking up as a fish. An entirely changed system of life arises: new water temperatures, altered currents or storms, a shift in the entire ecological regime. Crab and shrimp disappear for decades and enormous schools of bottom-dwelling pollock show up. Salmon runs jam rivers but sea lion colonies dwindle. Until, without warning, the gulf wakes again as something else.

Scientists can find plenty of clues to diagnose these transformations after they happen—far too many clues. Tiny changes easily become big ones, and tiny changes happen all the time. A fluke of winter weather could bring very cold air during an especially low tide, killing most of the mussels and, if the timing is right, allowing competitors like barnacles to take over the tidal rocks; and the mussels might not recover for years. And big forces can be subtle. Oddities in the paths of the earth and moon through space repeat the pattern of the size of the tides every 18.63 years, and somehow that affects the population of fifty-year-old halibut resting on the dark sea floor. Systems interact constantly, invisibly—the concentrically nested ecosystems, the interwoven currents from the far side of the globe, the swirl of wet winds and the mountain claws that catch them. Internal resonances switching frequencies, plus fishing, pollution, and climate warming: nothing causes the changes, or everything. Science calls the system complex. The unknowable giver of life answers no direct questions.

In Prince William Sound, spring used to reach an orgasmic climax when cloudlike schools of herring gathered to spawn in late April and early May. Females would lay eggs on kelp and rock along miles of shoreline, and males would spew sperm into the ocean to float freely among the eggs, white plumes drifting like curtains, streaming in the water from headlands. Birds, seals, whales, and people gathered for a feeding frenzy. Twenty years ago, after the *Exxon Valdez* oil spill, the fish became deformed and diseased and then the huge schools stopped returning. Years later unrelated herring populations in the gulf crashed as well. The cause, like so much else in this spinning, disintegrating universe, cannot be proven.

But enough herring remain so that, on one bright day in May, Diane Selanoff, a Chugach Native, climbs sure-footed in rubber boots from a skiff, over seaweed-covered bedrock near Tatitlek, in the heart of the sound, to pick fresh sprouts of popweed coated with translucent herring roe like tiny bubbles glistening in the sun. Her eyes shine and her smile refuses to be repressed, as happens when an experienced middle-aged woman feels an old joy well up. She says she'd like to lie down and eat it all. This is the food that sustained Diane and her single mother when she was a girl, beach food, when she was growing up poor in the village of Port Graham. She picks a piece for the bag and picks a piece to eat. Slightly salty bombs of pure freshness explode on the teeth. The animals long ago taught the people which foods would heal them, Diane says, and popweed was a good cure, eaten with this delicious seasoning of eggs or laid on the place that hurts when taking a steam. Diane smiles, works, and talks, mostly in English, but with a lot of Alutiiq words. They say in Chugach villages that when the tide is out, the table is set. Chitons, clams, mussels, and cockles, the seaweed and beach plants, the seabird eggs and octopus—it's all there at your feet, delicious and healthy, much of it ready to eat raw from the ocean. The sound seems to gently offer us everything we could need. In the sun, on the rocks, above water so clear the skiff seems to levitate, I smell the seaweed and herring spawn. Tiny intertidal creatures let out barely audible clicks in the drying air, a sound like scattered raindrops, over which the flow of Diane's low, steady voice continues.

She says, "The ocean is the source of all life for our people. Not only for our people. For everyone."

2

Killer Whale Culture and
Human Spirituality

Dropping to her knees, Eva Saulitis, a young whale biologist, cast a slip of paper into a fire on a pebble beach. She had just begun learning about Prince William Sound, but, like many others, she could sense spirits above in the misty spruces and down below in narrow underwater canyons, where great, warm creatures slid through green shadows. She summoned them with her burning words, calling to the killer whales. In the morning, Eva and another researcher would sometimes stride through forest and over meadows to the top of a mountain behind the camp, where they scanned miles up and down the passages of the western sound. Upon spotting the killer whales' black backs and high dorsal fins they galloped down the steep path to a little aluminum boat with an outboard motor to photograph and identify the whales. At night, on a cot in a canvas tent behind a storm berm of rocks, Eva worried that she wouldn't find the whales from the mountaintop next morning, or she'd make a dangerous mistake, or there would be something wrong with her, personally, that would doom the work. When she fed the fire, she incinerated her doubts, written on slips of paper, bidding the whales to show themselves. They usually did. Killer whales were easier to find in the sound then, a quarter century ago.

When killer whales' backs break the surface, the sea itself comes to life. Their skin has the liquid smoothness of calm water. A black bulge rises like the sea flexing a muscle. The dorsal fin comes next, a vertical saber that seems to rotate along the axis of the back. Now comes another back and fin, and another, different sizes and slightly different shapes, apparently swimming in formation, silent and without a splash, but quick. The pod is cruising its territory, circuiting home waters the way Native hunters do, surveying the homeland to find out what's changed. Eva and her colleagues know the pod right away and set about photographing each back and dorsal fin. These fish-eating killer whale families are the most stable in nature—including humans—with children that never leave their mothers. Eva can remember the first time she saw each whale.

The sea is another world that science encounters where the waves meet the air, or a little below, at the low tide line or with cameras or divers lowered from ships. From those moments we've learned something, but in little pieces, as if seeing through a distorted glass that can focus on just one narrow spot at a time. Often the glimpses only accentuate the immensity of it. These scientific challenges deal out discoveries grudgingly and leave much unknowable, but the quality of mystery itself helps in an investigation of human nature. The sea teaches humility even as it excites the imagination. It is populated by animals similar in many ways to our own kind, but who remain always intriguingly beyond reach. In its capacity to inspire, the world below the ocean's surface raises an essential question for understanding ourselves: why do we care about these creatures? Mere curiosity doesn't begin to answer it.

Killer whales emerge from that other world like spirits at a séance, focusing the attention of the researchers in the boat to a transcendent level, but giving only clues and cryptic suggestions about what their whole lives are like on the other side. Without looking at notes or photographs, the scientists immediately recognize animals, delight in new offspring, and wonder about those missing. Sometimes they're able to see flashes of a successful hunt, when meat-eating whales take a porpoise or a seal, or when a coordinated team of hunters kills a huge gray whale—then they can hear, as well, the raucous sounds of victorious whales celebrating underwater. But conclusions about the majority of the whales' lives come slowly, because these brief sessions—like murmurs from the departed—tell mostly of the moments when the two sides meet.

Could a whale be a spiritual being? For some people, the natural and spiritual world are one. Eva Saulitis requested help with her burning scraps of paper, not symbolically but because she hoped for an answer from the wilderness, so full of life and meaning. In the woods she performed the ceremonies of a personal creed, rites for spirits she sincerely believed were present. The indigenous people who lived here felt those presences, too, by other names. The spiritual and the real were inseparable facets of all animals, objects, and forces. In that universe, there could be a religion of inhabiting a place, a religion of the senses and of one's body acting in the world—eating, working, mating—and, for Eva, a religion of study, in the science of learning from the killer whales. She sought the whales with a need much deeper than that of filling a camera or field notebook.

Eva's adult need for connection followed a childhood of emotional isolation. She grew up speaking Latvian in the strictest of families, in upstate New York, dominated by her father and fed on his harsh, joyless Catholicism. She was a victim of history. Two decades prior to her birth Latvia fell to Germany and her father was drafted and sent to the front lines as a member of the Waffen-SS. He was an alcoholic when she was a child. Not knowing the man in any real sense, her adult conversations with her siblings were diagnoses about how he got that way, rather than nostalgic memories. He began speaking of the war only when a stroke damaged his brain, and then it came out as a twisted nightmare. Lying in a convalescent hospital, he thought he was awaiting prosecution for crimes against humanity and could only be calmed with assurances that the trial was "not today." He was pitiable, but it was too late to know the truth—to know if the war had imprisoned an innocent man or if he had authored his internal hell.

As a college student at Syracuse University, Saulitis saw the movie *Never Cry Wolf,* the dramatization of Farley Mowat's memoir of spending a scientific field season alone on the Canadian tundra, and in the process nearly becoming a wolf, running naked among them. She decided at once to be a wildlife biologist in Alaska. As a graduate, she worked at a fish hatchery in Prince William Sound and saw her first killer whales. She had to sit down on the dock to keep herself from jumping in with them. She volunteered to work with Craig Matkin, a commercial fisherman who had built a career as a self-funded killer whale researcher. On an early trip, a group of killer whales chased their Zodiac inflatable boat, lunging and snapping, leaping from the water. Eva was thrilled, screaming and laughing—transformed—but Craig was scared; in more than a decade of the work, he'd never seen whales act that way. Sometimes fish-eating whales, called residents, make eye contact and play gently with the boat, but these whales, meat eaters, called transients, were rough and crazy. Eva remembers the moment as a marker for her relationship with the whales. When she approaches them she thinks hard of her benign intentions, hoping to transmit her calm and obtain their permission. She takes out her cameras only after the encounter has been established, a sort of courtesy.

Modern anthropologists usually ask Native communities for permission before studying their cultures—their hunting and gathering practices, marriage arrangements, family and clan structure, their seasonal round of finding food and patrolling territory. Eva, Craig, and other killer whale researchers ask the same questions. Like human beings, killer whales occupy the whole globe—every latitude and every ocean. Like human beings, their hunting techniques, social arrangements, and vocal dialects vary, adapted to a habitat or molded by habit. Like people, too, the whales innovate skills and pass them on—the essence of culture—until their practices become so engrained that pods behave as if they are from different species despite being physically identical. In the northeast Pacific, including Prince William Sound, killer whales are either fish-eating residents or marine-mammal-eating transients, each half of the population with distinctive patterns of movement, mating practices, and genetics—they never mix. The residents stay in their home territory for life, in tight, matrilineal families, while the transients break up into smaller groups and roam. There's no reason transients can't eat fish, but in captivity they will nearly starve before doing so. Outside this region, however, this split among killer whales does not hold—it's a unique cultural division of these waters.

Culture accelerates evolution. A human tribe that learns to sew can occupy a colder habitat within a generation instead of waiting for biological evolution to provide a coat of fur. Scientists saw killer whales in British Columbia learn to catch too-fast white-sided dolphins by corralling them against the shore. Pods in Alaskan waters learned how to snatch fish from long-liners' gear and how to use the sound of the boat engines to recognize which boats tended to be easier pickings. The whales' ability to innovate adaptations and to share those skills with other pod members and calves allows them to live anywhere, whether feeding requires catching skates, leaping on the shore to grab sea lions, or working in a team to drown a whale. Whale biologist Hal Whitehead, of Dalhousie University in Halifax, found evidence that killer whales' cultural adaptation was so successful that it slowed their rate of genetic change: they had less pressure to evolve genetically once they could conquer new waters culturally. Human evolution similarly became more cultural and less genetic.

Killer whale culture extends beyond hunting. Pods sometimes leave plentiful food behind to tour unused reaches of their customary waters. They have certain special beaches where they rub their bellies. They pass on family songs, which slowly change through generations. When food resources exceed their needs, they play. Teen whales swim a little apart from the pod and romp, out of direct supervision of their elders. A National Geographic documentary shows a pair of whale brothers off Argentina trapping seals, eating their fill, tossing one seal back and forth, mouth-to-mouth and tail-to-tail, in a game of catch, and, finally, gently depositing a last seal pup back on the beach, unharmed.

Play is among the most difficult of biological activities to define. It can be a form of practice, or it can express purest freedom, whim, effort without purpose—a model railroad builder laboriously painting a layout, or a torpedo-fast pod of Dall's porpoises in Prince William Sound, romping in the wake of a boat. Or a killer whale tossing a porpoise it is about to eat high into the air. Eva and Craig's voyages to the sound could fit a definition of play: Craig began photographing whales for curiosity in the 1970s. When grants gave out, he took up fishing to pay for his whale-study habit. He never got a PhD because he didn't want to leave for that long. The work gained value in the wider world after the *Exxon Valdez* oil spill, when scientists realized his long-term data uniquely described individual animals from before the disaster, allowing comparisons to what was left afterward. But even now, these projects are firstly about being here with the whales, not gaining honors or padding a list of published papers. For Craig and Eva, the science serves the work, not the other way around. Their meetings with whales remain at the apex of life, the activity to which all else contributes, its purpose intrinsic—which could be a definition of play, or of worship.

The boat left port with hardly any food on board. Instead, at the dock we set up an empty chest freezer in the fish hold. The sound would provide halibut, silver salmon, snapper, cod and rockfish, blueberries and beach greens—plentiful wild plants harvested from just above the tide line and eaten like salad. While listening for whales on a dangling underwater microphone, Craig absently jigged a lure on the bottom, reeling in a big-mouthed, seaweed-colored greenling; he sheered a fillet from each side of the still-living fish and passed the meat through the companionway to Eva, who had a hot pan ready on the stove. Ten minutes later, lunch: white, stir-fried fish, greens from the garden at home, and brown rice.

They live in the Kachemak Bay town of Homer in a post-and-beam house they built on a forested hillside at the end of a dirt road. They plant a large vegetable garden and keep chickens and bees, and once a year they help a friend slaughter a cow fed from the grass and wildflowers on the hills east of town. They were both married to other people when they fell in love, in the sound, researching whales. Although Craig's two marriages had been to whale researchers, no one would have expected a relationship with Eva after eleven years of working together professionally. Their personalities and worldviews were so different. Her strong forehead and straight, strawlike hair are Old World; when she's thoughtful she can look grim, like a fatalistic Latvian peasant. But Eva brightens when she reconnects with the spirits of the sound: her eyes glisten and dart, a look I've seen on my youngest daughter when she's about to dance. Craig isn't like that. He's mellow, practical, and suspicious of religiosity. A child of California beach culture, he goes barefoot on deck even on chilly, wet days in the sound, pink toes against blue fiberglass. But the sound changes him, too: it focuses and intensifies him with purpose and belonging. Perhaps that connection to the sound connected them in love.

Eva is an accomplished poet and essayist. She wrote this poem, titled "The Lover."

Always the firm pressure
of lichened boulders,
always, the stubble and coarseness
of moss. Always the conversation
of water or bird, its complaint,
its news, its story-telling.
Always the damp, flexing limbs,
heady smells nose-deep in ground cover.
Always the body, willing, the welcome.
Always its gifts, corpses of small
seabirds, sometimes only the graying
of wings.

On the whale research cruise I joined, she was no longer performing ceremonies in the woods, although she said she was sharing with Craig's eldest daughter about Wicca-inspired spirituality. She was more confident now. She published a book of her essays and began teaching writing in Homer during the winter. Perhaps—and this is my thought, not hers—the need to connect through ritual recedes when you recognize how deeply, physically, a part of a place you can be. Ceremony could become as extraneous as a telephone on the pillow between lovers.

The lover in her poem was the land itself, not a person. "Yes. That's how palpable it is to me. It's like a person," she said. "I feel the presence of a conscious spirit in the natural world."

Anchored in a tiny cove on an island in the southwest sound, we paddled ashore in miniature plastic kayaks to pick blueberries. The drowsy green woods were fully enclosed from the sky except for big drops of gathered rainwater that clattered down from the long, sagging bows. The silence of deep moss rendered hypnotic the repetitive process of grasping one bright blue orb and then another and the gradual increase of the blueness in a plastic bag—the only contrast from universal green. The sound erases the rest of the world in a few days. Being is different here. Time smoothes, pulsing slowly with the tide, losing the quantized, mechanical tick it has in the city. Decisions in the sound are creations, not selections from a menu of choices. Cognition, or thought, is different here, too. It's continuous, not suited to boxes. Whole ideas grow up, long thoughts leading to unexpected destinations—unlike the flitting of city thinking, which is mostly reactions to questions, messages, lines and squares. From this perspective, that city life, if remembered at all, looks like a mechanical complex of herky-jerky activity, as incoherent as a hazily remembered dream. Both mental frames are real—urban or outdoor—but the continuity that arises in this environment makes it easier to feel connected to other living things.

Eva and I talked at length about how our thought processes adapted to Prince William Sound. She believed that a kind of thinking emerged from the place, the same way the Native people experienced its nature. Giving in to that pattern of thought meant becoming part of the environment. The answers she once sought from the spirits of the sound in ritual she now received from the experience of life there itself—as if each day offered meaning and one had only to listen calmly in order to hear.

"The Native way of thinking is a natural part of being out in that place," she said. "Your mind will start thinking that way, because the evidence is going to tell you that this is the way the universe operates, that you can ask for help in making a decision, and you will receive it. And if you don't find the animals, there is a reason. It's not because you didn't look or listen to the radio in the right place or whatever.

"It's about meaning. There's a meaning to the experience of whether we find the whales."

The meat-eating transient whales have become harder to find. They were poisoned by the oil spill and their flesh is contaminated with long-lasting industrial chemicals—pesticides, fire retardants, and lubricants from electrical transformers—which accumulate from industry and agriculture on more populated shores around the rim of the Pacific. Besides, the marine mammals they subsist on are scarce. Here and elsewhere on the northern Gulf of Alaska coast, many species of marine mammals, fish, crab, birds, and shoreline invertebrates have grown less common, often slowly and in patchwork fashion, changes that are scarcely noticeable except by those who were here before, a generation ago. When Eva first started counting the AT1 pod of killer whales and recording their calls, she could usually find them looking from the mountain behind the whale camp or drifting and listening. Seals were everywhere. Whales foraged for them all up and down the island passages. They would sneak through entrances just big enough for their bodies in search of seals hiding in tiny coves as smooth as ponds, leading the research team into concealed waters otherwise undiscovered. Gradually, as their overall numbers declined, most of the seals disappeared from those places. The whales no longer use them as hunting grounds. They have switched foods, moved where the remaining seals can be found, or disappeared. To Eva, a hidden cove can feel like an abandoned house, empty rooms haunted by the memory of the family that once feasted, sang, and played there.

"There is definitely a sense of emptiness to certain places," Eva said. "When you go to a place you used to see harbor seals, and they aren't there anymore, and you see that over and over again—they don't come back there—that definitely feels diminished.

"There's this feeling that someone used to live here and doesn't anymore."

The sound is heavy with mystery, full of secret places where life stews down in the mud, on fractured rocks, and within tide-pool universes, where I have watched a twenty-four-legged sunflower star chase a herd of fleeing hermit crabs like Godzilla pursuing extras through the streets of Tokyo. A brutally remote beach of rounded pebbles with miles-long piles of white clam shells and blue mussel shells—a narrow channel through it opens to a broad estuary with guard-tower rocks sticking up, topped with trees like pikes. Great blue herons stride with huge wings up a ramp of air, legs bouncing below on springs. A seal periscopes, its head pushing to the surface the way a kid looks up from under the covers after crawling to the wrong end of the bed. A little island arises from clear water with white stair steps of shattered granite like a Greek temple's, ascending from the shadowy kelp forest below to the mossy edge of the spruce forest above, where a bear has left an offering of half-digested blueberries on the top step. Thirty-five hundred miles of these hidden coves, bay within bay and fold within fold, life riding on the back of life in ever finer crevices, more than a person could ever absorb, all lie within the sound, uninhabited and alone. It was so much richer here once? The idea seems impossible. And can it be true that this life is being drained out, as if by remote control, from far away in space and time, without so much as a human eye to see it?

Prayer helps cope with the unknown and defends the self from disappearing in the face of infinity. Perhaps the infinitude of the oceans, wave upon wave, provoked the first prayers, along with birth, death, overwhelming love, natural disasters, and the other moments when we are forced to acknowledge our powerlessness. As a skeptic, I avoid beliefs that protect the human ego when simple, natural explanations work as well, so I was tempted to discount Eva's spiritual practices. However ancient they are, these rites could be merely manifestations of a biological urge to negotiate the unexplained. On the other hand, true skepticism requires fair consideration of all possibilities, and I have to admit that my heart, too, belongs to the sea. I've been remade. I feel the invisible force irresistibly drawing back toward the places whose spirits captured Eva. It's strong, and to chalk it up simply to beauty or aesthetics doesn't get even partway to an explanation. Besides, to explain away this power—killing off these spirits—suggests a horrifying result: neutrality in the face of the ongoing diminishment and destruction of marine ecosystems, including this one.

3.

Chugach Culture and
the Spirits of Nature

Killer whales usher the dead to the spirit world. In Chugach villages they tell stories of frail elders, awaiting their time, who passed away with the next approach of a pod of tall, black dorsal fins and became whales. One of Eva Saulitis's and Craig Matkin's whales is named for Mike Eleshansky, an old friend who predicted he would be leaving with the next pod that entered the bay. Jim Miller, a carver in Port Graham, carved a seven-foot pole for many deaths the village was grieving, faces of the departed looking from the holes of a kayak led upward by a pair of killer whales, traveling from the blue of the sea to the blue of a star-flecked sky. Over the winter the children in the village school had helped Jim make the pole and it was dedicated, next to the airstrip, on Memorial Day, 2007. Do they really believe the whales take away their dead? Jim simply answered with stories about what happened when people died.

Jim and I gathered driftwood from the bay around Port Graham, picking pieces of red and yellow cedar that had floated hundreds of miles from the south and east on the coastal current, hauling them down the beach, a man at either end, to where my aluminum boat softly crunched the gravel at the rising tide line. Driftwood logs lay partly exposed in the storm berm along the crest of the beach, silvery shapes sculpted by years of collisions between waves and rock, exposed like naked limbs emerging from the blanket of rounded pebbles, beach grass, and creeping pea vines.

Jim's eyes sought the shape of a potential mask or box within each log, finding flaws or potentialities I could not detect. As we walked, he told me about his childhood, son of a Welsh man and a Native woman of mixed tribal heritage, feeling abandoned when they separated for a few years and left him in Minnesota to live with a family friend, and of looking for a place where he would fit in as a young man, trapped by drugs and alcohol on the streets of Ketchikan. Now he is a counselor in a village weighed down by grief and alcohol abuse. Everything he talks about returns to the subject of healing.

Jim said all wood is sacred, but what's found is more sacred than anything he could buy, because the wood that washes up is a gift from the creator. By the same logic, Jim gives away most of what he carves. To carve a paddle, a spoon, or a soul catcher—a rounded, palm-sized object for a healer to hold—gives new life to a piece of wood for a time, for as long as it remains useful, before it is properly thrown away again. "The point is that everything has life and spirit. Connection. Connected to the trees and forest and understanding," he said. "Some people think certain land has more spirit, more connection, than other places."

The oldest way of thinking may be to believe, as Jim Miller and Eva Saulitis do, that meaning resides in beloved places and animals. Today law treats land and nonhuman creatures as commodities equated with money. Only certain outspoken believers declare kinship to nature. When they say killer whales escort the dead, they are not speaking metaphorically, for they see spiritual and material qualities together, and perceive people inside animals, too. It's a distinctly odd outlook these days, but once—and perhaps for the majority of human existence—it was a dominant ideology. To explore human nature, it's worthwhile to take that worldview seriously. We will follow the idea in the chapters ahead, considering what science tells us about why one would grant a sense of personhood to animals, and if such intelligences are, in fact, present in animal minds. But the question isn't only about who is right or wrong—about whether or not there is a thinking someone behind the eyes of a whale. First, in this chapter, it is also worth asking what it was like to live in that universe, the ancient one in which everyone believed the natural world had these simultaneous qualities of physicality and spirit.

Three centuries ago, when there were many more Chugach villages—when the Chugach people ruled Prince William Sound—they met each August for their largest and most festive celebration, the Feast of the Dead. Only the richest villages could afford to host. Emissaries would paddle off in two-seat kayaks to invite all the other villages, but returned saying no one would come. Host villagers then fed the messengers with practical jokes—clam shells filled with mud, for example—and laughed when they opened the shells and splattered themselves. There was drumming and shouting down the beach, around the next point, where the visitors were hiding. The people would cheer and welcome the visitors into the smokehouse. Each village then took an evening to dance, each person presenting his or her own dance, with comedy to cheer up the grieving and masks to make them laugh, which were unceremoniously discarded afterward. The feasting and gifts came after the dancing—the entire festival lasted a week or two—with gifts to commemorate those who had died that year, given with the statement that the dead person could not use these things anymore and a request to the receiver that "when you use it, remember him."

Killer whales gathered in August, too. Pods came from as far away as British Columbia, eight hundred miles to the southeast, to join in a great, apparently joyous social group in salmon-rich waters. Craig and Eva met them on August 1 for five years running, counting as many as eighty whales closely mingling. They theorized that because the whales never leave their mothers, they meet with other pods to find mates to whom they are not closely related, as villagers often do when they marry someone from another community (a village is like a tribe). In British Columbia, whale congregations sometimes start with a greeting ceremony: the whales from different pods form two lines about fifteen meters apart, face each other while pausing for up to thirty seconds, then submerge and surface in a closely intermixed group. The whale culture of the sound doesn't include a greeting ceremony. For unknown reasons, Craig and Eva haven't been able to find the congregation at all since 2004. As for the Chugach Feast of the Dead, it was lost generations ago.

Cultures change. If you accept that killer whales are smart enough to invent new ways of hunting and organizing socially, as seems indisputable, then it follows that changes in their mental processes could unpredictably alter their relation to their environment. Killer whale pods follow their matriarch, who can be eighty years old (their life spans and time to maturity are similar to our own). When a pod grows too large, an older female may lead her offspring and descendants off in a new group. That's the only way resident killer whales ever leave their mothers. The new pod evolves new vocalizations. Listening with underwater microphones, researchers have learned to recognize the repertoire of each pod and classify its relatedness to other pods' calls. They've identified clans and developed trees of distant family relationships by listening to the whales, similar to the family trees of human languages. Genetic studies confirm that they've got the families right.

The villages of Prince William Sound and Kachemak Bay were analogous to pods. The people here, now called Chugachmiut in their Alutiiq language, shared a culture that resembled that of the Eskimos from farther west and around the edge of Alaska to the north and also, but not as much, that of the Indians to the coastal east and interior north. The urge to banish grief with mirth is more like the Yup'ik or Iñupiat Eskimos than the more solemn Athabaskan or Tlingit Indians. Their stories mixed the animal characters of Tlingit mythology and the cosmology of the Eskimo. Their government resembled the Eskimos' as well—they had little. The men who early visitors identified as chiefs were probably just rich hunters, strongly respected family leaders. When such a leader decided to go to war against another village—as they often did—others could get in their kayaks and follow, or not, based on their own judgment. As members of a self-organized community, they would fight voluntarily. Warfare against other cultures and other villages—the same villages invited to the Feast of the Dead—determined who would control the rich hunting grounds of the sound's waters and mountains. Within a village, private property meant little. But each of the sound's eight villages protected its collective property, its territory of islands and passages and streams for summer fish camps. Those resources yielded wealth and power to village traders, owners of slaves, and givers of gifts.

Explorers of the sound—my own family, for example—can easily feel like the first to navigate an exquisitely branching complex of coves or climb to a meadow-encircled lake hidden behind a rampart of mountains. Railroad magnate Edward Harriman clearly had that sense on his expedition in 1899. Harriman had chartered a luxurious yacht for a glorified pleasure cruise with the day's most famous nature writers and scientists as celebrity docents. In the fjords of western Prince William Sound, Harriman's passengers named withdrawing glaciers for their own Ivy League alma maters—Harvard and Yale glaciers were substantial in size, those named for the women's colleges were relatively petite, but Harriman's own Columbia Glacier was immense. These little jokes are still engraved on the official charts and maps of the region. But those glaciers already had names, serious ones. It was just that no one who wrote books had ever bothered to record them. Every bight and islet had a name. Every mountain had been climbed and owned. Steep rocky islands had been refuges for women and children during wartime, reached by special removable ladders. Caves had been burial places for the mummies of the great. Salmon streams had been the subjects of wars. Stories passed down for generations placed magic little people among certain rocks on Hawkins Island, a spirit-crazed woman on Gravina Island. Every place meant something. Every place was full.

Harriman arrived after more than a century of decline for the Chugach, whose communities had collapsed under the pressures of invasion and disease, their losses accelerating with a flood of wealth seekers from the United States. Within a decade of his ship steaming through, government teachers were discussing consolidating the five surviving villages to make it easier to deliver health and education services, but no one recorded details about the Chugach culture until 1933, when only the villages Chenega and Tatitlek remained. As a summer project a Danish anthropologist interviewed a few Chugach people. Kaj Birket-Smith's primary informant was a dignified old man from the abandoned village of Nuchek, Alingun Nupatlkertlugoq Angakhuna, known by his Russian name, Makari Feodorovich Chimovitski, who was interpreted by his daughter, Shanuq, known as Matrona Tiedemann. Birket-Smith's young protégé from Bryn Mawr (a small-glacier college), archaeologist Frederica de Laguna, dug bones and relics around the sound. Birket-Smith's report was fragmentary and often contradictory, and the stories were not well told. He admitted there was no way of knowing how much he heard was truthful or accurate about Chugach culture, which had already been under siege for 150 years. De Laguna found clues to the past in tools, village garbage heaps, and fish camp sites, and determined the boundaries of territorial hunting grounds, but constructed no coherent story. The earth, the ice, and the rainforest rot had moved too fast to find much. But at least Birket-Smith and de Laguna's work survives: when village people such as Jim Miller or Diane Selanoff tell stories now, they're usually quoting those rare books, whether they know it or not.

The Sound, apparently empty to explorers, once was full not only of names but of spirits. Makari told Birket-Smith of these hidden inhabitants, stories that resonate with Eskimo cultures across the Arctic. Every object in the natural world had a spirit owner with a human shape: certainly the land and the ocean floor, the source of food, but also the sea swells and the high tide, the rocks, ice, trees, and leaves, the Aurora, even sleep—everything. One of Birket-Smith's informants had met these spirits from time to time and received their assistance. But humans and animals were something else: they had souls. They owned themselves, unlike the spirit-possessed world around them. And they were equals. Humans and animals exchanged forms. A man could, in reality, be a bear or a codfish. Likewise, a dog could be man. The sea otter had been a man engulfed by the tide while feeding on the shore on chitons (which cling to rocks like flexible snails), and his transformation explained why the otter's organs look like a human's. Consequently, a sea otter killed on land had to be accorded a human burial. An otter taught Makari his song, a profound secret that brought him success in hunting. Saying the secret names of the fire would help it burn.

The most fearsome and important souls were those of the killer whale and the raven, but the raven above all. It was the raven who brought daylight to humankind and taught people to make fire. The raven stories go on and on, and there may have been many more that Makari didn't remember or discarded for religious reasons (he was raised in the Russian Orthodox Church). Missing is the account of raven as creator of the world that is ubiquitous in other traditions.

The ravens themselves are still there, talking from the treetops with weird and expressive calls, which seem so rich in meaning but remain undecoded, the mysterious aural dimension of the forest. The ravens' language has foiled scientists with variety. The calls, more than eighty unique sounds by one count, vary by pairs and groups of birds, and ravens use different dialects in different parts of the world. Bernd Heinrich, the famous raven expert at the University of Vermont, concludes that raven calls are culturally determined. Heinrich finds evidence, as well, that ravens communicate with other species, leading hunters to their prey in order to scavenge a share of the kill, even if the hunter is a bear and the prey is a man. Makari said the Chugach people once knew which raven call meant a boat was approaching, and the direction.

Ravens' social relations are even more complex than killer whales'. In the winter in Alaska, unpaired ravens meet at dusk to communicate and play together, as many as eight hundred at a time, apparently establishing their plans for the next day, when they will overcome competitors by mobbing a food source. They seem to have same-sex friends. Their acrobatics and games of dropping and catching sticks leave a mess of needles on the snow and outlines etched in the powder by blasts of air, lasting shadows of long wing feathers. After these noisy assemblies, the birds steal away to rest alone, quietly, undetectable to predators. Once ravens are mated for life they no longer attend evening happy-hour gatherings, instead defending a well-defined territory together.

Corvus corax has used its intelligence to span climates, making ways to thrive in habitats ranging from the deserts of Mexico to the frigid shores of the Arctic Ocean in Alaska, where the birds knit discarded welding rods and cargo ties from the oil fields into nests of plastic and metal atop heat leaks from buildings. There they can incubate eggs in temperatures that are scores of degrees below zero. They're a creative force. And they watch. A researcher studying Arctic ravens, Stacia Backensto, lost the ability to trap them, even birds she hadn't seen before in areas she hadn't visited, because they recognized her and, presumably, warned one another. She took to disguising herself with an oil worker's uniform puffed up with pillows, a fright wig, and a fake beard and mustache (she later was scolded by the University of Alaska business office for spending research funds at the Party Palace in Fairbanks). Even at that, the ravens recognized her more easily than did the other humans on the oil field; yet we usually can't tell ravens apart or understand what they're up to.

Among Makari's stories, a strange, short tale described how Raven tricked a whale, probably a humpback, into letting him fly into her stomach. There he met a woman who asked him not to harm the heart of the whale, for doing so would destroy her home. Together, they harvested and smoked herring as the fish came washing into the whale's mouth. The Iñupiat (Eskimos of Alaska's north) have a similar story in which the woman in the whale was clearly the whale's kind and beautiful spirit. In both stories, the interloping Raven thoughtlessly destroyed the spirit, killing the whale. When the Chugach people came to the shore to butcher the whale Raven had killed, Raven tricked them into believing the food was bewitched and they ran away. Raven instead recruited the Steller jay and magpie and the three ate the whale by themselves.

Ravens, jays, and magpies are related, members of the family Corvidae, with crows, jackdaws, nutcrackers, and rooks. They are, together, among the most intelligent animals on earth. Superficially, it's interesting that the Chugachmiut recognized these birds' special qualities and promoted them to the top of their cosmology. At a deeper level, the mythologists' perception was more profound: that the animal masters were unpredictable and inscrutable, agents of chaos, capricious in their concern for mankind, like gods and like the ecosystem itself. The Chugach belief in the will of animals didn't merely project human thought, like the talking animals in animated movies. These other spirits were far more interesting, far harder to understand. The universe was full, rich with meaning to be newly discovered on every ocean journey. One was never alone, and never sure what these other voices might say.

Through the few words that come down to us, it's almost possible to imagine oneself into Makari's place. Returning from such thoughts to the mundane world feels like leaving a color-saturated movie for the gray sidewalk outside the theater. We commonly equate myth with fantasy or metaphor, setting it aside as an inferior kind of truth. But current scientific study of animal intelligence gives the Chugach view more credit than that. Something significant is going on in the minds of ravens and their corvid cousins. Science now obliges us to decide if a person could be inside; and to do that, we'll also have to decide what a person is. On that path, nature and human nature must be explored together.

4.

How Animals Think

The aloneness of being human is knowing only one's own thoughts. No one knows the content of another's mind. Despite craving love, respect, and connection, all that comes to us is the outward manifestation of thoughts and feelings that are themselves invisible. Without thinking about it, we mentally model what's going on inside other people, tracing their trains of thought, and we produce predictions—intuitions— about what they will do. Knowing a person well amounts to having accurate intuitions, a model that predicts that person well. Psychologists call this mental modeling "theory of mind." A talented politician or socially gifted family member can seem like a mind reader. But we're often wrong as well, deluded by our own wishes and prejudices and our need to be at one with others. We're always forced to choose between loneliness and trust.

No one knows a raven. Raven psychology may not work at all like our own, so our human mental models cannot reliably interpret their behavior. It's easy enough to apply human theory of mind to an animal, but if a dog seems remorseful after making a mess, that could be a conditioned response rather than a sign of insight—he may associate hanging his head with lighter punishment. Or maybe the dog's innate programming tells him to bow as he would before the dominant member of a pack. Field scientists watching corvids, primates, or whales in the wild can't overcome the many possible explanations behind any apparently thoughtful action they see, or the skepticism of colleagues who wonder if such observations reflect the unconscious bias of a human applying his or her own theory of mind. The public is willing enough to believe animals think the same way people do, but throughout the twentieth century lab experiments trying to find human mental qualities in animals were repeatedly debunked when the human influence was removed—including the horse that seemed to do mathematics, the apes that could form sentences, and the dolphins neurologist John Lilly gave LSD in the 1960s, believing they could communicate directly with people.

A breakthrough came when an English scientist started trying to think like a bird rather than expecting a bird to think like a human being. Nicky Clayton liked to get out of her office at the University of California, Davis, at lunchtime and enjoy the sunshine of the Central Valley, often alone because her American colleagues, work-obsessed, remained chained to their computers. She watched western scrub-jays—corvids, like ravens—stealing morsels from students' lunches and caching them. Corvid caching behavior already interested researchers of spatial memory, because the birds' memory of places can be so much better than our own. Clark's nutcrackers can make tens of thousands of caches and remember them up to six months later. What Clayton noticed on the Davis campus was that scrub-jays often left their food caches only briefly, returning a little later and caching the same food again somewhere else. What were they thinking?

Clayton claimed she could not study spatial memory like other researchers because she has no sense of direction, and therefore had to study a different aspect of memory—although, in truth, I suspect this explanation was part of the ditzy-blonde persona she projects as an ironic antidote to academic sexism. Instead of focusing on where the birds cached, she asked why they chose their hiding places. Clayton caught scrub-jays and brought them into the lab. Her great idea was to study the jays' intelligence in the context they use it—caching among their social peers—and to do so in controlled experiments that would rule out varying explanations of the behavior. Harder than it sounds. Nothing highlights the distance between the lives of wild animals and modern human beings like bringing birds into a building made for purposeful work and waiting for them to do something regular and predictable. While a scientist surreptitiously watches, pencil poised over data sheet, the birds flitter and flutter, their aimless behavior tedious and seemingly meaningless. But if the experiment is cleverly designed there is enough common ground between the two worlds that information begins to flow. For example, Clayton figured out how to get the birds to cache where she could keep track of them, in ice cube trays decorated with Lego blocks. She noted when they would move their caches, which suggested why.

The scrub-jays didn't like to cache when anyone was watching—even the presence of a TV videographer would interfere—and they would retreat to the darkest and least visible corners to avoid being observed. Later, if an opportunity arose, a jay who had been watched caching would wait for a moment of privacy to move the cache somewhere more secure. Nicky found their choices about recaching depended on who had seen them, how well, and the probability of that bird stealing the cache— whether the other was a mate, a dominant, or a subordinate. Were the birds using theory of mind to analyze the thinking of potential thieves? Maybe not: they could simply associate certain birds and caching situations with theft, recaching as a conditioned response. To rule out that possibility, Clayton and a colleague conducted an ingenious and arduous additional experiment. They hand-trained scrub-jays from hatchlings without giving them the opportunity to be thieves themselves. These birds didn't take precautions against theft by other birds. Without being thieves, they didn't recognize the possibility in others and were easy marks. Apparently, jays developed their cautious behavior not by being victims, but by projecting their own behavior as thieves on the thinking of other birds. Theory of mind.

Clayton's accomplishment marked a turning point in the study of animals' mental abilities. Others showed that New Caledonian crows spontaneously innovate tools, meerkats teach their young, songbirds use linguistic-style grammar, and African cichlid fish reason inferentially about their place in a hierarchy—an accelerating stream of discoveries revolutionizing thinking about animal minds. With the simplest tools, Nicky had turned a lock that had foiled generations of scientists, cracking the door between our lonely but social species and others like us with whom we share the earth. Presented with this laudatory point, she smiled coyly—it's not for her to say, but she's not arguing.

Nicky almost became a dancer instead of a scientist. As a child, her parents went dancing every weekend. She was interested in birds, fashion, and dance, and nothing has changed, as if she had reached her midforties without ever being disillusioned. While a full professor at Cambridge University, and famous in her field, she still goes dancing six nights a week—Saturday jazz, Sunday salsa, Monday ballet, Tuesday tango, Wednesday salsa, Thursday salsa again—and wears filmy Milanese dresses to work, with stiletto heels that shape her tiny body into a taut, birdlike arc. She recently helped an important dance troupe create a bird-inspired ballet for Charles Darwin's bicentennial. Nicky presents herself to the world innocently expecting to be accepted in the light spirit she intends, an intimate tango with one partner followed effortlessly by another, her unspoiled optimism carrying her speedily onward in each thing she tries.

Clayton and her husband, Nathan Emery, a solid bloke of an Englishman, met in California. The story goes that she marked up a paper he wrote about primates' social intelligence with corrections about how birds could do just as well. From that start, they became collaborators in social cognition research—Nathan wasn't likely to take offense at the edits anyway, Nicky said, as he was already smitten. She didn't want to stay in California, where people dress sloppily and work too hard (her first husband abandoned the state before she did). Back in England, however, Cambridge never offered Nathan a permanent position, so he eventually began a one-hour commute to teach at Queen Mary, University of London. Compared to Nicky's breeziness his brilliance is assertive, and I found his linear explanations of their work easy to grasp, but Nathan seems relatively earthbound. Nor does he love to dance, although Nicky remembers him saying he did when they were dating. Nicky has dashing dance partners. One owns an Iranian restaurant, which he opened up on a sunny bank holiday just for our party of three to have lunch—Nicky, a doting older research partner, and me—the proprietor hovering over our table and dispensing treats for which he wouldn't accept payment. Nicky enthused how lucky she was to be surrounded by "all these lovely men."

Clayton's taste for complex social situations suited her work. Negotiating such relationships was probably the original purpose of intelligence. Psychologist Nicholas Humphrey first pointed out in 1976 that nothing in the life of a gorilla requires a big brain except getting along with other gorillas. Once you know how to harvest a piece of fruit you can do it again without a lot of thought. But to obtain the advantages of living in a coordinated group—advantages including protection from predators, control of productive food sources, and opportunities to mate—a gorilla needs to be able to cooperate, working out personal needs in relation to the group and the potentially changing wishes of other individuals. The ability to form a theory of mind about other animals doesn't only help prevent theft, as Clayton found in the jays. It may be even more important for building connections, resolving disagreements, and sharing, as the most social animals also tend to be the most intelligent: primates, dolphins and toothed whales, elephants, parrots, corvids, and humans.

Scrub-jays may be inveterate thieves, but Clayton and Emery's team also observed them passing food to one another through the mesh of their cages. Immature rooks and jackdaws—black, crowlike corvids—use food sharing to develop complicated networks of affiliations, including alliances between birds of the same sex, which we would call friendships. The process probably helps them select the best possible lifelong mate. Jackdaws pair up at an early age and never stray, even for a secret dalliance, even if their union is sterile. The importance of that choice may explain why, as immature birds, jackdaws are the most avid food-sharers of all. A jackdaw will even pick up a worm another bird has dropped and give it back to him or her.

Jackdaws are all black like ravens and crows, but with long, curved beaks. Although I'd never seen one, they seemed familiar to me when I saw Nicky and Nathan's flock at the aviary near Cambridge, living versions of etchings and illustrations from English children's books. Nicky told me one day that I'm a jackdaw because I met my wife in high school and we're still together. For dinner that evening she picked me up in her muscular blue convertible. Back home Barbara, my wife, was packing lunches for our kids and teaching second-graders; I was whizzing by Sir Isaac Newton's and Charles Darwin's digs on a balmy evening with a dancer in a tiny black dress and silver jewelry. Nicky and I are exactly the same age. We looked as if we were on a date as we walked into a fashionable Italian restaurant. She was enchanting, flattering, open to my impertinent questions. I held my notebook under the tablecloth to avoid breaking the mood.

Thanks to those yellow sheets of notes I can remember much of what she said, and how, and I can return to the table in my mind, seeing the waitress's knowing but respectful smile and the quiet, unhappy couple seated in the back corner, near the bathroom. Now that I'm back in Alaska, looking out my study window on winterbound birch and spruce trees—and a magpie caching in the snow—I call on particular images to reassemble that dinner in Cambridge, points of reference pinned to the bulletin board of my memory around which my imagination strives to post other, less definite pictures, sounds, and words. But it's never certain which of these elaborations are real and which are manufactured for me by my mind. My notes have often corrected memories for me throughout my career and I've interviewed subjects whose recall of significant life events turned out to be completely illusory. Since I didn't write down what Nicky's jewelry was made of, I have to admit it might not have been silver. Brain scans show that processing of memories of moments and events takes place in the same place as imagination. It is called episodic memory and works differently from the memory of facts or skills that we recall without having to think about when or how we learned them.

Nicky's first major breakthrough with the scrub-jays, before the work on theory of mind, had come when she demonstrated they have episodic memory, too.

Nicky's jays remembered not only where they hid foods, but when, what kind, and who was watching, all elements of remembering a moment. Moreover, they used the content of their memories flexibly. The jays kept track of the perishability of different foods—nuts that last a long time or worms that spoil quickly—and returned to each cache before its expiration date. When Nicky and her research partner, psychologist Tony Dickinson, cleverly presented the birds with new information that a particular food would rot faster than expected, the jays adjusted their schedule, abandoning caches now anticipated to be too far gone.

The jays planned for the future, too. Given access to a pair of potential cages, one in which they had reason to expect food and one that usually had none, they provided for themselves by caching in the enclosure where they would be more likely to be hungry if locked in. A student of Nicky's, Dean Alexis, conducted an analogous experiment with children and found their performance didn't match the birds' until age four or five.

Does that mean that birds can imagine? Do they mentally travel in time, reconstructing scenes from the past and assembling hypothetical situations in the future? Somehow, that's a spookier notion than believing they can think about other animals' mental states: creativity and nostalgia define so much of our sense of ourselves. But Nicky isn't contending that corvids experience the world as we do. Her experiments succeeded because she investigated the jays' mental capacities as their tools for adapting to their environment rather than by trying to get them to behave like people. By the same standard, she now insists her results cannot be interpreted as finding little humans inside the birds. Theory of mind and episodic memory may mean something profoundly different in a corvid's head, something we can't quite imagine. If so, the gulf between us and other animals may span as broadly as it did before. Our own theory of mind may not be adequate to allow us to see as a jay or a raven sees.

On the other hand, her knowledge of animals' intelligence makes

Nicky something of a vegetarian. She eats hardly anything—usually she skips dinner, subsisting on a good glass of wine at a pub with Tony Dickinson after an evening of dancing. When she does order a meal it's often an appetizer of calamari. She eats marine invertebrates because she needs the protein and she reasons that their simple nervous systems put them somewhere outside the moral circle of intelligent animals. But at one of our meals together, I mentioned that some species of octopus seem to be craftily intelligent. They've been caught sneaking around aquariums at night to eat fish from different tanks, returning home so that in the morning keepers don't know what happened. Octopi open jars, they drop rocks on enemies, and they interact playfully and emotionally with their keepers—their skin turns bright red when they are angry or excited. Nicky didn't seem to want to hear about that while eating her calamari. She admits she's only consistent within limits—she can't give up her leather designer shoes.

It's not unusual to equate personhood with intelligence. The thought of eating an animal capable of complex thinking makes some people squeamish, as if it were close to cannibalism. Philosopher Peter Singer's work first inspired the animal rights movement in the 1970s with the logic that the right to life comes with the mental ability to value one's own existence through time, a basic form of self-awareness. That moral dividing line could be defined as presence or absence of theory of mind, episodic memory, emotion, ability to feel pain, or another standard chosen by an ethicist making his or her own rules. It's a peculiarly Western idea, descended from Enlightenment concepts exalting the individual— "I think therefore I am," in that statement's popular interpretation. But the uncomfortable corollary of this line of reasoning is devaluing organisms that lack complex cognition, including the overwhelming majority of animals, as well as human beings not capable of higher thought— babies or the developmentally disabled, or people who have lost mental capacities due to age or dysfunction.

As I said at the end of Chapter 3, the question of whether an animal can be a person requires asking at the same time what a person is. If intelligence equals personhood, then many animals are better qualified than some human beings. But the question seems much deeper than a simple equivalence of thinking ability and a right to life. The Chugach people, who deeply esteemed the minds of animals, nonetheless feasted on their flesh. The Chugachmiut weren't lonely in the natural world. For them it was full of relatives, gods, and adversaries, animals other than human beings who were people, too. They acknowledged and honored the gifts of life from the animals they consumed with special rites and with hunting practices that protected the animals' spirits and habitat. Offering such respect sustained the connections that anchored them as part of a complex network of living things. They knew what makes a person, but it's an answer that might not be found in a laboratory.

5.

Thinking and Being a Person

What you want to believe should be examined most closely because, of all the things you know, it is the most likely to be wrong. If you don't care whether a belief is true or not, you're probably unbiased and can trust your initial judgment. Danny Povinelli, of the University of Louisiana, wanted to believe that chimpanzees have theory of mind and can analyze the thoughts of others, as people do. He first got excited about the idea at age fifteen when reading that chimpanzees might share our concerns for self and existence. As a researcher, however, he felt a growing internal tension over the years and many experiments. His chimps could do some mental tasks easily, but they had great difficulty with other situations that seemed similar but required a step beyond the visible. Finally he had to accept that a new explanation fit better. He said, "I realized I had to do a profound reorganization of my thinking, and it was extremely difficult."

Povinelli asked if a kind of reasoning other than a theory of mind could explain the results of the experiments. Clearly, Nicky Clayton's birds and his chimps were capable of reasoning—they couldn't possibly interact with their peers as they did using associative learning alone (that is, through conditioning, as when a trainer elicits a repeated behavior with a reward). But Danny pointed out that animals might reason about what they could see without thinking abstractly about the invisible content of another mind. To understand this, think of it in human terms. If you meet a woman wearing a SAVE THE WHALES campaign button, you don't need to theorize about her thinking or motivations—you can tell just by looking at her that she will support laws to protect whales, and likely hold a constellation of other environmental and political views that are usually associated. Likewise, a jay might not need to think about the internal mental state of another bird to know the bird is a thief. Past behavior might be enough.

What about the hand-raised jays who never suspected theft because they'd never done it themselves? Povinelli said that could simply be how the birds learn about stealing, knowledge they can then apply to the *behavior*—not necessarily to the *thoughts*—of other birds. Perhaps stealing is part of their species' developmental program. In an experiment of his own, Povinelli placed chimps in a room where they could gesture for food from a woman who had a blindfold over her eyes and another with a blindfold over her mouth. The chimps made their begging gesture equally to each person. Unlike children in the same situation, the chimps didn't immediately grasp that a person with eyes covered could not see them, presumably because they couldn't imagine what being blindfolded is like.

Nicky and several other scientists point out that blindfolds and cooperative food sharing are unnatural for chimpanzees, so the experiment is another example of a scientist asking an animal to act like a human being. Besides, just because Povinelli's chimps didn't demonstrate a theory of mind in that particular experiment doesn't rule it out. Povinelli counters that no animal has passed the same test in any setting. The discussion gets heated on all sides. I got in the middle of a misunderstanding between Povinelli and Clayton, which Danny concluded with an e-mailed apology to Nicky that he signed, "All hugs and kisses, Monkeyboy." He likes her and respects her work, but believes she has fallen into a trap along with the rest of the field, seduced by the human tendency to theorize about other minds—in her case, about corvid minds. He said, "It's evidence for the incredibly powerful nature of the human mind to re-create anything it sees in its own image."

More is at stake than an experimental protocol. Povinelli's work was taken up by Christian creationists as proof that Darwin was wrong and that mankind is special, made in God's image. Povinelli was dismayed, but he was the one who brought up God in the first place, writing provocatively of "ghosts, gravity, and God" as attributes of a uniquely human style of thinking. Take gravity first. Our common sense tells us it is a pulling force, drawing objects down to Earth. But Einstein remade gravity as a curvature in the fabric of space-time pushing objects together. Einstein was right, but, even knowing that, our intuitive physics still insists gravity is pulling us downward. By analogy, Danny puts himself in Einstein's position, advancing a nonintuitive model of animal minds that are able to reason, but only about the visible. Our minds export our own broader mental abilities onto animals as we do to ghosts and God, two more intuitive byproducts of our uniquely abstract thinking processes.

I'm not sure if either Danny Povinelli or Nicky Clayton is correct. More scientists have accepted Clayton's evidence than have adopted Povinelli's distinctions. The influential Michael Tomasello of the Max Planck Institute in Leipzig moved from the Povinelli camp to Clayton's point of view, but then advanced a new dividing line between animals and humans: that only humans can form shared intentions with one another. Observers of animals in the wild contest that division, as they did the others that fell before it—for example, the disproved idea that only humans can use tools. Where will this dividing line finally end up? Or will it dissolve entirely into a complex shading of differences between animal and human minds? Povinelli could be right that the entire field is deluded and its paradigm of mentally sophisticated animals will collapse; or maybe he's wrong and one more experimental result will seal the conclusion that corvids or primates can reason about the unseen. Or maybe the debate will go on without end, because people on either side want too strongly to believe their own position—perhaps because they need to believe, or are constructed to believe.

In that case, we're talking about the wrong animals, and what we're really interested in is the nature of human identity. Nicholas Humphrey's original 1976 paper on the social intelligence hypothesis—the idea that intelligence was an adaptation for getting along in groups—went on to speculate that social thinking naturally led to assigning mentality to animals and phenomena, expecting fairness from the weather or the spin of a roulette wheel. Kaj Birket-Smith, the ethnographer of the Chugach people, made a similar point, writing, "How does one unconsciously, 'instinctively,' regard one's surroundings? The whole world is reflected in the picture of the observer. The child who strikes a chair against which he has stumbled, does so without metaphysical speculation as to whether it has a soul, but solely for the thought that the child lives, and therefore everything else must live. It is the same with the Eskimos. Every object, every rock, every animal, indeed even conceptions such as sleep and food, are living." Setting aside Birket-Smith's condescension toward the Eskimos, he may have had the instinct part right: our intelligence may be built on the foundation of seeing life in everything around us.

A psychologist at Stanford University devised an ingenious experiment to find out when this sense develops. Susan Johnson created a fuzzy green blob capable of blinking and beeping, but without a front or back or any clue about which direction it was oriented toward. She could surreptitiously move the blob on a table using a magnet. A twelve-month-old baby would witness the blob blinking and beeping while an adult talked nearby for about a minute. If one assumed the blob were interacting with the adult, then the side facing the adult would be interpreted to be its front. As the experimental trials turned out, the babies did reason that way, because when the blob rotated, they turned to look in the direction that the inferred front of the blob would be facing—the child glanced to see what the blob was looking at now. To pass the test, a baby had to reason about what the blob could see even though the blob didn't have eyes or any other features.

Johnson believes we're hard-wired by evolution to see mental states in other animals and inanimate objects. Finding intention in other beings would help anyone survive in a natural setting, whether to hunt or avoid being prey. The system works in adults whenever we watch an animated film—animators easily bring to life a lamp, a toaster, or a geometric shape and give it intention, emotion, and personhood by a few simple motions. Whether we discard these impressions as illusions or imbue them with spirituality depends on our beliefs and understanding about what's real. To put that idea in context, imagine a husband who is forced to watch his beloved wife as her brain slowly dies—as we'll see in the love story in Chapter 6. Until the end, he perceives her as the same person he always loved. Is that an illusion, or is he correct in believing that she was always more than a brain?

Looking for a soul in a raven has come to mean looking for an introspective conscious mind, a thing we feel, in intellectual culture, is the essence of a person. But people aren't the sum of their thoughts, like software packages running on a computer called a human brain. That's at least as much an illusion as projecting human thoughts on a green blob or a toaster. I doubt anything that happens in a Cambridge aviary can take us back to the misty shores of Prince William Sound three centuries ago, when the ravens could take off their coats of feathers and become men. We would be better off trying to become ravens. Or, to put it better, to become animals at home in our own bodies and therefore akin to ravens, whales, and the sea. We're human from the moment we're born, blank of experience, and immediately identify our mothers and bond with them. We remain human when age and illness erase our minds at the end of life. We're equally ourselves when we don't think of ourselves at all. The thinking "I" who is writing this, always self-conscious, remains fundamentally alone, and will never find empirical data proving a raven or a killer whale has its equal right to life. But that's not me. I'm much more, built with truths in blood, meat, and bone as well as mind. As an abstract mental entity I may be alone, but as a physical man I am not.

6.

The Essence of Carol Treadwell

In the video, Carol Treadwell's eyes look serious, concerned, a bit bleary, but not really afraid; however, some of the clues to her emotions are missing, because a blue fabric drape obscures her eyebrows and forehead, where her skull has been opened and her brain lies bare to the room, pale neural flesh under a netting of fine, black blood vessels, all overlain by numbers on small slips that a surgeon has placed to help remember what goes where. A national television audience watches. The surgeon probes an area of the brain near Carol's tumor while an assistant shows her picture flash cards. The place where words are kept is in the way of their work and they need to know exactly how close they can get. They expect her to lose the ability to speak for three weeks, but with a mistake she could lose language permanently. Carol remembers what a dog and a cat are, but later she can't recognize a spoon or a kettle on a flash card, so the surgeon backs off. The cards are segregated by living and nonliving things because the brain organizes them differently. Carol says, "An elf. No. It's called an eagle." A smarmy TV reporter from NBC asks Carol if she wants to send a message to her husband, Mead. She says, weakly, "Just that I love him, and I'm doing great."

Seven months later, as the Treadwells' influential and well-off friends overflowed Anchorage's downtown Catholic cathedral, Carol's eulogist declared that she had always been fearless—not just courageous. No one could know that except her. But anyone watching the *Dateline NBC* documentary could see she somehow defeated outward signs of dread as successive surgeries and a rapidly growing tumor dismantled her mind. If the cancer was a foe, she treated it with airy contempt. In the documentary, Carol's surgeon at UCLA Medical Center, Linda Liau, has to tell her, a week after the surgery with the flash cards, that tests have confirmed her cancer is now the fastest-growing type, meaning her survival is extremely unlikely. Mead's head hangs and he seems to hold himself together by writing notes. Liau appears heartbroken. Carol, who didn't lose her speech in the operation after all, tells her doctor, "Be happier."

"I'm sorry," Liau says, struggling.

"Life could be worse," Carol assures her.

Carol had wondered if NBC would run the program if she died before the new television season. By the time it aired, she could no longer stand or speak except for a few rare words and her body drooped and inflicted strange pains, sleeplessness, and difficulty eating. She watched at home in Anchorage with an expressionless stare. The Treadwells' house is a fashionable modern box on the downtown shoreline, looking across Cook Inlet to the mountains, one of the best addresses in town. As the broadcast hit time zones across the country the phone rang incessantly and hundreds of electronic well-wishes flooded in by computer; the next day Federal Express deliveries piled up continuously with fan letters, advice, and inspirational books. Mead had to spend twenty minutes on the line with a faith healer someone on the other side of the world had hired for $2,000. Millions of people were moved by Carol's optimism, but she was beyond comprehending the adulation. Between these interruptions, Mead frantically worked the phone looking for a cure, networking to doctors through his extraordinary political contacts—he had already recruited the head of the National Cancer Institute to consult on the case. Family and friends viewed his pursuit of vapor-thin hopes as brave optimism or frantic denial, but either way his work was in keeping with Carol's will to live.

This mythic optimism, which became so heroic, must have come across as at least a little conceited when Mead, Carol, and their Ivy League friends were young, in the early 1980s, and looking forward to launching careers in business and politics with the help of their families and social allies. Why shouldn't the privileged be optimistic? But their romance did have a quality of adventure and, for Carol, required the courage to abandon corporate life in Manhattan for Alaska, where Mead was already a well-known Republican political operative. She had dark hair and quick eyes, wicked humor, and a sense of irony that allowed her to navigate a party and create laughter. Mead, on the other hand, was a brick of sincerity, dressed in pressed L.L. Bean duck-hunting garb, with a broad, boyish face and bangs. The gossip columnist in the *Anchorage Daily News* took to calling him Tread Meanswell and mocked his self-importance, but what may really have bugged his detractors—and perhaps attracted the fearless Carol—was Mead's unself-conscious enthusiasm.

Besides, he was an Alaskan man, a frontier ideal, even if he had come from Connecticut and had a Harvard MBA. When their romance was budding Carol visited for ten days and he took her to the Prince William Sound town of Cordova to participate in sea otter research and to his friend Michael McBride's one-unit wilderness lodge, in the mountains south of Kachemak Bay. Carol had been all over the world, but hardly anyone has been to a place like Loonsong Lodge, all by itself on a clear, deep lake encircled by enormous, impossibly steep mountains and an unnamed glacier, a place that can be reached only by helicopter or small plane except when winter freezes the rivers, and then only with difficulty. In fall the air is clear and crisp and yellow birch and cottonwood leaves shine like sparks against the dusky green of tall spruces. McBride himself found the place when he was caught in a storm on a mountain goat hunt above the tree line and needed shelter; he bought fifteen acres there, the only private land for many miles in any direction, a hidden outpost among seaside mountains with high pocket valleys of heather and ice—the same mountains that catch the Pacific storms and make them into an engine of life. One winter Mead brought a scientist here who was studying the sort of people suitable to send on one-way trips into space (Michael and his wife, Diane, seemed to fit the profile). Mead had helped Michael install a huge bubble-shaped window in the master bedroom, which they lowered by helicopter in a crate suitable for a grand piano. Mead and Carol gazed from that window on the lake that fall, on the sky, mountains, and ice, as they fell in love during their days alone. They would go on many wilderness journeys together—including a week on Wrangell Island, north of Siberia, in a cabin they shared with a polar bear in a box—but the time alone at the McBrides lodge became the touchstone of their relationship.

Two months later, Mead and Carol saw the play *Shadowlands* in London's West End, the largely accurate story of the love affair between C. S. Lewis, an Anglican writer and author of the Narnia children's fantasy novels, and a Jewish American divorcee named Joy Davidman. The natural landscape Lewis imagined for the Narnia books, including *The Lion, the Witch and the Wardrobe,* could, in its wilder stretches, be the mountain peaks and secret meadows of Prince William Sound or Kachemak Bay around Loonsong Lodge, and its talking animals were like tea-sipping versions of those Makari told of from Chugach legend. As a child whose mother died when he was seven, Lewis had been seduced by the magical power of spirits in the woods. But as an adult, he became a priggish and self-satisfied Christian. According to William Nicholson's script, Joy Davidman rescued Lewis from sterile self-righteousness. She came to him as a fan, but beguiled him with her vital personality and sexuality, breaking him free to be born again as a real man, capable of love and passion and able to feel real loss and pain, as he never had allowed himself to do after his mother's death. In the play, as in life, Joy died of bone cancer shortly after transforming Lewis, leaving him shattered but, in another sense, newly whole as a human being.

The play so moved Mead that he proposed to Carol over dinner later the same evening. The cosmic coincidence of the play and the Treadwells' lives had several dimensions: Mead's father died when he was a child, Carol was a devout Catholic, and her brain cancer may already have begun growing when they saw the play, a dozen years before her death. And *Shadowlands* was playing in Anchorage when she did die. But Mead and Carol didn't need lessons in suffering. Their son, Jack, was only nine weeks old when he died in his crib on the eve of his christening. At another point in their marriage Mead's speculative business ventures nearly ruined them. Carol turned their home into a bed-and-breakfast and put on a yard sale that cleared $1,000. Mead came home from a business trip to find Texan tourists walking around the living room wearing his old shirts, and the mortgage paid.

Mead's investments eventually worked out. An Internet company he helped start held a successful initial public stock sale. He bought into cellular telephone frequency licenses, which became attractive to a large telecommunications firm. He and Carol entertained the cell phone executives on their patio overlooking the inlet on a July evening after a tough day of negotiations. Later, very early in the morning, Mead woke to hear Carol making a strange sound in bed next to him. She began panting through her teeth. He turned on the light to see her eyes darting in every direction, her muscles twitching, and foam coming from her mouth. Five minutes later the seizure continued as Mead went to the door to admit an emergency medical crew from the fire department; but as he returned, Carol was awake in bed and wondering, placidly, why a dozen suited fire fighters were pouring into her bedroom. Her manner didn't return to normal that day, either because of the seizures and exhaustion or the antiseizure medicine, even as they learned at the hospital that she had a brain tumor and they began calling family around the country. She seemed never to emotionally react to the news.

Mead already knew how brain damage could change a person. In March 1981, he had been waiting at the White House for a meeting with press secretary James Brady when Brady was shot in the head in an assassination attempt against President Reagan. Although surviving with his frontal lobes largely gone, Brady's lack of inhibition shocked Mead when he visited him after a partial recovery: a cagey press briefer and witty man had become a fount of filthy jokes and inappropriate sexual remarks. A brain injury can damage senses or memory, or the ability to form intentions, or it can change a personality, making a deep, thoughtful person shallow and inconsiderate. Brains can function without memories: one patient could masterfully conduct a choir and adore his wife without being able to remember a single event or fact; others retained theory of mind without being able to imagine the past or future. The capacity of mental time travel fades even in normal adults as they pass middle age: both memories and fantasies become less detailed, less vivid, flatter as we grow older. Most of us never have to reckon with the physicality of our own brains, instead self-imagining a single, unified mind, an entity capable of introspection and revelation, perhaps lit from within by a soul. But every memory or trait exists as a particular piece of matter at a place somewhere in the head—Carol's memories of the mountain lake at Loonsong Lodge, for example, or her irony—each divisible from the whole, vulnerable, and not necessarily essential.

What is essential? What makes the person? Mead found out over the fifteen months of Carol's disease, from the period when the cancer seemed in stasis to its sudden growth, into the period of consulting experts and finding experimental cures, the three operations, each taking more brain, the tumor branching out into new areas and growing with lightning speed, displacing the healthy tissue inside her skull. Through it all, Mead said, "She was never not Carol. That was the amazing thing about it." She didn't stop being Carol when she began forgetting words, choosing wrong ones, or finding elaborate paths around the words she couldn't find: she refused a pickle one day by saying, "In the theater of my mind this pickle has no role." She didn't stop being Carol when she couldn't remember her children's names. Nor when she tumbled into the bathtub and cried with frustration that she could no longer walk to her own toilet. Nor when she looked out on the world with quizzical, half-surprised interest, as if not sure why she found herself where she was. She didn't stop being Carol when she stopped talking entirely, stared blankly into space, having thoughts no one could fathom, or perhaps none at all. Or when the house filled with family and friends waiting for the end and she could communicate only by blinking or by the expressive content of her eyes, as she lay on the sofa in the sunroom at the ocean side of the house, gazing out the big windows onto the sea. Certainly she didn't stop being Carol when her nurses proposed to stop feeding her, as eating was difficult and there seemed no point, and she kicked her legs and grunted angrily, still refusing to give up. Nor on her last day, squeezing friends' hands. Nor when she said her last word, gazing into her six-year-old daughter's face, just the child's name, "Natalie."

"We obviously saw capacities go away, but the person never went away," Mead said. "During the last couple of weeks you weren't able to get much reaction sometimes, but you were still there talking to your wife and she was still in there. Mom was still in the house. . . . If you're in love with somebody, you can look right past the silent stare, if that's what you're getting sometimes.

"As things went away, I know what she cares about. She cares about living and she cares about loving. The people she loves are here with her," he said. "I would not let, in my mind, Carol be anything less than she always was."

Carol died after midnight but well before dawn one morning in the fall. She was lying on the couch in front of the sunroom windows, under a full moon that shone on the water. Mead, sleeping upstairs, sat up in bed and felt her tap him on the shoulder and say "Goodbye." The nurse attending Carol called him downstairs a little later, and by the time he was by her side she was still and waxy. The family gathered and said the Lord's Prayer.

A disinterested reader isn't obliged to accept Mead's word in these matters. He'd hardly be human if he weren't biased. But the point is that we're all biased, because everyone who loves eventually loses the beloved. We're all endowed with this paradox of an organ that allows us to travel in time and project ourselves into the minds of wives and ravens, but that also is material, ultimately made of food and destined to become food again. Disease deconstructed Carol's brain; but, as Mead pointed out, all adults lose brain every day, as mental experience narrows imperceptibly but inexorably with age. And sometimes age also brings wisdom, a consolidated core of knowledge that explains many things at once, as if the mental instrument were enhanced while diminishing. From a distance, we're purely paradox: a physical being, by nature divisible and perishable, from which emerges an indivisible person.

Down in the heart of living, on the other hand, the human essence isn't so difficult to grasp. Amid the extremes of death, birth, sex, or the forces of the ocean, mind and body can collapse into one, a unity of nature and spirit. At these times, it's clear that the essential quality of oneself does not depend on a single organ or even a single organism. We are born in society with others of our kind and with other living things, and with forces that seem alive—the tide and the rain—and we grow within that matrix like cells in the veins of a tree, subsisting on the nourishment of our neighbors. Mead knew Carol, and he wouldn't let her die. Likewise, he fought a plan to clear-cut the forest around Loonsong Lodge. We gain meaning as part of a system of life and give meaning to the places and people we care about.

Connections make the person. That's the truth that emerges when all else is stripped away by disease or, simply, by a moment of clarity. And these connections aren't only to other human beings. A sense of kinship to animals or a link to certain places can lend a sense of identity and self more durable than any personal relationship. Connections to wild nature are essential to the human animal. Wild places fulfill ancient capacities for belonging. They help make sense of one's own existence. And wilderness matters not only when it is useful or close at hand. To be whole, we need the sea and its animals, even a dark kelp forest under storm waves on an autumn night in the Gulf of Alaska, where no human holds sway. The spirit of vast, unruly creativity ultimately connects everyone.

7

Connected or Alone

A scientist at Oxford University has invented a way of scanning a brain while paring it away in very thin slices, the first step to downloading the content into a computer, where the thoughts, memories, and emotions of the former person could live on forever. I doubt it's possible for a person to live in a computer, because the content of the mind grows from the peculiar biology of the body, but someday a new species may arise from such an experiment, a hybrid of a machine and the mental shadows lifted from a body that has died. What would it be like to be that creature, able to remember breathing, birthing, and loving, but no longer able to do anything but dream? Existence itself would be a dream. Perhaps computer minds could experience constant orgasms or swim freely among whales. One might be able to do anything, like an avatar in one of those online virtual worlds. This fantasy has some familiarity for anyone living partly in cyberspace. And so it seems likely that, just as people feel today, computer minds would crave authenticity above all else, the unmediated contact with the sensuous, unpredictable real.

I doubt the ancient Chugach people ever felt that craving. Were they more real? Certainly, the brain-shaving future represents a way of living that is the opposite of their lives. On one side, an individual could imagine himself entirely disconnected from the body, existing in isolation as the boiled-down essence of a self-conscious ego. On the other side, a culture taught that the connection of mind and body extends the self to include the animals, waters, and forces of the natural world. In real life, a person could live anywhere on a continuum between those extremes, neither entirely egoistic nor fully unified in body, spirit, and community. But, for a moment, let's consider only the extremes. I'd like to know what "I" means, and, with that meaning, understand how it fits us into the rest of the world.

A voice of commentary seems to speak in your head, like the narrator of a film, saying, perhaps, "After I read the next sentence I will get a snack." That sort of internal monolog, a key quality of consciousness, often asserts that it is the rational core of the self. It seems in command of reasoning and intention, the boss of brain and body. Re-creating such a monolog in a computer would presumably amount to success for a brain-shaving machine. That voice lives in a world of ego. All objects are separate from itself. The monolog's sentences begin with "I" as the subject, followed by the verbs of its actions, and finally the objects upon which it acts (the snack I will get). This self is its own ultimate good, and it is ultimately alone. It can get along with others, negotiating and trading with them according to their respective power and wealth, but it always places people, like things, as objects in its sentences: only the self can be "I."

If you were to build a society around this individualistic self—as Western culture began doing during the eighteenth century's Enlightenment— you would need a way for these self-interested executive beings to work together without robbing or killing each other. An all-powerful monarch could force them to get along, or they could enter into contracts to establish rules for a safe and productive playing field. The U.S. Constitution is such a contract. Americans' rights come from those written words, subject to amendment or revocation according to a specific process, and applicable only to those who are party to the agreement. Currently, rights apply to born or naturalized adult citizens, but in the past they were only for males, whites, nonindigenous people, and so on, and still you need proper papers to prove who you are in order to have rights. This arrangement represents a philosophy now called legal positivism, which says law derives authority from the agreement of those who are party to it.

A society built on a legal positivist contract will not necessarily destroy the biosphere—its members might reach an agreement to prevent their own demise—but it is unlikely to leave significant resources in the control of free members of other species. Those species have no part in the contract. This is an urgent point. The chief scientist for the Nature Conservancy, Peter Kareiva, has argued that a fully domesticated biosphere is already inevitable, and that the best role for conservationists now is not to fight that outcome, but to direct the final dismantlement of wild nature thoughtfully, without waste. We would keep preserves for study and we'd protect natural systems that provided certain needs better than technology yet could manage.

Kareiva told me over a beer that he personally values nature differently—to him, it has a special, transcendent goodness—but in his work he aims to speak on a basis that the broader society can understand, the basis of materialism. People will pay for special places and charismatic animals because ecosystems provide material benefits, such as cleansing water or air, or because they contain objects of aesthetic and psychological enjoyment. But as human needs expand, such arguments are bound to fail, because they put a price on the survival of undomesticated animal lives against other goods in the marketplace. People have been generous at times in setting aside resources for wilderness. Forgoing those resources amounts to buying nature from the economy. But with each generation the negotiation starts over again. The price of nature rises as the planet gets smaller. One generation might lose only half, but the next loses half again, and the next half of that. An environmental victory on economic terms lasts only until there is more money at stake; since the economy always grows, the threat always escalates.

Against such a downward trajectory there is this hope: human nature isn't really that way at all. The egotistical internal monolog of consciousness is not the center of the self and social systems need not use it as a basis. Science more and more finds the mind emerging as a manifestation of the body's systems, not a free-floating entity. The monolog may be no more than a baton passing between parts of the brain where real work is being done, biologically, in the intuitive workings of experience. Much of our most important thinking occurs without awareness, followed seconds later by the conscious mind concocting reasons for what we've already decided. Brain scientists stimulating the parietal cortex have produced in their patients a conscious will to act that the patients sense as having initiated themselves.

Once mind and body are reconnected, the connection to the rest of the world is easy enough to see. It's unavoidable for an animal like us. We're physical, social beings, defined by the strands of affiliation that tie us to the world. The "I" of a person is always complicated, almost a "we." Every sentence in the thought monolog emerges in the context of many other story lines, and none writes itself. The ego's subject-versus-object isolation constantly entangles with other influences: the grammar of life is undiagrammable. In the mixture, some objects, parts of ourselves, gain intrinsic worth beyond any relative valuing in the economic marketplace. For example, the love for a child or mate, or love for a place or an ecosystem. The divisions fail completely at life's extremes, such as at Carol Treadwell's death. Humanity intermixes mind and body, self and other, both of our own kind and other living things, and of the connections rooted in rock and water.

This other, more ancient idea of human nature spawned its own law, which is different from legal positivism. The natural law rests on what is commonly called justice. It exists outside ourselves; it is universal; it can be perceived by human minds, but cannot be changed. According to Cicero, the Roman philosopher,

> Of all the questions which are ever the subject of discussion among learned men, there is none which it is more important thoroughly to understand than this, that man is born for justice, and that law and equity have not been established by opinion, but by nature ... for law is the highest reason implanted in nature, which prescribes those things which ought to be done, and forbids the contrary.

Cicero means by *nature* its magnificent forces and bounty that contain and feed us; the essential similarity of all men, even from different countries; and the tendency of people everywhere to gather together for common benefit. For proof, he points to virtue—the many selfless acts people do, such as the ethical treatment of one stranger to another even when far beyond the reach of punishment or social reputation. A sailor going to the aid of a vessel in distress, for example, and bringing it to safety rather than plundering its cargo. The goodness of such virtuous conduct, Cicero said, is unaffected by others' opinions of whether or not it is good; it is good in itself, like a tree or a horse. One knows virtue by respect for its honorable quality, while bad actions engender opposite emotions. All points that seem so simple as to be archaic, if not naïve, except that two thousand years later, Cicero has found solid scientific support. Psychologists and neurologists find that emotions such as disgust do drive moral choices far more than mental calculation does, and economists consistently find people prefer cooperation over personal gain, even in experiments with real money at stake.

For a connected human self, one who loves good, like Cicero, the continued existence of other untamed species is not just another want that can be priced alongside new cars or steak dinners. Wild nature is indispensible for our survival as integrated beings, necessary for our human dignity, and an essential fiber in the web of who we are. To destroy the oceans and the rest of the natural world from which our selves and goodness partly derive would be a shameful act, analogous to casting aside our bodies and living in computer chips. Many people feel that way. For them, the prospect of the end of wild places carries not only sadness but also a sense of personal loss.

Human physicality relates us to the natural world in a way that renders living oceans and wilderness intrinsically valuable, like family or other things that shouldn't be sold. So says one tradition. Another defines values more relatively, putting prices on all species according to individual wishes. On the coast of the Gulf of Alaska, for more than 250 years, people with each worldview have claimed the same ground and water. That conflict continues, over culture and over this shoreline, a microcosm of the world ecosystem and the people who will decide its fate.

Part II

THE ENLIGHTENMENT
COMES TO ALASKA

8.

How Captain Cook Saw Alaska

In the spring of 1778, Captain James Cook looked for a harbor along the sandy, stormy, mountainous coast of the eastern Gulf of Alaska, searching for a place where he could tip the larger of his two vessels, the *Resolution*, so his carpenter could fix a leak. That eastern coast offers little shelter and Cook couldn't make much sense of the maps and descriptions of earlier Russian explorers. On May 12, as storm winds rose from the east, an opening in the shore six miles wide beckoned to the north. Cook steered that way, hoping for shelter and maybe a shortcut to the Bering Sea and then the Northwest Passage, naming the cape he passed Hinchinbrook as a thick fog descended. The ships felt their way around the end of Hinchinbrook Island into a long, wide bay relatively protected from the wind—now called Port Etches—and dropped anchor in the first cove encountered, barely an indentation with a steep shore, now called English Bay (I use all modern names here).

John Gore, the *Resolution*'s first lieutenant, took a small boat to hunt for dinner near some tall rocks that stand in the middle of Port Etches, but before he could get there two large skin boats of Natives approached and he fled back to the ship. But these people seemed friendly enough, floating alongside the *Resolution*—one danced and sang, stark naked, while others wore long coats of fur and offered lengthy speeches the Englishmen could not understand. They tossed a staff aboard the ship, decorated with feathers, that seemed to be an offering of peace. Captain Cook sent it back with a piece of iron wire attached, which a Native man appreciatively made into a bracelet, tossing the staff back again for Cook, as well as a gift of a well-made raincoat of waterfowl skins. Without boarding, the Natives paddled away about two hours later, suggesting by pantomime they would be back again the next day. They were likely from Nuchek, like Makari, as it lay just across Port Etches. But the village never was seen by Cook, who departed when the fog lifted the next morning for deeper in the sound.

Two worlds had met, curiously and without alarm, utterly different in experience and outlook, but not as different from each other materially as both are from us. Ordinary people in both cultures killed the animals they ate. Both obtained heat and light from flames they tended. Neither had a toilet or a telephone, or engines for propulsion, or help for dying children. Cook's ships may have intimidated with higher technology and more powerful social organization, but the Chugach people made better boats—Russian fur traders adopted their skin craft rather than wooden boats—and they knew better how to live richly in the place where they found themselves. The people of Hinchinbrook Island accumulated great wealth. One of their otter skin coats could have sold in Asia for more than a British sailor could ever dream of earning. Cook himself grew up poorer than a Native child would in the sound, born in a clay hut with a thatch roof in the north of England, son of a day laborer, with only two siblings who survived childhood out of six born.

At forty-nine, however, as he sailed into the sound, Cook's previous explorations had already made him world-famous. Other explorers had visited Alaska and crossed the Pacific, but Cook first pinned down the ocean scientifically on charts and in logs and images, fixing the great unknown into a replicable grid of published knowledge. True science was being born in Europe's Enlightenment, and Cook emerged as its brave and methodical hero. After his death he grew into a mythic symbol of the power and individualism of the new age. In life, he was simpler: a dogged pursuer of accurate physical facts who, by measuring the world, expanded the culture he represented. That role provided no shared point of reference with the people he found in Prince William Sound. The moment of his arrival began a conflict between two realities.

Cook's early education was rudimentary, but he learned enough to try out as a grocery clerk in the seaside town of Staithes, which led to a teenage apprenticeship on a ship carrying coal along the North Sea coast. A decade working up through the ranks in that trade brought him an offer to command a large merchant ship, which, incredibly, he turned down in order to volunteer as an able-bodied seaman in the Royal Navy, a position more commonly filled by the press gangs' kidnapped draftees. Cook's personality is there, in that decision, which his contemporaries must have thought utterly irrational—to quit the top of an easier, safer, and more profitable service in order to join at the bottom of one known for danger and brutality. He must have believed he could rise again, this time on a taller ladder.

In two years the admiralty promoted Cook to master, the rank of the Royal Navy's highest noncommissioned officer, responsible for all of a ship's workaday details. In that role he sailed to Canada during the French and Indian War, where he met a military surveyor and other men with technical knowledge and learned celestial navigation, trigonometry, and the use of optical instruments to make accurate maps and charts. The quality of his charting brought him command of a small ship surveying the coast of Newfoundland, where more exceptional mapmaking put him in line as a reasonable candidate to command a much more important scientific expedition. That voyage, sponsored by the British navy and the scientific Royal Society, bound for Tahiti, would carry gentlemen scientists to measure the transit of Venus across the face of the sun in order to learn the sun's distance from the earth.

Cook accomplished his assigned task; learned from his passengers more mathematics, astronomy, biology, ethnography, and geography; and charted much of the South Pacific, returning to England as one its most successful explorers, now promoted to captain. His two succeeding voyages, during which he mapped the Pacific and the Southern oceans and studied the people there, made him among the most important explorers of all time, the man who completed the map of the world. In the decades after his death great and minor artists enlarged Cook's heroism to godlike proportions—for example, Samuel Coleridge, who followed his Antarctic journals phrase by phrase in creating "The Rime of the Ancient Mariner."

The discovery that most impressed Cook's contemporaries, coming on his second voyage, was empty water. Europe's geographers believed a vast and luxuriant continent of rich people lay at the southern end of the world, hardly farther south of the equator than Spain was to its north. Cook sailed south of the Antarctic Circle among the storms and icebergs of the Southern Ocean, circling that end of the world, his sailors, without gloves, working lines encased in ice. The empty water amazed Europe. Oddly, from today's perspective. The more amazing fact now is that so many learned people believed in a fantasy continent in the first place. Classical philosophers had planted the delusion by speculating that the bottom of the planet would need extra weight; then a careless translator added to it, misinterpreting the writings of Marco Polo to place the lush Malay Peninsula far to the south; and finally hopeful sailors contributed with a few ambiguous sightings of islands and cloud banks. On that evidence, cartographers confidently drew the southern continent on world maps.

We think differently today. We can hardly conceive of "theoretical geographers." Cook gave this pragmatic view to us: to see things as they are. He went and looked, he measured carefully, and he wrote clearly and simply of what he found. He was lucky, too, as the first explorer to carry navigational technology that allowed him to fix exactly where he was when he encountered land or the lack thereof, including the chronometer and new mathematical astronomy to find longitude. Nothing seems more obvious today than to use technology to record reality objectively—that's how encompassed we are in Cook's paradigm.

When he reached Alaska, however, the simplicity of deductive reasoning began to fail Captain Cook. It's easier to prove what doesn't exist than to understand a complex reality that does. Cook hadn't expected to find the southern continent, but he did believe a Northwest Passage led over the top of North America, and he put great weight on an imaginative Russian map he thought would lead him to it. Some writers say he was fatigued, ill, or even going mad on this final voyage, others say he was behaving normally; so many have reinterpreted Cook that the man himself is lost in their many prisms. His own journals show, however, that as the fog lifted from Port Etches and he sailed into the heart of Prince William Sound, Cook's curiosity failed him. The Nuchek people who had planned to meet him that morning chased behind in their boats but couldn't catch the *Resolution* and the *Discovery*. Although they were unlike any Native people he had met before—their decoration and boats looked like Greenland Eskimos in a book he carried—Cook was more interested in making distance to the north in pursuit of the route shown on the map. And although passing through an inland sea so broad it took an entire day to cross, Cook noted only the weather, not the myriad islands encircling the distant foreground, the passages between them, scores of miles long, that opened to view like corridors as he passed, or the outermost circumference, white mountains surrounding all. George Vancouver, who was aboard the *Discovery* that day and returned sixteen years later as a captain himself to fill in the map, wrote that Cook ought to have been especially interested in this extraordinary place, but instead he was uncharacteristically negligent, taking little of the usual geographic information—the location of islands and rocks, the shape and topography of the shore, the latitude and longitude of land and ship positions, compass readings, or tides—even missing an important island Vancouver himself remembered. The resulting chart was sketchy and some important information it did contain was wrong. The logs of officers other than Cook were more detailed and vivid. It's as if Cook, researcher of the world, suddenly was overwhelmed by what he had found and turned away. In his journal he was later halfway apologetic, exaggerating the sound's enormous size by way of excuse.

Across the sound to the north the ships rounded Goose Island and entered the twenty-mile-long fjord of Port Fidalgo, anchoring in its nearest of many branching harbors, which Cook named Snug Corner Cove, a deep V of smooth water penetrating into an amphitheater of steep little mountains. They were in the territory of Tatitlek, whose present-day village lies eight miles across the port. A storm that had kicked up in the afternoon made Cook glad he had found such a protected spot, but the weather wasn't bad enough to stop three Chugach men from approaching in two kayaks to inspect the visitors' wooden ships. The English were seeing kayaks for the first time, too.

The next morning a couple of large skin boats came out, tried to trade, found they had more of value to exchange than the Englishmen, and then pursued and seized one of the *Resolution*'s boats while under the gaze of a hundred sailors. The English, quite surprised, threatened force, where-upon the Chugachmiut returned the boat with friendly gestures. They then paddled to the *Discovery*, where most of the crew were belowdecks eating, climbed aboard, and overcame the men on watch, holding them at knife-point while grabbing anything handy to toss into their boats, including a rudder. The English rushed up the companionway with their cutlasses and took back the ship without anyone getting hurt, and again the Tatitlek Natives retreated, nodding apologetically. Which didn't stop them from a third attempt on an unoccupied boat. Lieutenant James King wrote, "The old chief with the greatest composure beckond us to lay down our Arms; & he to Mollify us began to dance & sing, stripping himself Nak'd & throwing round him a garment with pieces of hoofs that made a noise as he shook it about, & afterwards continued trading with us, as if nothing had happened."

Cook hadn't encountered such audacity in all his travels and he was bewildered, as anyone is who finds himself under an unfamiliar legal sys-tem. Chugach villages held most property collectively: whatever lay within their domain, their islands, peninsulas and bays, belonged to them to-gether, to be protected by force (as modern people have armed police to protect their property). Even after a hunter harvested meat it remained collective property: the village shared equally in any whale, sea lion, or other animal, except for indivisible parts, such as sea otter pelts, which had their own special property rules. Individuals did own personal tools, clothing, and such, but that kind of ownership didn't mean so much. Tak-ing from a neighbor concerned only the two people involved, to be settled simply by the return of the item taken. The community's interest was in peace, and maintaining harmony usually outweighed the desire to punish. Only women could own houses and large boats—things used by many people, as opposed to kayaks, which could only fit one or two—and Cook noticed that women did command, although Makari said husbands steered their wives' boats into battle. The *Resolution* and the *Discovery* and their boats fell neatly within the Tatitlek villagers' rights to take, if they could keep them.

Cook understood none of this and interpreted the incident in light of an entirely different code. The important point to him was that the Natives seemed to have no fear of guns, suggesting they had not encountered Europeans or Russians before. If so, he was the discoverer of this territory and could claim it for England. Legal positivism. The concept had developed, conveniently, with other intellectual changes of the Enlightenment, as sailing nations were accumulating colonies around the world. When Christopher Columbus found America in 1492, Europeans in power believed all the earth belonged to God, leaving it up to the pope, in 1493, to decide which nations would control the New World and bring the gospel to its people. After a little negotiation, Spain got the North Atlantic, the Pacific, and the Americas, while Portugal would own Africa, the Indian Ocean, and all that. The precedent stood until challenged in the court of public opinion in 1608 by a Dutch lawyer, Hugo Grotius, on behalf of the Dutch East Indian Company, which had seized a Portuguese ship in the Strait of Mallaca, off Java—waters the pope had awarded to Portugal. Grotius argued, among other points, that the pope had given Portugal what he didn't own, because the God the pope represented had already granted a quitclaim deed to the physical universe. "Our Lord Jesus Christ when he said, 'My kingdom is not of this world,' thereby renounced all earthly power," Grotius wrote. "If the Pope has any power at all, he has it, as they say, in the spiritual realm only. Therefore he has no authority over infidel nations, for they do not belong to the Church."

Without a religious basis for the law, Grotius cited the same authority for justice used by Cicero. He turned to the idea of natural law, the sense of right and wrong that he supposed God had built into all people. When it came to empty land or other things that no one owned, he argued that natural law followed the Roman legal tradition of *res nullius,* known in the schoolyard as "finders, keepers." Quoting Seneca, Grotius compared finding a new land to taking a seat in a theater—until taken, seats are open to all, but once you're sitting in one, it's your personal place. Unless you get up and leave. Possession matters, too. Slipping out briefly to the popcorn counter doesn't relinquish your original claim on a seat, but let it go cold for too long and someone else could legitimately grab it.

All this assumes, of course, that the seat, or the land, is empty. If the Chugach people had known what Cook had in mind, they surely would have fought harder, the way they fought for the kind of property that really counted.

Cook wanted the sound as a potential passageway for ships, nothing more. His men heeled the *Resolution* over in Snug Corner Cove for the carpenter to remove the metal sheathing from the hull and stop the leaks. But in all the idle time while that work was going on, he didn't bother to send someone ashore to look at the Tatitlek village. No literate person recorded much about their culture until Birket-Smith arrived 150 years later—as distant in time from that first encounter as we are from the American Civil War.

Cook remained fixated instead on a Russian map published in 1774 by Jakob von Stählin, which showed all of western Alaska as a collection of scattered islands. Cook thought he could sail through those islands to get to the Arctic and hoped the sound was a channel. But Stählin's map was largely fanciful, in the tradition of the imaginative cartography that Cook had worked doggedly to dispel in the Southern Ocean. The space where Alaska lies had been filled by many such daydreams over the centuries. Somehow, the once skeptical Cook now believed. He needed a map. Faced with an intricate coast and limited time to find a passage, his strategy lay in simplifying the task—cutting away its complexity—stripping it, as it turns out, of much of the real interest. The trade-off of reductionism: exchanging the knowledge of the whole for focused knowledge of one discrete part.

The *Resolution* and the *Discovery* sailed west from Snug Corner Cove for the wide opening to the north that might be a passage, but were stopped by rocks and currents near Bligh Reef, the same obstruction the tanker *Exxon Valdez* hit two centuries later. (The reef got its name from the nearby island, which Vancouver later named after the *Resolution*'s unpleasant master, William Bligh, after Bligh had become a captain and was set adrift in the south seas by Fletcher Christian and his mutineers aboard the HMS *Bounty*.) From that spot, two possible passages were in view, one north and one east. Cook sent Lieutenant Gore northward with two boats and Bligh eastward with two more. Bligh's route turned out to be Tatitlek Narrows, leading back around toward Snug Corner Cove. Gore, on the other hand, found a wide, deep channel whose shores rose straight to mountaintops, like an enormous hall, which he could not see the end of after a full day on the water. He excitedly believed he had found the path. Another, two-day journey was planned. But a junior officer who had gone with Gore wasn't so sure the opening didn't dead-end (in fact, it was the fjord of Valdez Narrows and Port Valdez). When the morning came, the wind was fair to sail south, out of the sound, and Cook took it. Gore's route, Cook reasoned, might or, more probably, might not be the way through, but so might innumerable other routes in the sound, to explore all of which could take the entire summer season.

A week later, Gore's enthusiasm for finding a passage north overcame Cook's better judgment and the ships entered an inlet 150 miles long and 20 to 60 miles wide—large enough to show on any globe—which exploring required sixteen days riding and fighting ferocious tidal currents, threading mud banks and grounding on them, and giving up hope of success before the end, a waste of time, perhaps of no use to anyone, just a "trifling point of geography" that Cook left behind in irritation without even giving it a name. The admiralty later named it for him, Cook Inlet. The inlet nearly touches Prince William Sound, but differs, with a washed-out flatness and a bigger, paler sky, unlike the sound's gothic catacombs of dark-treed passages and deep, lived-in waters. Glacier-milled rock flour colors the inlet's water milky—the eye can read it as solid—a swirling tableland that melds into the shore's silty banks, themselves half liquid and some-times quicksand. Cook inferred from this dirty shade, from the inlet's enormous tides and the continental mountains around it, that no passage to the north led through here. But Gore, whose will was stronger than his intellect, insisted on being sure, and Cook knew the theoretical geogra-phers would never rest unless Gore's notion were disproven. So they sailed up, drifted up on the tide, past Port Graham, past Kachemak Bay, past the volcanoes, the Forelands, up to where the inlet branches in two arms—one, where Anchorage now lies, viewed with small boats by Bligh, and the other, the silt-clogged cauldron of Turnagain Arm, partly observed by King.

Here, after giving up exploring the inlet, at a place he called Point Pos-session, Cook put Lieutenant King ashore with two boats of men to claim the land for England, barely willing to wait to sail onward, to the Arctic and its impassable ice, and then back to Hawaii and his death at the hands of Natives there. The point is low and flat, a strip of tape edging the anonymous gray vastness of the inlet, which shimmers like spilled water-color paint. A fit stage for the absurdist drama of King's rite of posses-sion. A dozen Natives met King's two boats on the beach and, after the usual demonstrations that neither side intended harm, tried unsuccess-fully to communicate with him, another officer, and James Law, the *Dis-covery*'s surgeon. King writes, "After some time we returned to our Party & performd the Ceremony of taking Possession, by hoisting Colours &c.; & by drinking his Majestys health in good English Porter, by us, as well as by three of the Natives who repeated what we said; & what we did not expect, were fond of the Liquor; they had also the empty Bottles; we contriv'd to place a bottle that the Captⁿ had given us (with a Parchment Scroll in it) not in a conspicuous open place, for that the Natives would find, but under some rocks by the Side of a Stuntd tree, where if it escapes the Indians, in many ages hence it may Puzle Antiquarians." (No one ever did find it.)

The Englishmen walked up a hill to obtain grass for their goat, alarm-ing the Natives, who grabbed their weapons and put on their wooden ar-mor, scurrying, hidden among the trees. King was intimidated and decided not to try to visit their village, but the tension eased as he walked back toward the beach and the two groups again fell in together. Dr. Law bought a dog from one of the Natives in exchange for two buckles. Then, without warning or real reason, he shot the dog in the head. King watched the reaction of one old man, who "Eyed the Dog, his companion, & us alternately, & without speaking a word retir'd sideways as did the others; & when they got some way from us they quicken'd their pace; nor would all our Shouting or laying down of Arms avail, not one would come near us."

It's often given as an excuse for European and American theft of Native American lands—and the ownership of slaves, as well—that one cannot judge those actions out of the context of their culture. But Cook, his officers, and many literate people of the eighteenth century were well aware of the ethics of taking indigenous property. Francisco de Vitoria, an important Spanish theologian and jurist, asserted Native land rights as early as 1532. Grotius's natural law held that all lands, if occupied, must already have been explored by those living on them, and so none could be possessed by right of discovery. He also discounted the right of possession by war because making war without cause was unjust—it was really theft by another name. When Cook left England on his first voyage, the president of the Royal Society, James Douglas, fourteenth earl of Morton, advised him to treat the Natives as "Lords of their own country," for "They are the natural, and in the strictest sense of the word, the legal possessors of the several Regions they inhabit."

Cook himself came to regret his voyages' curse on Pacific peoples, largely because of the venereal disease his sailors spread—almost everyone but Cook himself was guilty, it sometimes seems—shattering paradise on island after island. He also blamed his crew for the new practice of some Maori men to violently force their wives and daughters into prostitution in exchange for bits of iron. "We debauch their Morals already too prone to vice and we interduce among them wants and perhaps diseases which they never before knew and which serves only to disturb that happy tranquility they and their fore Fathers had injoy'd. If any one denies the truth of this assertion let him tell me what the Natives of the whole extent of

America have gained by the commerce they had with Europeans." So Cook wrote in his journal, intended for publication. But his editor deleted that passage before contemporary European readers could see it. To include those words would have put readers in Cook's unenviable position, of perceiving himself as the carrier of a destructive contagion and themselves as the unthinking biological agents of disease.

Besides, a new kind of law dismissed concerns such as those Cook expressed for the Natives and their ruined paradise. When Grotius had set aside the earthly authority of an external God, he turned instead to human nature's internal sense of justice as the basis of law. But Thomas Hobbes, writing forty years later, advanced another version of human nature: that it consists of crude self-interest, that the strong always seek to dominate the weak, and that competition and conflict are the essential condition of life. Without an all-powerful sovereign to impose a system of justice, mankind always clawed violently for survival and supremacy and nothing but destruction and terror could result. Hobbes wrote, ignorantly, that Native Americans "live in that brutish manner" of anarchy, except "small Families, the concord whereof dependeth on naturall lust." It could not be unjust to conquer such people, or for any state to make war against another, since justice came from laws, and no ultimate sovereign existed to impose laws on kings. Hobbes wrote:

> To this warre of every man against every man, this also is consequent: that nothing can be Unjust. The notions of Right and Wrong, Justice and Injustice have no place. Where there is no common Power, there is no Law: where no Law, no Injustice. Force, and Fraud, are in warre the two Cardinall virtues.

Hobbes's view of humanity prevailed over Grotius's natural law and introduced the subjective individualism that is the rationale for the legal system that we live under today. On a battlefield of savage conflict, members of a social group grant rights to life and property to one another through a mutual agreement, a kind of peace treaty. That is the essence of legal positivism and a theme of the political science that led to the drafting of the U.S. Constitution. The only right everyone is born with, by this thinking, is the right to self-preservation—the right to fight for your own life, not to expect help from anyone else. Those excluded from the law's alliance—including all indigenous peoples—remained exposed to the state of war, without legal defense against theft of their land or freedom. That kind of law allowed the taking of America and remains today as the justification for the conquest of the final hidden coves of the wild ocean. The powerful are free to take creatures and habitat that remain unowned, and what is already property can be wasted or consumed as owners see fit.

Cook's doctor was perfectly within his rights to shoot the dog. And the old man was wise to run.

9.

Rethinking the Moon

A reader might ask why intellectual history matters. After all, few people without a bachelor's degree have ever heard of Grotius or Hobbes. Among those who do know these writers, few call them to mind when deciding whether to recycle a plastic water bottle, throw it away, or toss it in the ocean—or voyage to remote Alaska shores to gather up bottles thrown overboard by others. Nonetheless, our habits of thought guide these and other choices in the channels created by our tradition. We're taught how to see the world, how to divide the reasonable from the unreasonable—and those lessons had to be created by someone. For example, Hobbes's idea that human beings are, by nature, competitive. The world economy is built on this idea. But is it true? Or is it a model Hobbes and his heirs advanced that became reality through our customs? Like children growing up in the corridors of a single building, it's up to us to consider if our world could have been designed differently.

One of the challenges of examining how we think is the burden of working from within the same mental framework that is itself the subject of the inquiry. Take, for example, the word "truth." Before the Enlightenment truth came from sacred and classical texts, tradition, social hierarchy, and the imagination. But Cook and other scientists of the seventeenth and eighteenth centuries instead chose to see things as they are—to take truth from nature in a way that anyone could perceive the same way. No more theoretical geographers. No going back. Some do try, but the retreat into faith, subjective truth, produces discretionary beliefs useless in measuring the physical world, which is the only world we're still certain of sharing.

This afternoon I skied from a dim tunnel of snowy birch trees to the shore of Cook Inlet, where open water rippled orange with the sunset between ice sheets speeding silently northward with the incoming tide, and the sky was luminous, gradations of deep blue, and the crescent moon was as green as a brushstroke. Children often believe the moon follows them through the sky. Humility comes with the realization that it does not. A sliver of moon can turn the inlet into a river, driving heavy ice that carries away ships. This power, seen for what it is, shrinks a child from the godlike center of the universe down to a midsized mammal, helpless in the cold water. A trade of magic for fact. But the trade is a good one, because in the process, the universe becomes infinitely larger and more interesting, unfathomable but intelligible.

Chugach people used the tide to tell time. The moon was a powerful spirit, a man named Kaliq who went into the sky over his frustration at not being able to get up early in the morning to hunt. We don't know if Kaliq affected the sound's tides—their growth with the full and new moon—but it's hard to believe such a dramatic effect would go unnoticed. Ancient people around the world who did notice the dance of moon and water often explained it by animal or human analogies, anyway—the planet or moon themselves breathing, for example. Philosophers beginning with the Greeks advanced partly physical theories, and by the time of Saint Thomas Aquinas, in the thirteenth century, every writer explaining nature had to address the tides, in some cases producing mechanisms that now seem bizarre. Systems based on the earth standing at the center of the universe were bound to fail. Anyway, theory was irrelevant to practice: sailors had handy rule-of-thumb methods of predicting the tides based on the phase of the moon that worked far better than anything the philosophers could offer, as Chaucer mentioned in *The Canterbury Tales* in 1386. But the old truths didn't give up easily. In

1600, church leaders objected to an astronomer's contention that the sun was a close star circled by the earth—they burned him at the stake.

Galileo also was arrested, but not killed, in 1633, after he advanced the first model of the tides that somewhat resembled reality. A series of intellectual giants who followed produced tidal theories using the new astronomy, including Kepler and Descartes. The great breakthrough came in 1687 with Newton's *Principia Mathematica,* which defined much of our mathematics and physics, directed in significant part at solving the problem of tides. Other geniuses refined and extended Newton's work—Euler, Bernoulli, Halley, Cassini, Laplace—and extensive campaigns of empirical measurements buttressed and supplemented their complex mathematics. By the late nineteenth century the math had become too hard and time-consuming to do by hand and governments built mechanical computers with hundreds of gears, belts, and dials to work out the equations and make tide predictions, machines that grew to the size of large rooms and continued in use until supplanted by electronic computers in the mid-1960s.

I think now of the tide book that lives in my back pocket all summer, the ink wearing off its cover, the paper conforming to the shape of my body, and the columns of figures on each page, dates that approach with promise, actualize, then recede in memory—the times of high and low, marching forward about forty-eight minutes a day, and the size of each tide, listed in feet above or below mean low water, breathing in and out twice a week as the flood and ebb do twice a day. The tide is the ordering principle of life on our beach on Kachemak Bay—the size of our outdoor space, our ability to come and go, the animals we can find, food we can harvest, times we can swim, row, or walk. This book is our oracle, never wrong, but leaving out so much: the waves, the sky, snow, rain, and wind, or the warmth and stillness, and our emotions, cares, what we'll want when the water changes direction. No—sometimes wrong, but rarely. There was the time, in the spring of 1964, when my wife's family (she was an infant at the time) was bewildered by water flowing into their camp far up among mossy trees of western Prince William Sound—the great earthquake that shifted the region that March had sunk the land itself and they were present for one of the sea's first incursions. The exception proves how these mathematical calculations synchronize our lives.

Although science became central to living, the mystery never evaporated. *Apollo* astronauts left mirrors on the moon that allow precise measurement of its distance by shining a laser from Earth and calculating the time it takes that light to get to the mirror and back. After measuring through a full 18.63-year tidal cycle, scientists established in 1989 the rate at which the moon is slowing down in its orbit as a result of energy spent in the work of moving the oceans. But even with that level of knowledge, tidal predictions remain beyond the scope of pure theory and mathematics. Tides are influenced by the shape of the shore and the sea floor, the water's depth, and resonant frequencies within bays and inlets that act like the pipes in an organ; and influenced, too, by changeable factors, including density differences in the water and pressure differences in the air. Equations describe the tides, but getting the predictions right for any particular place still requires long-term measurements of the rise and fall of the water. Scientific truth differs in that way from the religious truth it superseded. Instead of a God omniscient of everything, large and small, but inscrutable, this kind of truth yields practical rules through effort and the shared work of brilliant minds, rules that approximate reality but must be constantly amended and refined and can never fully dictate the details of life, in which infinite degrees of freedom confound any prediction.

Does that sort of truth help us understand ourselves? Walker Percy, the southern Catholic novelist, thought not:

> Every advance in an objective understanding of the Cosmos
> and in its technological control further distances the self
> from the Cosmos precisely in the degree of the advance—so
> that in the end the self becomes a space-bound ghost which
> roams the very Cosmos it understands perfectly.

This is a bigger version of the child's loss of the illusion of the following moon—he's concerned about something larger than trading magic for fact. Percy critiqued materialism as the modern self attempting hopelessly to reconstruct itself by accumulating meaning through things—treasuring, for example, eighteenth-century antiques as a tangible link to something real, which he said no eighteenth-century citizen would have felt about sixteenth-century furniture. Writing in 1983, Percy's own point was something of an antique, carrying on a rearguard fight begun in 1726 by Jonathan Swift with *Gulliver's Travels,* a book conceived as a satire on the self-importance and shallowness of Enlightenment science and politics. Swift imagined mankind as a low and sordid animal elevated only by humble obedience to universal truth; Percy described a futile modern search for a self which, without God, is a vacuum. They were on the same side, but the literature they produced reflects the basic difference in the cultures in which they lived—each with a different approach to exploring what it means to be a person.

English literature before the eighteenth century wasn't about self or personal development at all, and when it seems to be, that's usually because, as modern readers, we bring our presuppositions to writing that was created in an entirely different world, one that predated the entire idea of psychology. Instead of creating internally unique individuals, writers made characters that reflected types and classes, fulfilling roles in a morally certain landscape. Human nature was known. Personality was determined by the mix of the four humors, a prescientific medical theory dating from the ancient Greeks. The course of a narrative or a play reflected allegory, classical themes, or another established path, not surprises or revelations—Shakespeare's audiences knew his tragedies would end with the main characters dead and his comedies would end with weddings. These works were born of a spiritually unified universe similar to the Chugach people's. When Swift sent Gulliver on his strange adventures, he was confident readers would understand the satire because the folly of human ego so obviously fell short of the God-centered order in which they lived. Later readers, in an age that exalted human ego, missed all that, and relegated the book to children's nurseries as a weird fantasy. A lot of literature from before 1800 can be hard to understand today, but that doesn't mean writers have gotten better; readers themselves have changed.

Our minds instead were trained on novels, like Walker Percy's, that seem to portray reality as if through a window—whether the view belongs to an all-knowing narrator or the mind of a character. Human nature emerges from the story, the place, and its unique characters; a novel can be fantastical, but if not believable in terms of folk psychology, the story doesn't work. Allegorical statements or heavy symbolism break the illusion. This elevation of personal experience dates to the Romantic poets—Wordsworth, Coleridge, Keats, and so on—writing in the first decades of the nineteenth century, a time that idolized Captain Cook. They encountered the world directly, individually, and found divinity in beauty, in sentiment, or simply in their own will, as Percy Shelley did when he blasted through social convention and extolled values fit for a hippie: equality, atheism, free love. Their poetic process of self-exploration prefigured the psychoanalytic view of the mind, with the unearthing of unconscious complexes and archetypal dramas: people looking for spiritual mysteries within themselves. Now electronic screens tell our stories and realism's demands are higher yet. Still, customary stories prevail. We continue to retell the story of the individual searcher in every form: murder mysteries, detective stories, any tale that depends on finding something hidden, personal journeys of self-discovery, final epiphanies. Everyone's a Captain Cook, sailing the world alone in search of something real.

Stories reflect the intuitions of their time—the sense of rightness that writers and readers find in an imagined situation. Culture trains us in these intuitions, each of us in our own way. Psychologists probing the roots of morality have found that our judgments come from feelings—the sense of revulsion or dread that arises when contemplating a heinous act—and that our explanations of those emotions as ethical decisions usually come after the fact. These emotional forces, these intuitions, work slightly differently among different groups of people. Experimental surveys find two clusters of Americans: those who react more strongly to stories relating to fairness and caring; and others who have stronger emotions about loyalty, authority and respect, and purity. The echo of an old argument. Presumably, Shelley would have been in the fairness-caring group and Swift one of the loyalty-authority-purity people.

Which is better? Maybe Walker Percy was right, and by seeking God within, by seeking to *be* God—a futile search—we've contracted a disease that hollows us out, leaving an emptiness endlessly fed by material wealth and hedonistic pursuits. If so, then freedom, equality, and individualism are viruses, like the syphilis Cook's men carried to the South Pacific, breaking down the social and spiritual unity that kept us healthy. But, whether a virus or a value, the rise of the individual was a supremely economic innovation. Adam Smith was founding economics, writing *The Wealth of Nations,* while Cook was off exploring; Smith's book was published in 1776. Market economies rely on the freedom to trade, on equality that allows merit to be rewarded, and on the power of the individual to make choices— for ambition, for acquisition, to be what he or she chooses to be. Competition brings all into play—the impulse Hobbes said was at the human core. As Smith wrote:

> It is not from the benevolence of the butcher, the brewer, or the baker, that we expect our dinner, but from their regard to their own interest. We address ourselves, not to their humanity but to their self-love, and never talk to them of our own necessities but of their advantages. Nobody but a beggar chuses to depend chiefly upon the benevolence of his fellow-citizens.

Individual freedom proved powerful in many ways. For scientists, individual minds free from dogma found new truths in reality. Poets followed the scientists' path to new truths in the imagination. Politicians created mass democracy when freed from religion as the basis of authority. Adam Smith's self-interested economic individual—always rational, always seeking to optimize his own well-being—created unimaginable wealth in the nations that first embraced the Enlightenment. I cannot argue against individual freedom. I would never give it up. Even if everyone did renounce freedom we couldn't return to a unified society of traditional roles and beliefs. The journey to self-definition goes one way only, like the exchange of magic for fact made by a child who recognizes the moon doesn't really follow him through the sky. Besides, freedom is indispensible to prevent the horrors of the modern totalitarian state.

On the other hand, if individuals must compete as Hobbes and Smith imagined, then the cost of freedom may be unbearably high. Earth is finite in size, not large enough if all its inhabitants strive to be their own gods of power and wealth.

10.

Competition in Men and Snails

A blind boy listened intently to the radio for news of miracles, trusting in the announcer's promises that complete faith in Jesus would transform his cold, lonely existence in a Dutch boarding school into something warmer. Geerat Vermeij was only five years old. At night he slept in one of a long row of oak beds in an unheated dormitory. At meals he sat at a long table under the eye of a strict house mother. Children were kicked and whipped for violations as minor as tardiness after a weekend at home. Inmates were allowed no personal possessions, so the pebbles that stuffed Geerat's pockets were his only belongings. During recess periods he would hide in a corner of the schoolyard counting and sorting them, avoiding other children his own age as well as the taunts of older bullies who roamed the grounds. He remembered the entrancing vision of glistening, rounded pebbles on a gravel path along a canal where he had once walked with his mother, before his eyes were surgically removed at the age of three.

An inescapable odor of diesel exhaust settled over the school grounds one day, convincing Geerat that he would never escape imprisonment in the narrow boundaries where teachers had trained him to move by sense of touch. His mother, Aaltje, had the same fear, imagining Geerat as an adult still at the institute at Huizen weaving baskets and making brooms. Despite her husband's job in a candle factory and Holland's general post–

World War II poverty, Aaltje's self-image was as a member of the gentry. In the family's apartment in Gouda she kept fine china and antique chests left over from an elegant past. She was intensely ambitious for her two sons. Escape beckoned with an offer of work and a house on a farm in America—in New Jersey, a state that allowed blind children to attend regular public schools with sighted children. The family made the expensive, irreversible journey, but the offer turned out to be a hoax. The parents took demeaning jobs and lived in shabby housing. Aaltje's china and chests were smashed in the move. But her ambition survived. Geerat excelled in school, driven by her hopes, and driven by his own sense of obligation for the sacrifices his family had made on his behalf. And driven, too, by his fascination with the smell, sound, and touch of nature, especially the seashore. He went to Princeton University, where he amazed his professors with his skill in, of all things, finding and identifying seashells.

I met Vermeij at the University of California, Davis, where his large office contains an extraordinary research collection of shells. Finding a snail shell missing from one of dozens of metal drawers, his hands flashed among hundreds of tiny boxes containing thousands of seemingly identical snails the size of beads. A touch here and there shifted him rapidly across the room to open other drawers for the misfiled shell, which he finally found and dropped back into the correct box with a tiny click. His jerky movements suggested a nervous flurry, but were intended to be decisive, like his breakneck pace crossing the campus and climbing its stairs with his white cane. Vermeij found long ago that a big, fast stride helped him blast through patronizing doubts and his own shyness. He had waded among mangroves, corals, and algae-slick rocks on the remotest shores of the Pacific, Atlantic, and Indian oceans, north to the Bering Sea, and south to Australia. He had been pierced by a stingray, bitten by a moray eel, and captured by an Indonesian island chief. On Guam he turned over a rock and calmly identified a stonefish by touch—the most venomous fish in the world—before carefully replacing it.

"I do believe in taking risks in every way," he said. "Intellectually as well as physically."

At the Huizen Institute, children attended Sunday school each week in a nearby town, where a teacher asked the boys and girls to contribute a penny apiece to a collection to "give back to God what he has so generously given to us." One Sunday, as little Geerat was handing over his penny, it occurred to him that God couldn't possibly want or need his money. The teacher's explanation that the donation was symbolic and would be used for a good cause wasn't adequate. Geerat concluded that he had been misled, and that if this claim about giving money to God was untrue, or merely symbolic, then the other stories about God might be fabricated as well—including the tales of miracles that had given him so much comfort when he listened to the radio. He began stripping away what he knew to be true for certain from what he merely had been told and wanted to believe. In the process he discarded God, as much as he needed him at the time.

In 2004, Vermeij took another intellectual risk on the same path: he tried to complete the Enlightenment's unfinished business, connecting the strands of natural history and human nature. His book, *Nature: An Economic History,* carefully and lucidly explained the laws that govern the rise of life and its meaning, laws without which we would not exist. Those laws, he maintained, remain operative in our daily behavior, although we call them "economics" rather than "ecology." The two words mean the same thing. Moreover, by observing the fate of simpler ecological systems, Vermeij forecast our fate on Earth.

Seashells told him most of what he needed to know. They record the history of life. Shells tell the story of the genetic code their owners carried, the animals' development and geography, and how they responded to the traumas of changing climate, geology, and ocean chemistry. And single shells tell the story of individual animals, too—how they grew, whether they dominated or were preyed upon, and even their close calls, inscribed in calcium carbonate: a repair in a shell can recall a snail's escape from a crab in an ocean that disappeared a hundred million years ago. As a naturalist, Vermeij read shells voraciously, whole libraries of them that had lain unopened for thousands of millennia. And not merely in a metaphorical sense: he interpreted individual shells to understand how they worked, their innovations and weaknesses, and the course of the occupants' lives, and he fit those meanings into the animal's ecological context and its place in the chain of its ancestry. He traced this evidence past the point of scientific obscurity and into the realm of papers written for a potential audience of just a few other people in the world. He liked knowing what no one else knew, including twenty genera and thirty species that he himself discovered. He believed that understanding nature deeply and finding its patterns required deciphering living mysteries to the very end of a coiled shell.

Life leaves behind few artifacts as articulate as a shell. Compare a shell to a poem. (Vermeij's work inspires us to look for similarities between natural and human products.) Each is the work of an individual, recording both the presence of the organism itself and its interaction with a place and the other creatures there. But this information stands at a remove from its creator, perhaps long dead. Only through the act of reading can meaning be extracted, revealing something, although never everything, about who lived and what happened. Rocks and other products of unthinking, inanimate processes also store information, but without purpose. Living organisms, on the other hand, are assembled by genes specifically to occupy the environments in which they live. Each birth is a proposed solution to the puzzle of how to survive in a particular niche. Good ideas are passed on because they work as plans for constructing the next generation. At base, life *is* meaning. And only meaning can beget meaning. Organisms create meaning when they act in the world—not only by producing offspring but also by leaving behind a record of themselves in seashells and poems. Either document is a form of immortality. But of the two, the seashell says much more: it contains more raw information, and its tale encompasses millions of generations' solutions to environmental challenges we don't even understand.

Mollusks solved advanced problems in fluid dynamics before the first animal emerged on land. In the fast-flowing suck of receding waves, a snail attached to a rock must resist both the drag of the water and the lift it creates as it swirls over the shell like air over a wing. An animal with a long, low profile reduces drag but exposes itself to increased lift. Limpets—snail-like animals that graze on rocks in the intertidal zone worldwide—address these physics with shells shaped like conical Chinese straw hats, with the steepness of the peak tuned to the particular wave conditions of each species' habitat. Another variable: for survival out of the water, during low tide, the shell must hold a water reservoir that provides cooling through evaporation from around the limpet's foot.

Upon close inspection, shell structures often suggest an adaptive purpose. Periwinkles are knobby to disperse heat but tall and narrow to avoid absorbing sunlight. Some tropical snails are shaped like drills to rotate down into the sand and are sculptured with grooves that act as ratchets to prevent them from unscrewing from the sediment. Shelled cephalopods (relations of the octopus, of which the nautilus is the only nonextinct species) swam in open water, submerging and rising by adjusting the gas pressure in a series of curved shell chambers. The shell is ingeniously corrugated and buttressed to make it strong enough to withstand changing water pressure while remaining light enough to make swimming possible. Scallops also swim. Thousands of eyes along the edges of the shell warn of threats, shell ridges provide strength for flapping to create jets of water, and an aerodynamic shape creates lift to pick the scallop up off the bottom as it moves. A yardwide tropical clam has a shell that allows it to collect energy through photosynthesis as well as by eating.

The labral tooth is hardly noticeable, just a little spine attached to the opening of some predatory snails. Vermeij studied it obsessively from 1992 to 2001. The tooth gives snails leverage in opening bivalves and barnacles, speeding up the process by about two-thirds in laboratory observations. Looking at snails in the field and in museum fossil collections, Vermeij found the tooth on scores of species where it had never been reported, because no one ever looked. What made his finding important was that in sixty of those cases, the tooth had evolved independently. Over the last 80 million years these sixty lines had invented—not inherited—this little bump. And in every case, the innovation had occurred in warm, productive waters in the western or eastern Pacific or Indian Ocean, and never in cold waters or where resources were not abundant.

Why? Why do some limpets in the tropics dig holes for themselves in solid rock, making them virtually invulnerable, but none in northern latitudes do so, even though northern limpets face harsher storms and rougher waves? Why does the elaboration of periwinkles' heat-dispersing architecture depend both on the latitude and on the number of similar species present? Why are exotica such as the drill-and-ratchet-shaped snail or the submarine-like nautilus or the photosynthesizing clam found only in warm, rich seas, while northern shores have far simpler and less varied species?

Vermeij was in a fourth-grade classroom in New Jersey when he first touched smooth and intricately detailed shells from warm ocean waters and wondered why they were so different from the crude, chalky shells he was accustomed to finding on the North Sea. His teacher, Caroline Colberg, had brought the specimens back from Florida and set them on a window ledge near Geerat's desk. What he felt amazed him. He wanted more. Classmates brought him shells their fathers had gathered during the war in the Pacific. Mrs. Colberg allowed him to pursue his question without regard to any curriculum and it drove his career: why are northern and warm-water shells so different? He groped across beaches all over the world for the answer.

On one hand, it seems obvious: cold-blooded creatures naturally flourish in warm water. An ecosystem powered by more energy can afford a deeper food web with more animals. The relationship of the total size of an ecosystem and its largest animals works out with tidy mathematics on isolated islands, where individuals scale to the acreage of the place where they live—explaining the fossils of island-dwelling dwarf elephants and hippos and perhaps even tiny people found in Indonesia. But animals at the top of the food web aren't simply bigger versions of their prey. The killer whale needs the concentrated energy of meat to power its fast swimming, its flexibility, its group cohesion, and probably its remarkable brain as well. Human beings developed our energy-costly brains only after moving up the food web to calorie-rich game. It's also true that the killer whale and our own kind are recent additions, roughly contemporary in our appearance on Earth. The fossil record shows the newest species tend to be the best built. Among mollusks, those from the north—the makers of those chalky shells from the North Sea, for example—are evolutionarily older. Animals from the poorest environments, such as clams living in cold, deep northern seabeds, are ancient, with thick, heavy shells made of material that requires constant maintenance. They can't move much and have few other defenses. They also have relatively few predators because, in their scarcity, they'd be tough to live on.

All that said, warm water alone doesn't explain why individual species should be more evolved. If living is easy, why change? Vermeij looked at differences in shells that had evolved in seemingly similar habitats. He found the key in their defenses. A mollusk can be attacked by being swallowed whole, by being extracted from its shell, or by having its shell smashed. Its potential defenses—larger size or a stronger shell, for example—require energy, and often other trade-offs, such as a heavier shell that slows movement. The most successful species finesse these limitations with innovations, such as smaller, safer openings, construction with thinner, stronger material, and shell sculpting that increases strength without increasing weight or uses smoothness to deny an attacker a good grip. Vermeij found that shells from the Indian and Pacific oceans tended to use better designs than shells from the Caribbean Sea and West Africa. Yet the well-defended shells from Madagascar and Guam were more likely to be found broken than the apparently more vulnerable shells from Jamaica.

Little was known about what kills tropical mollusks, so Vermeij and his wife, Edith, set out to find their key predators. They picked out a crab with a huge claw from a museum collection in Washington, D.C., and tracked it to its habitat on Guam. In a lab on the island they put living specimens of the crabs in aquarium tanks with the island's well-built snails. With frequent crushing sounds, the crabs produced shell fragments like those found on the shore—but some snails escaped to repair their homes. Here was a clear basis for natural selection to improve shell architecture: only those escaping snails would reproduce. Next, the couple tested the Jamaican shells. Rather than import live snails from the Atlantic, they brought empty shells to the crabs, inhabiting them with hermit crabs as bait. One hermit escaped and moved into a decorative shell hanging on the wall of the lab, but the devastation inside the aquarium was total: the big Pacific crabs made short work of all the weak Atlantic shells.

Vermeij returned to the natural history museums, where damaged shells from the fossil record completed the story. In the Pacific and Indian oceans, the competitive escalation of offensive and defensive weaponry had driven the evolution of snails and crabs like nations locked in an arms race. When snails adapted better shells, crabs responded with stronger claws. In the Atlantic, where mass extinction had happened more recently, progress hadn't come as far, and both snail shells and crab claws remained weaker. In northern waters, where the environment provided less energy, organisms couldn't afford so many expensive improvements—for example, to grow a larger or more complex shell—so the progress of evolution there was slower and the competitors had not reached the levels of sophistication found in the tropics. The sixty independent innovations of the labral tooth (discovered later) fit as a perfect example. When resources supplied the fuel, competition drove adaptation to more capable life forms.

Vermeij noticed the similarity to the human arms race in 1983, when Ronald Reagan launched his buildup of nuclear weapons and attempted to construct a missile shield in space, competing with the Soviet Union for ecosystem dominance. The United States won the race with innovation and superior wealth. As on the seashore, competitive success provided the power to dominate, taking control of more energy and resources that begat yet more power—and forcing aside lesser competitors to stagnate in leftover niches, as primitive clams are left to the cold deep or as the world's poor scratch out their living on arid subsistence farms. Winners grow, become larger, faster, smarter, and more flexible, like killer whales or corporations. Vermeij matched up the terminology of ecology and economics and began using them interchangeably, talking about the economics of a tide pool and the adaptations of competitors in the marketplace. The same concepts worked in playgrounds, in politics, or among computer-generated automatons, beyond biology: any collection of self-interested entities should behave the same way.

Competition drives evolution and economics toward complexity, diversity, greater capability, and the concentration of power, the one-way flow of evolution, economics, and history. But, of course, there is a limit. The energy in the system is finite. When dominant competitors can no longer afford to innovate, the ecosystem stagnates in a state of equilibrium. Now the character of the system changes. In the fast-evolving setting of excess wealth and competition, individual animals drive natural selection and improvement. But individuals don't matter as much in ecosystems at equilibrium. Life becomes cyclic, ruled by external forces rather than innovation from within. Vermeij noted that human individualism receded as well during those long spans of history when growth and innovation were dormant. He characterized economically stagnant societies as community-oriented, conservative, socially stratified, religious, and hostile to newness. The advance of science and culture depended on investing in ideas that might not work. That kind of investment requires excess wealth.

Equilibrium ends with major disruptions—mass extinction—that unleash a new competitive rush. Usually, the most dominant players in the old regime disappear. Vermeij wrote, "We continue to obey the fundamental principles of economics set down at life's beginning by inanimate, unintelligent processes, even as we invent startlingly novel abilities and institutions. Nothing in the historical record or in the arsenal of economic principles suggests that humanity can alter the directionality inherent in history even as we overwhelm the biosphere."

The physical space and sunlight available for nonhuman life forms is diminishing. Earth's surface is only so large. Out of the totality of plant photosynthesis—the net primary production that ultimately feeds all humans and land animals—we use a fifth to a third, by various accounting methods, from a high of 80 percent in south-central Asia to a low of 6 percent in South America. A world map showing where human use of primary production is greatest looks like a map of where people live: the world's poor use less per person, but are more numerous. If their consumption were to match that of people in the industrialized world, the total human take would increase by three-quarters. In the oceans, economically important fish species are depleted in 91 percent of coastal ecosystems globally. In some coastal waters only the primary producers remain—algae that's sickly because of lack of thinning or microbial life that dominates where whales, sharks, seals, crocodiles, and turtles once ruled. Figuring our percentage of take from primary production there wouldn't tell the story, because we didn't only appropriate the ecosystems' energy, but also destroyed its capacity to generate life.

Barring catastrophic disruption—from a pandemic or nuclear war, an energy crisis or climate upheaval—the scenario of a human economy that grows without ecological limit is one in which the planet becomes a food-making machine. Useful land is occupied by buildings and farms. The ocean is for food, energy, waste, and transportation. The library of seashells ceases to grow and begins to dissolve as the oceans turn acidic with carbon dioxide from fossil fuel emissions. Nondomesticated life survives in refuges—both the official kind and those that are simply places of no use to people—but these restricted ecosystems are frozen by their constricted size, always subject to extinctions within their narrow boundaries and pruned food webs, vulnerable to changes in climate, pollution, disease, or political whim.

"I'm afraid I just don't see any great outcome from this," Vermeij said, sitting in his university office amid his Braille notes and his shells. "It's going to look a lot like what we have today, but more so. A completely human-dominated Earth and biosphere where most of the diversity is in human artifacts and human occupations rather than biological ones."

11

The Russian Conquest of Alaska

Russian frontiersmen dismantled Alaska's marine ecosystems amazingly rapidly. The first to arrive at the unpopulated Pribilof Islands found rafts of sea otters so numerous they impeded vessels trying to land. Animals were easy to approach and kill. In their first year, a Russian expedition took 40,000 fur seal skins, 2,000 otter pelts, 14,400 pounds of walrus ivory, and more whale baleen than a ship could carry. Within six years, no otters were seen on the Pribilofs. Fur traders exterminated the region's most extraordinary marine mammal before Captain Cook arrived. The Steller sea cow grew up to forty feet long grazing on seaweed. A single pelt stretched over a wooden frame made a large boat. The tasty meat—7,000 pounds from one animal—could be eaten fresh or dried and used like bread. The thick fat layer provided oil to drink, cook with, or use for light and heat. In 1742 the first Russian expedition to reach Alaska discovered the sea cow; in 1755 a Russian government engineer noted sea cows were getting scarce and said hunting should be reduced; by the 1780s, the sea cow was extinct. All that's left are a few skeletons in museums and cartoonish drawings.

Russian American historian Hector Chevigny observed that ultimate frontiers draw psychopaths. Both the daring and the viciousness of the fur traders who began the job of denuding the Aleutian Islands go beyond anything familiar today to normally functioning minds. Launching

on the open North Pacific in wooden riverboats held together with animal skin straps—iron being unavailable in the wild Russian Far East—
they sailed with only the sketchiest idea of where they were or where they
might find land, and without much food except what they could catch.
The horrifying storms here sometimes sink big steel ships, but the Russian fur traders' boats were small enough to pull up on a beach. The lucky
ones landed on treeless islands midocean, ghostly mountains standing
like rocks in an endless river of fog, their dark, volcanic cliffs protruding
from undulating tundra, furious waves battering cooled lava into black
sand. Here they dug underground dwellings and stayed for years, accumulating skins that would sell for fabulous prices in China, slaughtering
Natives for domination or perhaps on a whim, using their few guns—
which, with the compass, were their primary points of technological superiority. After the people of the Aleutians fought back against Russian
atrocities, fur traders in 1764 systematically hunted down the men of
several islands, torturing their families for their whereabouts, and burned
villages, boats, and tools, leaving survivors to starve and freeze without
means of survival.

But this wasn't madness; it was rational self-interest. If these lower-
class Russians had stayed in the settled part of their own country they
would have stifled at the bottom of a stagnant system of caste and perva-
sive bureaucracy. Like American frontiersmen going west, Russians bound
for the Far East could win wealth and freedom if they were tough enough.
Their work explicitly fulfilled the program of Adam Smith's free market
and Thomas Hobbes's warlike state of nature. Their brilliant empress,
Catherine the Great, a German, had read deeply all the Enlightenment
philosophers and adopted a radical policy of laissez-faire capitalism, especially in Alaska, where she refused to grant a government monopoly that
might have brought the trade under more effective control; nor did she
provide meaningful enforcement of her own laws, which nominally protected the Natives. Competing merchants could extend her empire into
North America without government expenditure or effort. As if starting a
huge experiment, the Russians—and the Americans who followed—

unleashed this new free-market philosophy on virgin territory in Alaska. The subsequent story of the stripping of Alaska's coastal life neatly demonstrates Geerat Vermeij's theory about the inexorable rules of power and competition, which are the same in ecology and economics. The newcomers spread as an invasive species, top predators reordering the food web; and they also came as entrepreneurs and warriors. In fact, only the vocabulary differs to distinguish one description from the other.

Within two years of Catherine's 1769 decree freeing merchants to travel and compete, populations exploded in eastern towns. Ambitious operators accumulated capital from stock sales and loans and left for Alaska with larger ships and plans to set up permanent posts. The most important of these businessmen was the shady and dishonest Grigory Shelikhov, who landed on Kodiak in 1784 with an armed force and took the island in a bloody, unprovoked battle, massacred Native prisoners, and made hostages of one boy from each family to assure their parents' obedience.

Shelikhov's company, and the others that invaded—fighting one another as well as the Natives—brought the concentrated power of organization, numbers, and superior technology, displacing the freelancing fur traders. Shelikhov's manager in Kodiak, Aleksandr Baranov, obtained manpower by enslaving indigenous men as hunters and fighters, enforcing his orders with the lash, which he knew was an even deeper cause of shame in their culture than in his own, and sometimes flogged his victims to death. Leaving villages without their leaders and providers, fracturing their communities, Baranov systematically destroyed southwest coastal societies, cutting far deeper than the more random violence of the early fur traders. As the advancing front of marine mammal extermination progressed eastward from the Aleutians, Baranov used his Native forces to seize new land and waters, attacking resisting Native villages and competing Russian traders with ships accompanied by swarms of as many as seven hundred two-man kayaks, plus supporting crews in skin boats. The kayakers died by the hundreds in battles and storms. Ultimately, Baranov pushed as far south as California and built towns in

Sitka and Kodiak larger than any on the American West Coast. He even tried unsuccessfully to make a deal with King Kamehameha of Hawaii. And his hunters slaughtered most of the valuable fur-bearing animals on the great arc of American land and islands spanning the north edge of the Pacific Ocean.

Baranov devoted his early years in Alaska to winning control of Prince William Sound from the Chugach and from a rival Russian company. English ships came to trade with Natives there as well, including parties led by former officers of Cook's, who had learned the value of sea otter pelts when they landed in Canton on the way home from Cook's final voyage. Baranov entered the sound in 1792 in a skin boat with an obsolete little cannon on the prow, accompanied by a fleet of ninety two-man kayaks. The size of the force must have impressed the Chugach villagers, but in the fifteen years since Cook's visit they had learned to deal distrustfully with foreigners and bargained at length, insisting in one village that Baranov marry a Chugach girl. He agreed, although already married with children to a woman back in Russia and married in all but name to a Native woman in Kodiak.

Like other Russians in North America, Baranov had adopted Native food, technology, family, and even clothing, because he had no choice—tenuously linked to home, focused on profits and used to hard living, the men were materially poor. When a well-armed English trading ship, the *Phoenix*, sailed into the sound where Baranov had camped, near Nuchek, its officers at first took him for a Native. Speaking German, their only shared language, the Englishmen invited Baranov to dinner. He likely hadn't tasted bread in several years. Their table probably would have been set with fine wine, and dinner would be followed by cigars. Baranov was deeply impressed. When he built his own ship, he even named it the *Phoenix*.

Chugach people like to say they never were defeated in battle by the Russians. However, most have Russian names—Kompkoff, Selanoff, Totemoff, Evanoff—and the Russian Orthodox Church is at the center of each village community. Traditionally, village decision making was nonhierarchical, but indigenous church leaders called lay readers became chiefs, and each village still has a chief. Russians moved people around—whole villages—for business reasons, mixing the different cultures of upper Cook Inlet, the Aleutians, Kodiak, and Prince William Sound, so in a few generations what it meant to be a Chugach Native was no longer very clear. Marine mammals that had provided food, clothing, tools, and spiritual connection were driven to the brink of extinction. The main Russian fort, at Nuchek, declined along with the sea otter, becoming a filthy, ramshackle trading post and source of booze, which fur buyers used to seduce hunters into dependence on the cash economy rather than their traditional and more successful subsistence practices. Death came in waves: epidemics, scarcity, conflict. Skills and stories went with the dead. The grief is unimaginable. Alcoholism overwhelmed entire villages, behavior soaked in shame, responsibilities forgotten, children wandering ignored, seeing what they shouldn't have to see. So much forgotten that today, from the Aleutians to the sound, no one knows anymore exactly what's Russian and what's Native: traditions, ways of thinking, the all-important *banya*—the steam bath, the heart of togetherness, that time when nothing is hidden—the Russians might have brought that, too. Native archaeologists dig for clues to their own forgotten cultures. Dancers make up steps, imagining what might have been.

Strands of Russian ways and traditional indigenous ways were wrapped with another strand, the strand of the grief itself, the kind that comes too fast to work through, that runs too deep, because not only have the teachers died, but the universe they inhabited and taught is dying as well. Grief passed on generation to generation, the way descendants of Holocaust survivors or war veterans pass on their pain, tied up with anger in a hard knot—how do you please an angry father or learn about joy from a joyless mother?—all woven into the culture along with the other, better strands that give strength and nurturing. Children in such homes can learn from their parents that they're lost and that the cold, hard surface of unprocessed grief is the right way to meet the world. This is the heritage of Russian genocide—so Jim Miller calls it, the Port Graham carver and counselor I introduced at the beginning of Chapter 3, whose life's work now is to loosen the knots and separate out the strands tying his village's twelve dozen people to their ancestral losses, the alcohol, the physical and sexual abuse. Get rid of the bad traditions, bring back the good ones, create new artistic designs based on the ones in museum catalogs of the objects that collectors took away.

I met Jim in Tatitlek, a village in the sound where Chugach teachers and youth gathered for a cultural heritage week to pass on the practices of the past. Miller had learned carving from books because no one in his village knew how anymore. Now he taught by example, leaning over his own work in a folding metal chair in a garage near the school, joking with the teenage students, catching their attention by mentioning how much his art could sell for, inspiring them by talking about the spirit in a piece of wood that is carved with love as a gift for a particular person. He wanted them to find purpose. To have purpose, to feel something good coming from the work of one's own hands, something valuable—that, he believed, was a path to healing. He said, "I see art as a way of reclaiming some of these things. Self-respect. Respect of elders, family, and culture."

We rode out to the Tatitlek dump on four-wheel-drive all-terrain vehicles looking for eagle feathers he could add to his carving, a mask. A stream ran through the heaps of garbage, eagles wheeling overhead. Wind plastered plastic shopping bags to the chain-link fences. The young man who takes care of the dump gave me huge slabs of fresh halibut, as he'd caught a couple of hundred pounds more than he needed. People often give me fish when I visit Chugach villages. Jim Miller explained that those who like to fish just keep fishing until everyone's freezer is full—something good that survives from old traditions.

Jim sat one day that week with Patrick Norman, Port Graham's chief, in the Tatitlek School cafeteria as the culture week students went back to their classes. Patrick chuckled at the sight of village children grabbing Costco bottled water as they passed—illogical, he said, since "Tatitlek has the best water in the world." Patrick wants the young to learn their Alutiiq language. In Alutiiq, names for animals are descriptions, not sounds arbitrarily attached to an object—Alutiiq is good for naming. "By looking at it in the English way, we've forgotten how to name new things," he said.

"How you going to make new songs without new words?" Jim said.

New Native cultural learning is taking hold in Chugach young people, even if not in their parents. I met plenty of teens who were proud of their Alutiiq words and their artistic and hunting skills. They go back to Nuchek in the summer, near where Cook and Baranov landed, Makari's abandoned village, which their people now own, and where they have built a big summer camp. A young man who in the winter attends an Anchorage high school and plays on its football team told about going to Nuchek in the summer to learn how to take a seal, shooting for the nose or back of the head so as not to force the air out of the body and make the carcass sink, and how to butcher it so every part was used.

Broken hearts can last for many generations, but healing happens, too: on the beach in front of Tatitlek, around a bonfire on a cool, clear May evening, stomachs full of crab and seal, teenage boys jumped off a high dock into the frigid ocean. Couples walked where the smooth pink light reflected at the water's edge amid the scent of seaweed, the clicking of barnacles, living creatures at their feet and in the sky above their heads. If the damage had stopped with the Russian genocide perhaps the grieving could end now. But that was only the start.

The American Conquest of Alaska

Russia sold Alaska to the United States in 1867 without ever occupying, conquering, or even exploring the vast majority of it. The Russian American population peaked at 823 people, all on the coast, leaving large swaths of Native Alaska untouched. A drunken and undisciplined U.S. Army detachment assumed control in the former Russian capital of Sitka, sexually assaulting Native women and meting out retribution for perceived crimes by individual Tlingit Indians with indiscriminate killings and burnings of entire villages—acts similar to the first Russian fur traders' behavior of a century earlier. After ten years the army withdrew its forces and a Tlingit war party prepared to destroy the tiny American community of Sitka, deterred only by the intercession of another clan in defense.

Like the Russians, Americans never displaced most Alaska Native peoples. The conquest was ecological rather than geographic: they took the food. New Englanders and Californians slaughtered whales for baleen and walruses for ivory on the western and northern coasts, wasting hundreds of millions of pounds of meat and inflicting starvation on the Iñupiat. Fur traders acquired the last of the otter pelts and depleted other valuable furbearing animals. As wild furs ran out, fox farmers appropriated islands as natural enclosures, especially in the sound, the foxes running free and fattening up in part by eliminating nesting geese. Salmon canneries began cropping up next to rivers in southeast Alaska in 1878 and quickly spread up the coast, competing without regulation, competition sometimes causing gunfights—Alaska had no civil law at all until 1884 and little practical law enforcement for decades more. An entrepreneur could steam in from San Francisco or Seattle with his equipment and Chinese manpower, block a stream with a barricade to scoop up every fish, and maybe make back his investment in a year. Preventing any fish from escaping to spawn wiped out the run over time, but there were always more rivers with seemingly unlimited fish farther along the coast.

Running salmon fill a stream, virility surging toward climax, pure muscle and will, so thick the gravel bed beneath disappears. All that's visible is fish flanks moving, overlapping, refracting in clear, flowing water. A river of flesh. It happens fast, with the incoming tide, for the few weeks when these particular fish home in on their particular natal stream, bringing back energy collected across the broad ocean. No one could resist catching one. You can grab one in your hands, flashing in cold water, coming up with a jerking, wild silver life you can barely contain, a meaty bolt of lightning, so rich to eat and so full of health, as if the food could transfer the vital spirit of the fish itself. Among all creatures, spawning salmon were meant to be eaten; they present themselves to the streams at the end of their lives as if for that purpose. Thousand of streams, hundreds of millions of fish, billions of pounds of food—all free for the taking.

The cannery owners must have felt like they had won the sweepstakes, the prize to grab as much cash as they could hold. Conservation wasn't on their minds. They would have needed to leave behind only a fraction of a salmon run to spawn each year for the abundance to continue indefinitely. But they didn't plan to stay indefinitely. Competition ruled out long-term considerations. Like found money, Alaska salmon were yours only if you grabbed them first. A cannery operator who abstained from fishing to allow for the next year's harvest might not be in the same business when the fish came back.

The best private economic decision might be to destroy a salmon run—or wipe out a marine mammal population—if you could thereby obtain a profit quickly and invest it somewhere else, in Alaska or on Wall Street. Unsustainable practices often make sense when you're free to move and take your profits with you. Our economic lives depend on this fact. Nothing made of plastic or metal or manufactured and shipped with fossil fuels is sustainable. Look around you. We buy these things as cheaply as possible—technology, vehicles, energy—knowing we will discard them after we've exhausted their value. A man setting up a cannery on a salmon stream made the same economic calculation.

Besides, short-term depletion of a fishery might not hurt business even in the medium term. As fish, furbearers, or whales become scarce, their value rises on the market. The incentive to invest in harvesting increases as fish and animals become fewer but are each worth more. On a graph, the two lines split: fishing intensity rises as the resource turns toward zero. Alaskan salmon fisheries, despite becoming monopolies before 1900, followed the economic pattern of putting in more effort to chase fewer fish through the first half of the twentieth century. Whalers drove a sequence of species to the edge of extinction from the 1860s to the 1960s, despite being large-scale, economically sophisticated companies—that was how their business model worked. Globally, we're on the right-hand side of that curve today. Total fish catch peaked in the 1980s, but fishing effort has continued to grow with improved technology. Fishing fleets scour deeper and more remote waters, and, as predators disappear, vessels fish for organisms lower on the food web. As tiers of the ecosystem are eliminated—beginning with the trophic level of the predators, then the fish they used to eat, and so on—each succeeding level increases in economic value, transitioning from abundance and low value to scarcity and higher value, inciting increased fishing effort until collapse and a move down to the next lower ecosystem level. Sea urchin depletion radiated outward from Japanese sushi restaurants and spread worldwide, reaching new stocks in places where sea urchin take wasn't regulated because they had never been targeted before. Now fishermen are targeting even lower on the food web: jellyfish and zooplankton such as krill.

This is mankind competing in the state of nature as imagined by

Thomas Hobbes and as linked to natural history by Geerat Vermeij. But we expect government to look out for the long term. Hobbes envisioned a Leviathan—an all-powerful ruler—as the protector of the common interest that any rational person would prefer to pure anarchy. John Locke and the American founding fathers adapted that idea to create a constitutional democracy in which institutions controlled by law and popular consent could manage the competitive space. Yet the International Whaling Commission allowed the extermination of the whales. In territorial Alaska, the canned-salmon industry, consolidated in the hands of major industrialists, easily manipulated the political system using the same methods corporations use in Washington today: lobbying, capture of regulators through revolving-door jobs and cozy relationships, campaign contributions, and bribery. For the first forty years, the salmon industry enjoyed almost no regulation; when harvests crashed around 1920, another forty-year period began, of weak, industry-controlled regulation, until Alaskan statehood in 1959. By then the catch had declined to levels not seen since the canneries first opened. From the earliest years, fishery management avoided restricting catches, focusing on increasing salmon numbers by building hatcheries and reducing predators—killing seals, whales, trout, eagles, and other birds in immense numbers. Hunters in Prince William Sound claimed a bounty for each seal nose they turned in. Fishermen legally shot killer whales and sea lions to protect salmon into the 1970s.

Before the Alaska purchase, Chugach villages had owned their salmon streams, effectively limiting harvest since village needs were limited. During the run they lived on the banks, catching fish and drying them for use through the winter. They took a substantial but sustainable portion of the resource: consumption averaged nearly five hundred pounds of salmon per person per year. When canneries fished out the salmon runs, starvation struck Interior Alaska Natives on the Copper and Yukon rivers. The loss of salmon and herring all but wiped out the Eyak people of Cordova, Aleganik, Katalla, and the Controller Bay area. The last Native speaker and full-blooded Eyak died in 2008.

Within the sound, the Chugach survived the late nineteenth and early twentieth centuries by fishing the scores of dispersed salmon streams. Also, the decline of food resources was offset in part by a decline in the numbers of Chugach people through successive epidemics. Village visitors often found almost everyone sick. The Russians and then the Americans had created a cash economy in Prince William Sound with the trading post at Nuchek and smaller stores in the other villages. Natives needed money to fund raging alcohol addictions—a storekeeper in Chenega boasted of obtaining more furs by encouraging drinking—and to buy food to replace some of their former subsistence, and to obtain tools and derby hats. Chugach weapons finished off the sound's otters and they fished for the canneries and salteries.

By 1904, Prince William Sound bustled with activity. Towns had been born called Valdez, Fort Liscum, Ellamar, LaTouche, and Orca and, just east of the sound, Kayak Island, Chilkat Point, and Katalla. Scores of mining camps, canneries, and salteries occupied bays, islands, and mountainsides in every corner. Steamers ran back and forth like city buses and connected new towns to outside cities. Rail lines were mapped and cleared in the oddest places. Whites far outnumbered Natives, rushing to new gold and copper finds, canneries, and fox farms, and the villages shrank— Nuchek, Chenega, Tatitlek, Kiniklik, Eyak, and, to the east, Chilkat. Copper miners from Ellamar walked three miles to Tatitlek, bringing alcohol for easy sex. "After the parents are drunk the children are left to the pleasure of the whites," the Tatitlek schoolteacher reported. A Valdez doctor reported in 1908 that half the sound's Natives had died in the previous two years of tuberculosis, influenza, and hereditary syphilis. Coming on top of several previous epidemics, the sound's entire Native population dropped below 200. The doctors and teachers decided to consolidate the villages. Only Chenega and Tatitlek survived, their population down to 160 when Birket-Smith arrived in 1933 to record their fragmented culture.

But by then, they had the sound to themselves again. The mines had played out and the towns were abandoned except for Valdez, which had a road inland, and Cordova, with canneries. Fur farmers had left when prices dropped, leaving cabins back in the woods of tiny islands to collapse slowly under heavy snows. A few canneries continued to operate, including one at Port Nellie Juan supplied by Native fishermen from Chenega, but others stood empty on beaches, slowly rotting until the decks gave way, the walls washed off into the sea, and only the pilings remained, homes for barnacles and mussels. Place names are left on maps like water spots. My family looked for Golden, up Port Wells toward the glaciers, where in 1913 we could have patronized a drugstore, a post office, a restaurant, road houses, stores—but now it's a lovely cove protected by narrow islands of angular bedrock and backed by the smooth, swampy grass of a gently rising meadow. Nature has camouflaged the rusted iron and timbers of steep mountainside mine workings so one can again feel like the first to ever enter the narrow lagoon where once a tram and other heavy equipment hauled ore and men. The sea otters and salmon have come back, and the whales.

Nothing comes back the same, however. Nothing is ever restored. The story of ecology goes in one direction only. Finite energy that's spent can't be recovered; every death and birth creates a new system for distributing the new energy that arrives. Nor would we ever know it, if the ecosystem were restored, through infinite improbability, since we've no idea what we would have found within these waters before, and don't, in fact, know what's there now. The sea otter example is a classroom lesson for ecology teachers: otters live in kelp forests, using the algae's fronds as anchors, and feed on sea urchins, which crop the kelp and keep the forest healthy, providing shelter to many other organisms. Otters dead, urchins overpopulated, kelp forest destroyed, food web gone. Although otters were protected in 1911, they remained so scarce that even by 1968 few people had ever seen one. Otters were gone so long that, now that they're back, booming in the last twenty years, it's been discovered that they don't live only in kelp forests. When they're numerous, as in Kachemak Bay, they also live over muddy sea bottoms, and even float right out in the middle, in deep water—what are they doing there? How is this system supposed to work? A lot of otters or not so many? Does it make any difference that the sea cows are gone, or that the great whales were removed and now some species are returning? Having wiped out the large mammals, perhaps the numbers of many species will oscillate for generations.

Herring once mobbed the length of Kachemak Bay in huge spawning balls, so thick in places that oars pushed them aside when rowing. In April 1926, biologist George Rounsefell saw acres of herring roiling the surface of Halibut Cove Lagoon, where flat water sits in a bowl of encircling mountains: "about 50 belugas (a species of small white Arctic whale) were raising havoc, and thousands of sea gulls were scattered everywhere. Cormorants, murres, surf scoters, and divers were there in tens of thousands, and scores of bald eagles were circling about." For the school to pass by, exiting the lagoon, took half an hour. Herring flesh is a conduit carrying energy from plankton to predators, along with a few other critical forage fish species in Gulf of Alaska waters: sand lance, capelin, smelt, juvenile salmon, and pollock. Herring accumulate calories during plankton blooms to store as fat for the winter; their thick schools feed the species that Rounsefell saw, plus humpback whales, sea lions, harbor seals, adult salmon, and other fish—herring is an all-purpose food.

The fish were so numerous that the Russians named a large bay for them: Seldovia, where the region's dominant town developed. When World War I disrupted Scandinavian fish supplies a herring gold rush exploded there and along the Kachemak shore, in Halibut Cove and Tutka Bay. Salteries sprouted on the beaches, packing the fish in barrels with salt—twenty or more of these factories operating in a single year. Fish that didn't fit buyers' standard size requirements were dumped in enormous piles on the shore. The stink of rotting fish at times wafted eighteen miles across Kachemak Bay to Anchor Point. After a decade of the rush, the waste had fouled the spawning grounds and the runs died out. They have never recovered; far fewer herring come now, not enough for a commercial fishery. Ted Otis, who is studying what happened to Cook Inlet herring for the Alaska Department of Fish and Game—the runs on the west side of the inlet failed in the late 1990s, too, for unknown reasons—believes the stocks of fish may be too isolated to recover quickly. He said, "We get impatient, and it might not take place in ten years or twenty years. It might take two hundred years. Or it might not happen."

Ecologists are still figuring out what it means to cut an important forage fish species from a food web. It takes major computing power to solve the mathematics of the food web relationships that create ecosystem resilience. Each forage fish species is like a pipe bringing energy from plankton to top predators. In a diverse food web, when one forage species runs into trouble, such as a bad year for reproduction, top predators can turn to another food—switching from herring to sand lance, for example—until conditions change and they switch again. Predators tend to prune the most productive forage species, preventing temporary winners from permanently displacing the others. When all are present, there's always a fallback food, a surviving channel to move energy up the web. Taking away a part of the system might not matter for a long time, until that pipe is needed again. The consequences may come too late for anyone to perceive the real cause—too late to notice, for example, that beluga whales that once chased herring stopped coming into Kachemak Bay during the springtime more than twenty years ago.

Few people remember the belugas in springtime. Hardly anyone remembers there was a herring fishery here at all. Each generation remembers only the abundance of its own youth, not the life that's already gone. We're born into plenty and learn regret with age, as what we treasure slowly dies—the trees lost to parking lots, the ocean emptying. Our children don't yet know this loss. They think they are born into plenty, too.

13.

Evolution, Free Will, and Hope

An illusion protects us: the illusion that those who depleted fisheries and drove marine mammals to extinction didn't know what they were doing. It's not true. Even Russian America had voices of conservation. The herring fishermen in Kachemak Bay were told they were overfishing, but instead of restraining themselves they tried to stop the Natives from eating roe on kelp. Members of the salmon industry recognized fish were rapidly diminishing before 1900 at the same time they were saying salmon were inexhaustible. Anyone under fifty years old today was weaned on environmental warnings. We're sick of the nagging—even as we see the natural world changing irrevocably. My family recently visited Thomas Jefferson's home, Monticello, and came away saddened that so brilliant a man couldn't extricate himself from slavery because of his own weakness—his need for the lifestyle he enjoyed—despite knowing he was wrong. The conversation stopped dead when my teenage son, Robin, pointed out that we will probably be looked back upon the same way.

Social nicety depends on the illusion that our ancestors were blameless. It preserves our self-image, too. The rude alternative, Geerat Vermeij's prediction, is that humanity cannot overcome the laws of ecology, cannot decide against stripping and domesticating the oceans. He writes delicately, but essentially says we can't control ourselves—we can't choose against our consumptive nature. We do seem practically incapable of making some choices, even as individuals. Only death can stop some people from destroying themselves with alcohol. As groups, our ability to choose is weaker yet, because group decisions involve persuasion and relationships. As a species, it's hard to think of any choices we've consciously made. Nearly everyone wants peace. Yet war rarely ceases.

Such an uncomfortable prediction would come from a man familiar with social rejection. Loneliness tested Vermeij from the days at the Huizen school, when he sorted his pebbles rather than playing with the other children in the schoolyard. He learned there how badly communal living arrangements can work. His social isolation continued as a blind child with poor parents in a new country. American culture never suited him. The schools were too easy, the children too lazy. He wasn't offended so much by the informality as the lack of seriousness. The upheavals of the 1960s further manifested the country's decay and degeneracy. He hated America's rock music, violence, and anti-intellectualism. Vermeij detests pop culture to this day. At Princeton, he was intimidated by privileged students, free-flowing booze, and the caste system of selective eating clubs—his father, a vinyl salesman, earned half as much as his least affluent friend's dad.

At each stage of Vermeij's education, resistance cropped up because of his blindness. He was forced to prove himself again and again. Although he excelled at Princeton, graduating in three years, he encountered skepticism when he interviewed for a graduate school position at Yale. Attempting to prove him incapable, a professor sprung an unexpected test during a visit to the mollusk collection at the Peabody Museum, asking Vermeij to identify a pair of similar shells. "It was a dirty trick, because they wouldn't have done that to anyone else," he said. "But it was also reality, so I just dealt with it." He passed the test.

By then most of the women Vermeij liked were already engaged or married, and he had yet to begin his first romantic relationship. He believed his blindness was a near total barrier. But he fell in love with a graduate student employed as one of his readers, Edith Zipser, who had luxurious curly hair and a clear, bell-like voice. He recalled, "By then I was really thinking quite seriously that if I was ever going to get a girlfriend, this is the time. It was not going to happen later on. But in a way I wish I had more—I mean, I don't regret a thing—but it would have been nice to have had a bit more experience while looking around." Edie abandoned her own career in molecular biology to help with Geerat's research.

In 1975, Vermeij had barely gotten tenure from the University of Maryland when Edward O. Wilson published *Sociobiology: The New Synthesis,* in which Wilson proposed an evolutionary basis for social behavior in ever higher realms of the natural kingdom until, in the final chapter, he posed biological roots for what humans do. The chapter is as insolent as a precocious child. It shook people. Wilson predicted that the empirical tools of the Enlightenment would, by 2100, bring humanity total self-knowledge as machines made of flesh, stripping us of spiritual hope. I was a teen when the sociobiology phenomenon happened in popular culture and I've got it all mixed up in my mind with disco music, pheromones, and a movie about computers taking over the world. People started finding evidence of natural selection in everything. For Vermeij, the slipstream of that cultural event caught his career and he never fully escaped.

A few years later, Stephen Jay Gould became a scientific and cultural star by deflating the sociobiology fad, eventually landing himself a guest spot as a cartoon on *The Simpsons.* A mega-hit scientific paper he coauthored mocked the foundations of Wilson's theory and steered evolutionary theory in an entirely new direction. Gould said sociobiologists were just making up plausible tales when they explained everything with adaptation, like Rudyard Kipling's *Just So* stories, in which it's explained that the rhinoceros has floppy skin because of too much scratching. Just because a man wears pants, that doesn't mean his legs are adapted to fit in them. Gould believed adaptation did not drive most evolution. Instead mass extinctions and the luck of how they affected different species brought beneficial characteristics into play. An aspect of an organism that developed for an unrelated purpose, or for no reason at all, suddenly became newly useful. By this thinking, evolution was neither progressive nor repeatable: if the story were rerun, it would come out entirely differently. Individuals didn't matter, species did. And, critically, human behavioral traits were ineligible for interpretation as adaptations—they were lucky leftovers, too.

Creationists grabbed hold of Gould to counter Darwin at the same time as Ronald Reagan was pushing aside disco sexuality and environmental pessimism. Without adaptation and progressive evolution in the picture, God had a shot—what else was all this lucky evolution but providence? Gould opposed teaching creationism in the schools and claimed to be an atheist, but his theory does neatly sidestep Wilson's stark anti-spiritual conclusions. Gould even wrote a book to make peace between the truths of science and religion, calling the two realms "non-overlapping magisteria." Gould's universe could be consistent with humanity choosing to save the oceans. Freed from the ecological laws of natural selection, anything might be possible, including people going against their self-interest for the good of the biosphere. Any unknown potential could lie within, installed by chance, or whoever is finally in control. That subtext, the quality of mystery, must have helped make Gould's books appealing to his broad readership.

While Vermeij was groping his way through tropical tide pools and recognizing the importance of competition in individual animals, his ideas were shunted aside by Gould's worldview, in which only species matter, not single organisms. Study of adaptation isolated him professionally as his foreignness and blindness had isolated him socially. At a conference, "adaptationist" was used against Vermeij as an epithet. He was a naturalist when abstraction and computer modeling were predominant in biology. He said, "I have always felt I am a minority crying in the wilderness."

Since then, loads of evidence supporting the importance of adaptation have contradicted Gould's basic ideas (as he died in 2002, we don't know how he might have responded). Biologists saw adaptation happening in real time during career-long measurements of the size of the beaks of finches in the Galapagos Islands: when periods of drought made larger beaks advantageous, natural selection produced them. A diverse group of fish called cichlids that live in Africa's rivers and lakes adapted amazingly quickly into an incredible variety of ecological niches—more than five hundred species in the ten thousand years since Lake Victoria was last dry. It turns out that adaptation by natural selection can progress so quickly as to make other evolutionary mechanisms irrelevant.

Gould's emphasis on chance cannot explain the mounting examples of convergent evolution: Vermeij's sixty independent lines of snails developing the labral tooth or Nicky Clayton's corvids, whose brains, so little like primate brains, can think just as well. Likewise, the killer whale shares its sharp black-and-white markings with the Dall's porpoises it lives with and feeds upon, but both are more closely related to other species than to each other. (But the purpose of the markings remains a mystery, a solution found by nature for a problem that is not yet clear to scientists.) The cichlid fishes, evolving quickly in unconnected lakes with dissimilar genes, nonetheless often look and behave nearly identically, right down to their shape and detailed coloration. Only adaptation to similar environmental problems could bring so many disparate heredities to the same solutions. Human beings must be shaped by our environment, too. We are born survival machines, like all the other species that avoided extinction, not just lucky. By this thinking, our path might not be predetermined—indeed, its variety seems infinite—but, like a Calvinist's unavoidable heaven or hell, our destination would already be set by the universe in which we grow.

No one wants to hear that our fate is sealed. Or, perhaps worse, that our inborn will for striving, of which many are proud, will itself deprive us of the cooperative capabilities needed to conserve much of the biosphere. Vermeij's work at the hard details of biology and evolution made him the oracle of this unwelcome prophesy. I listened as he taught a freshman class of eighteen students at UC Davis, "Evolution as a World View," a lively and fascinating lecture that touched on the labral tooth, the flightless birds of the South Pacific, and the Davis public schools, all topics related to the forces of competition or its lack. He passed treasured shells among the students. Few were taking detailed notes. One student slept, his head lolling back. Vermeij couldn't see that, but at the end of the period he had to notice that no one asked a question. He later said, "I find it very difficult to motivate students, and I find it very depressing that so many students seem to be beyond motivation."

It's ironic that a man of such charm, daring, and intellectual grace would repeatedly find himself an outsider. Vermeij has paid a price for insisting on his unwanted truths. He also suffers from the dissatisfaction common to true competitors, always stretching, never settling. On a balmy evening with Edie among the fragrant vegetable stands of the Davis farmers market, I watched him focus his intense concentration and patience on evaluating a squash. Forced to fight for every inch he has gained in life, he seemed intent that even this task be completed perfectly.

Vermeij is clearly correct about how competitive systems work. The escalation of stronger crab claw and harder snail shell continues in tropical tide pools now as it has for millions of years. But a link may be missing in extending that story into a prediction for the planet's fate. People sometimes have enough. They can feel fulfilled and turn their efforts to cooperating and helping others. I see them devoting themselves to issues much larger than they are with no hope of their own contribution making much difference—knowing, indeed, that their contributions might be washed away in a momentary corporate shrug. Edie drives Geerat around Davis in an early-model hybrid car. Hybrids emit less carbon but cost more to own in total, even after fuel costs (at least at that time). So why would a powerful ecosystem-eater buy one? Thomas Hobbes, Adam Smith, and Geerat Vermeij don't have a quick answer for this question.

Why do we conserve at all? Sometimes saving resources can be explained away as rational self-interest. A farmer could preserve a forested lot for the future use of his own offspring. As Vermeij points out, cooperation can be the most effective form of competition. But we'll see in the chapters ahead that some large decisions in conservation history don't easily fit the evolutionary model of self-interested beings driven by their own benefit or even that of their heirs. Men sacrificed greatly to preserve the forests of Prince William Sound without any expectation of that place being a part of their lives, or even with a plan for any human use. We will examine their times, motivations, and psychology to understand why.

On the surface, their decisions resemble those of some of the American founding fathers who, unlike Jefferson, responded to their own words about liberty by freeing their slaves. They couldn't end all slavery by those decisions any more than you or I can personally prevent the destruction of the oceans, yet some felt the need to act, as some do now when they pick up trash from beaches. That drive to do good, emerging from a deep human preference for fairness and connection, is as real a part of us as the desire to compete.

It's not that Vermeij's theory is wrong in predicting the behavior of self-interested actors, it's that the nature of the actors themselves is in doubt. We may be better than we know.

Part III

HOW CONSERVATION HAPPENED

14.

Taming the Gold Rush

Other men trying to cross the White Pass beat their horses to death, but Will Langille made it without losing any on the way. The trail climbed the coastal mountains, from southeast Alaska to the Yukon Territory, where, in the fall of 1897, thousands of Americans without outdoor experience hoped to reach the summit, build boats, sail across a series of lakes, float down the Yukon River through ferocious rapids, and arrive at the Klondike River and Dawson City, a journey of more than four hundred miles, and there, somehow, get rich finding gold. Ships arrived daily at the head of Lynn Canal fjord, ninety miles north of Juneau, and dumped passengers, horses, and tons of equipment and food on wilderness beaches. Langille built a fire for a party who didn't know how. Campers swindled and stole from each other—someone took twenty-five sacks of oats Langille had already moved over the pass—and there was no law but the vigilante committees that punished suspected thieves with summary execution. The trail collapsed under incessant rain and the pounding feet of men and animals relaying thousands of tons of supplies one pack at a time. Dead horses piled up and rotted, killed by exhaustion, fallen from rocky precipices, or mired down in bottomless mud holes. As if drugged by greed, men left the animals where they fell, grinding their carcasses underfoot while plodding onward. A Canadian official who climbed the pass at the time described horses "sometimes in tangled masses filling the mudholes and furnishing the only footing for our poor pack animals on the march—often, I regret to say, exhausted but still alive, a fact we are unaware of until after the miserable wretches turn beneath the hoofs of our cavalcade."

Langille, among the earliest to arrive, gathered a crew to fix a blockage in the trail, but the men worked only for a morning before wandering off. A flamboyant newspaper correspondent recruited him to lead a crew of five hundred men with $2,000 worth of tools and blasting powder to improve the trail; however, only a dozen workers showed up. After spending a day blasting some bad spots, Will turned to getting his own group and their three tons of material over the pass, across the lakes, down the river. In his letters, he bragged of losing none of the two hundred pounds on his tall frame, of carrying heavy packs others couldn't handle, steering his boat through rapids safely instead of portaging through the woods like the less courageous. Jack London followed a few weeks behind Langille and they wintered in the same little community of cabins on the Stewart River, but the sort of hardships London made terrifying in his fiction Langille took in stride—after sleeping outdoors at −44 degrees Fahrenheit he wrote home to his mother, "To you I suppose it seems like something terrible to be sleeping out on the snow with the 'quick' down that low, but we had plenty of blankets and kept warm, the worst thing cooking, the dough freezes before you can get all kneaded, and everything of metal sticks to your fingers."

Will's exceptional qualities—his heartiness, charisma and physique, his contempt for the weak, the incompetent or dishonest, his wit and vividness as a writer, his lack of self-doubt—all contributed to accomplishments that made him Alaska's original and possibly most important conservationist. It was Langille who first recognized the need to conserve Prince William Sound. Before coming to Alaska he had joined campaigns to protect Washington's Olympic Peninsula as a forest reserve and prevent the elimination of the Cascade Range Forest Reserve in Oregon. His directness made him a steamroller at times, "mildly terrifying," in the words of a work superior, and sometimes intolerant and childishly boastful, but directness also gave Will a clear-eyed view of the waste and fraud that were the constant motif of the gold rush period. He was too honest to fool himself into believing Alaska was inexhaustible. Instead, his travels ultimately persuaded him that every acre of it should be withdrawn from willy-nilly exploitation and conserved for managed, sustainable use. Along the way, Langille's work sparked a conservation struggle that protected a vast region and helped change presidential politics.

Few among us can stand as a simple example of selflessness, the sort of hero whose story would demonstrate the hypothesis that people are capable of profoundly cooperative environmental behavior. Will Langille is close to ideal for the purpose. He went to Alaska as a gold seeker and an adventurer, but he learned from the place. He came to see, long before his better-educated contemporaries, that even those forests without commercial value had inestimable value in themselves, as intact ecosystems supporting animals and people. He acted on those ideas to Alaska's enduring benefit. And then he moved on, never to reap any benefit for himself.

Will's mother influenced him. Sarah Harding Langille had emigrated from Nova Scotia to the Hood River area of Oregon to homestead with her carpenter husband, James. She raised her three sons for a decade in a cabin at the foot of Mount Hood, where Will roamed the woods for days and nights through his teens. He attended school only to the eighth grade, but absorbed Sarah's literary sophistication—the boys nicknamed her Tant' Sanna, after a widowed farm boss of similarly broad build in a strikingly modern feminist novel of 1883, *The Story of an African Farm,* which they read together one snowy winter in the cabin. She legally separated from their father and in 1891 took over management of a failed wilderness lodge he had supervised building, the Cloud Cap Inn, high on the flank of Mount Hood. Tant' Sanna made the inn a popular destination for important visitors from the East Coast through the strength of her intelligence and her hospitality at the evening fire, and the mountain guiding skill of Will, then twenty-three, and his younger brother, Doug. Built of giant logs and local rock on a 6,000-foot-high ridge, it was a starting place for challenging climbs 5,000 feet higher to the summit, which

Will and Doug pioneered on various routes and led 150 times between them. Will installed huge cables over the inn's roof, ostensibly to keep it from blowing away, which he obtained from a ferry that once crossed the Willamette River at Stark Street in Portland, and he strung a telephone line to town to make business easier.

Famous conservationists and artists, scientists and the heads of the federal land agencies visited and joined the Langille boys' treks. Will and Doug had taught themselves botany and geology on their explorations, and they picked up more from their guests, spouting their new knowledge back to the next group. The mountain became an open-air university, "The inspiration that awakened better things in our young lives," as Will recalled. In 1896, four members of a learned commission from the National Academy of Sciences visited to report to President Cleveland on the merit and purpose of the new forest reserves, which encompassed Mount Hood. The cold and imperious Charles Sargent led the commission; he also founded Harvard's Arnold Arboretum and was the world's leading expert on North American trees. Doug, in his overalls, corrected Sargent on his classification of a Noble fir, to laughter and chaffing by the other commission members, and Doug pointed out a new variant of Engelmann spruce that Sargent later introduced to science. Those two days in the woods led to lasting connections. The boys became informal local agents of the reserve, and in 1900 one of the commission members, Henry Gannett, director of the U.S. Geographical Survey, hired Doug to conduct an examination of the Cascades in central Washington.

The next summer, in August 1901, Gifford Pinchot met Gannett in Portland while touring the West and, presumably on his recommendation, went two days later to Cloud Cap Inn with a traveling companion, Frederick Newell. Pinchot directed the Department of Agriculture's Bureau of Forestry and was a friend of Vice President Theodore Roosevelt (Roosevelt became president the next month with the assassination of William McKinley); Newell was soon to be head of the Reclamation Service. Doug took them on a spectacular but arduous twelve-mile hike over ridges, through thick forest and across rushing streams, over meadows of ripe strawberries, thick with blooming wildflowers. Pinchot was a legendary sportsman—he beat Roosevelt in their boxing and tennis matches—but after eleven hours on the trail with Doug he fell exhausted into a bed of fir bows. The next day, Doug showed off evidence supporting the need for the forest reserve—damage by sheep and fires—and dragged Pinchot and Newell on horseback over a 7,500-foot ridge, and then on foot across a glacier with crevasses and an ice wall, over terrain Newell thought fearfully steep, and back to the inn. Pinchot had found the kind of man he needed to protect the forest reserves—tough, knowledgeable, likable, and, of utmost importance to Pinchot, a good writer. That fall and for several winters Doug went to work in Washington at Pinchot's side, returning to the West in the summer as an inspector, routing out rampant corruption and winning over pro-development loggers and miners to the conservation cause, which he and Pinchot both did with astonishing success.

Will had been gone four years when Pinchot visited the inn. He had left on a boat north only thirteen days after the gold rush's starting gun— the arrival of the steamer *Portland* in Seattle, in July 1897, with a ton of Klondike gold. Apparently restless after scores of summits of Mount Hood, nearing thirty, and with serious financial difficulties—like many during those depression years—he hoped to make money as well as have adventures. But he realized within weeks of his arrival in Dawson that the gold prospects had been exaggerated (although the convulsive peak of the rush was still to come the next year). He was homesick and hoped that Doug would follow him. He sent back botanical samples, worked at various jobs, but never made the money he needed to take a break back in Oregon. In January 1900, he scrounged the means to drive a team of four dogs clear across Alaska to Nome, but he had no more success there, prospecting over the tundra a month at a time without aid of blankets or food except what he could catch. Not until November 1902 did he finally give up and return to Oregon. A telegram from Pinchot reached Will there in March 1903, asking him to elaborate on thoughts Doug had passed on about conservation in Alaska; Will responded with a brief letter, and Pinchot soon hired him to go back north to scope out lands for new forest reserves. In less than a week's time Will traveled to Washington, met Pinchot, and left for Alaska.

A previous expedition had picked out some islands in Southeast Alaska for the Alexander Archipelago Forest Reserve. Will surveyed that enormous, complex region by sail and oar in the summer of 1903. In 1904, Pinchot dispatched him again with simple instructions to inspect the entire coast of Alaska from where he had left off in Southeast, across the Gulf of Alaska to Kodiak Island and the Alaska Peninsula, including Prince William Sound, Cook Inlet, the river valleys and inland, and lands on Norton Bay east of Nome on the Seward Peninsula, and not to miss the central Interior up the Susitna River—essentially, as it worked out, to explore most of Alaska—and report back within a season.

In Will Langille, Pinchot had found perhaps the one man in the world capable of such a journey, which no one would dream of attempting today, and also able to gather the facts and report them expressively—found fortuitously, for Pinchot had urgent need of the information. A political sweet spot lasts only so long. Roosevelt's popularity stood at a peak with his landslide election in 1904 and Pinchot had become the president's close confidant and playmate—rowing and talking philosophy or wading fully dressed through a D.C. canal in the middle of the night—and their conservation program rolled on, with the creation of the Forest Service and consolidation of all the reserves' 86 million acres under Pinchot's control in 1905. Pinchot wrote, "We seem to win what we want at every point." But within two years, the sweet spot would be gone: the Congress hostile, the Forest Service fighting for its life, and the president's power to create national forests by executive order under threat. Pinchot couldn't predict those events, but any smart politician knows opportunities are fleeting. Langille had to work fast.

Will voyaged into Alaska on his own with an expense account of $2,500, a camera (with which he was already expert), surveying equipment like Captain Cook's, fountain pens, and lined paper. No boat, no assistant, no support of any kind within thousands of miles. He worked his way north and west, from town to town, by taking steamers, through the Southeast and past the wall of glaciers and mountains of the St. Elias and Fairweather ranges, until he reached Kayak Island, Cape St. Elias, where there was a town called Kayak—now long forgotten—and there chartered a small motorized launch. An improbable land rush had seized the area—the Bering River and Controller Bay area just inside Kayak Island, and the booming community of Katalla, on a shallow slough, where oil had been found from seeps, and where docks and dancehalls had sprouted with the dreams of competing rail lines webbing across the map and into the mountains. It was a terrible place for a town, with potential harbors blocked by towering surf or raked by frigid glacial winds, their bottoms shifting with flowing silt, and, inland, swamps, rock, ice, and mountain peaks. The entire 250-mile-long region, overlain by phony mining, oil, and coal claims, looked to Will like one more massive Alaskan land fraud in progress.

Will entered the sound in his boat, visited the villages and towns—Nuchek, Valdez, the places Cook had seen—tramped up a mountain on Montague Island for a picture of snowy meadows between patchy strands of dark forest, and the crooked tracery of the rocky coves penetrating the land at Port Chalmers, and little islands beyond, "bits of emerald which are a pleasant contrast to the surrounding wastes of snow and ice." His report put across the sound's beauty and recorded its natural history and social conditions. Will found information where he could: after seeing the poor showing of island fox farmers' gardens, he judged the sound unsuitable for agriculture. Nor did the commercial value of the timber impress him. The sound's forests deserved conservation for the future use of local people, but at first look he thought it was too late even for that, as the smaller islands had all been taken up for foxes and speculators had staked claims to immense tracts on the mainland. The white population didn't care about the future. "No one seems to feel any interest in the matter of forest preservation," he wrote. "The citizens, generally, are more or less transient, and they dismiss the subject as being of no interest to them."

Langille's boat ran low on gas and he couldn't buy more, so at Ellamar he left it behind and boarded a steamer bound for Unalaska, in the Aleutians, there to change ships for Nome and his exploration of Norton Bay. He wrote friends a hilarious account of his six-week fiasco there, in pouring rain and wind, with an incompetent companion, a terrier that got tired and had to be carried, and a runaway packhorse—although the adventure could easily have killed him, for he chopped his foot with an ax and couldn't walk, it became infected, streams flooded, and he could travel only by building makeshift rafts and floating downriver, finally to be helped by Swedish fishermen in exchange for the horse. Next by steamer back to Unalaska, stranded, taking another boat all the way back to Seattle to catch one north, more time spent in the sound, finally with good weather, and by October Will was in Seward, where he planned a reconnaissance of the Kenai Peninsula prior to surgery on the still-troubled foot and the writing of reports.

On that journey, from late October though November 1904, Will's thinking evolved in an extraordinary way. He traveled on his own, walking from Seward to Kenai Lake, fifteen miles, there buying a dory and rowing the S-shaped glacier-carved corridor to Cooper Landing, another twenty-five miles; he walked from there to the gold towns of Hope and Sunrise, forty miles through the tundra mountain valleys, likely dusted at that time of year with snow, and back again; floated down the Kenai River to Cook Inlet, eighty miles through canyon, lake, forest, and wetland—where the bright yellow and rusty red refuse from autumnal cottonwoods and birches shows like mosaic tiles on the streambeds of clear backwaters—to the picturesque Russian village of Kenai; there he put the dory on a steamer to Seldovia, and in Seldovia hired a Native boy, and together they rowed on, exploring forty-mile-long Kachemak Bay, and hiked from the future site of Homer over rolling hills of head-high grasses, always brittle in early winter, to Tustumena Lake, forty miles, and back, with side trips to fill in the rest of the geography. The towns were tiny—just twenty-five non-Natives in all of Kachemak Bay—and the forests often thin, fire-ravaged, and slow-growing; the canneries were outside-owned and self-contained; agriculture showed little potential; mining yielded disappointment, Will frequently finding "abandoned workings and unused hydraulic outfits—the derelicts which mark the scene of wrecked hopes." But the magnificent wildlife, the woods as they stood, and the humbling and exhilarating scenery and variety of the place, including its glaciers, "these living, moving ice tongues"— all this had value.

Will wrote, "It seems a far-fetched idea to seriously contemplate forest preservation where there is so little apparent need of it and so little to preserve." But he added new considerations. "Here the living forest, though small in size, is the product of many years' growth, which when destroyed does not seem to thrive under the civilizing hand of mankind, and so slow is this growth that the seedlings of today will be of little use at the end of this century. . . . The forest cover in its primal state is also very essential to the prolonged existence of living game, which represents the best of its kind and, if cared for, will be a source of revenue to the inhabitants and pleasure to the world for many years to come." Pinchot and his colleagues in Washington had yet to popularize the new meaning of the word "conservation," as the sustainable use of natural resources—that happened the next year, after approval on a horseback ride with Roosevelt—but Langille had already gone beyond the concept. Sent to Alaska to find valuable timber to conserve for responsible cutting, he had instead found land without much economic worth and decided to save it for its own sake. People fit within his vision, too, as part of an existing community of life worthy of protection. His idea called for maintaining the integrity of an ecosystem, not just avoiding waste while logging a forest.

There's much more to Will Langille's story—that winter he snowshoed and mushed twelve hundred miles, past Ship Creek and Eklutna, along the Susitna, the Yentna, Skwentna, and Talkeetna rivers, over the Alaska Range, down the Tanana to Fairbanks, and up the Chena, and in March another journey back south to the Matanuska, the Nelchina and Tazlina to Copper Center, thence over the Alaska Range northward again, up the Gakona, down the Delta to the Tanana, and back to Fairbanks, where, receiving Pinchot's telegram on May 1 directing him to Wrangell, in Southeast Alaska, Will walked two hundred miles to Circle and caught a boat up the Yukon, presumably finding his way via the new White Pass and Yukon railway to Skagway, a reversal of his original arrival in Alaska, to catch a boat. He continued such trips, becoming a nearly full-time wilderness traveler, crossing Alaska back and forth, finding out what everyone was doing, seeing more than anyone had seen before, until, leaving the Forest Service in 1911, he went to scout forests in Brazil.

But first the story in Alaska turned a corner: Langille's reconnaissance was done. Pinchot grabbed his opportunity, the political sweet spot, and following Langille's advice, put before Roosevelt an executive order creating the Chugach and Tongass national forests in 1907. With additions Will and his colleagues later recommended, Langille received charge of the entirety of Southeast Alaska in the Tongass; and, in the Chugach, Prince William Sound, the Copper River Delta and Katalla area to the east, and almost the entire Kenai Peninsula to the west, including Kachemak Bay and Kenai, all the way up Cook Inlet to Knik Arm—including the land where Anchorage now stands. An area, 26.7 million acres in all, roughly the size of Pennsylvania.

The pattern of conflict for the next century of Alaska politics established itself in Chugach National Forest over the next few years—generations of Alaskan development dreams supposedly frustrated by federal conservation. There were two kinds of dreams. In the simplest variety, common until quite recently, Alaska's perpetually enthusiastic economic boosters, the greedy and the ambitious, made themselves easy victims for any serious-sounding businessman who tapped into their grandiose collective imagination, even with purely fraudulent plans for railroads, mines, factories, or brand-new cities. (What's really remarkable is that gold rush credulity persisted in Alaskan politicians for an entire century.) The town of Valdez was born of such a fraud when perhaps six thousand would-be prospectors were sold on a virtually impossible route to the gold fields over a mile-high glacier and down a raging river; the lucky ones gave up in shame while many died of exposure, accidents, scurvy, or starvation in the wilderness. Promoters raised millions selling building lots in towns with no reason to exist and stock in railroads that couldn't possibly lay track: one railroad would supposedly run through the mountains of Kachemak Bay from Bear Cove to the Fox River, where there isn't so much as a footpath even today; other lines would impossibly track the sandbars of Controller Bay on miles of trestles defended by vast breakwaters, where crossings were only ever accomplished by wind, surf, and the occasional bear. Sixteen oil and coal companies at Katalla issued more than $100 million in shares (a dollar then was worth more than twenty now), but never produced more than dribbles of oil for local consumption.

The other kind of dream was similar, but, instead of being advanced by a scammer, it had a sincere promoter as deluded as his supporters. The Alaska Central Railroad, for example, bound from Seward all the way to the Yukon River: the rails never made it anywhere people wanted to go, and the line scarcely had an economic hope even if it had been completed, as the territory emptied in those waning days of the gold rush.

The kinds of big projects that were more than dreams usually were owned by large industrial interests with Wall Street backing. The trans-Alaska oil pipeline occupies this category, and the rich copper mines at Kennicott, two hundred miles up the Copper River. In 1906, J. P. Morgan and Simon Guggenheim formed the Alaska Syndicate to buy into that find, as well as a steamship line, Prince William Sound canneries, coal prospects, and nascent rail lines starting out from Valdez and Katalla— the latter to access coal, the former copper, but neither of them pursued for long. Will Langille visited Valdez in 1907 and wrote a wry account of the town's frenzy over stock sales for a railroad to compete with the Alaska Syndicate's; the conflict between the two companies erupted into a fatal shooting that summer in Keystone Canyon. (The syndicate's competitor, H. D. Reynolds, was arrested for mail fraud in Massachusetts the next year.)

Will Langille endured fury and threats when Pinchot and Roosevelt froze oil and coal lands and created the national forest. All the financial failures were blamed on the national forest. The forest killed the Alaska Central Railroad, somehow. After Roosevelt left office, Alaskan developers, the territorial legislature, and the nonvoting delegate to Congress,

Judge James Wickersham, fought hard for the abolition of Chugach National Forest, and the Forest Service did shrink it, leaving room for Anchorage, among other cities. Many Alaska conservatives still rant against such federal arrogance as Pinchot's.

I imagine Langille dealing with the frantic boomers with the same bluff front he presented to the men who abused their horses on the White Pass, men he had tried to rally into cooperation to improve the trail. When they would not join even for their own benefit, he left to march on in his own way. He sized up the opposition to conservation in Alaska in a report written shortly before his departure from the territory in 1911:

> Since 1898 Alaska has been cursed with unscrupulous promoters and their visionary "get rich quick" schemes, who have elaborated upon and magnified every mineral discovery and every possible resource, until it was conceived to be a land where wealth was to be had with little or no effort. Glowing prospectuses have deceived large investors and coaxed from their hiding places the savings of the poor, to be wasted in some unfeasible and fruitless endeavor from which it was never possible to receive returns.
>
> Conservation is, to some extent, delaying development but not actually injuring Alaska. It is hurting the speculators who cry out against it. It is giving the people a chance to consider its possibilities, and ultimately legitimate investments will be made and judiciously expended in the development of Alaska's real resources, which are of worth, and the growth of empire will go on slowly but steadily, and not the least of these resources are the forests which it is great wisdom to retain in the care of the Forest Service.

In his predictions, Will Langille was many decades early. Some of Alaska's elected leaders still view the future with gold rush eyes, pinning hopes on extravagant possibilities, idealizing the individual competitor, blind to the qualities of the place that cannot be improved upon. Will's central idea was more than beyond its time—it's still politically unsettled—his call for all the resources of Alaska to be put under conservation-minded scientific management to assure rational, sustainable development. He lived at the dawn of a great national debate, as conservationists such as Pinchot and Roosevelt sought to end the government's backing of the scammers and greed-addled prospectors—the laissez-faire policy of the public domain, Hobbes's state of nature, everything open to the first taker. But saving Alaska already made sense to some people. As Will's supervisor wrote in a cover letter to his recommendation back to Washington, "I realize of course that at the present time it will probably be impossible to take action . . . but I feel that Mr. Langille's suggestion is at least worthy of consideration."

Such thoughts should encourage anyone wondering if broad cooperation for the environment is possible. Roosevelt and Pinchot tried to halt the pillage of the American West by the rich and the strong with a principled stand for the future and the public good. Will Langille's Alaskan journey shows human nature can include the ingredients to make that kind of change happen. As we'll see in Chapter 15, the personal qualities that made these accomplishments possible were more than a fluke—those same ingredients are in everyone.

15.

The Anatomy of Sharing

Suppose you find yourself in this unusual situation: a person you don't know, can't see, and may never encounter again is given ten one-dollar bills to split with you, the only rule being that if you refuse your share, then neither of you gets to keep any of the money. This happens only once—no second chance, no communication between you, completely anonymous. How many dollars would you need in your pile to go along with the division of the money? The economics of rational self-interest predicts you will accept anything more than zero—a single dollar bill—because some money is better than none, and refusing will get you nothing. Chances are, however, that you will not accept anything less than three or four dollars, and that most people in your counterpart's position will anticipate that and will offer a split close to even. The experiment is called the Ultimatum Game, and it has been repeated many times at universities all over the world, at low stakes and high, with similar results, the receiver getting an average of 42 to 48 percent of the money. Numerous more complex experiments with real money come to the same conclusion: many of us prefer fairness and other social benefits to our own personal advantage.

So Hobbes and his Enlightenment followers were wrong to find only selfishness and competition at the core of human drives—people also cooperate, even when they are not compelled, when reputation isn't on the line, when they are beyond the reach of a contract to reciprocate. Experience said it had to be so. Otherwise we would find no explanation for a man like Will Langille, in his lonely epic to set aside millions of acres of forest for no other purpose than to maintain its ecosystem intact. Nor would we be able to understand people who help others in a thousand ways that define our personal relations every day.

I remember as a young man reading that science had proved that true altruism did not exist, and trying to understand what that could mean about the people I knew. Economics and biology had flourished for decades with a purely competitive model of nature. Since I left college, new experiments have overturned those proofs, emerging with the satisfying simplicity of a naïve question: what if you created a situation that tested the basis of human motivation? In college computer labs and on the dirt ground of indigenous villages, games for money produced clear results, replicated and retested, and spawned a new science that is still growing, linking economics, psychology, anthropology, and biology, all based on the idea we already knew was true: that we want more than what's good for ourselves alone.

If a single thought started me on the path of writing this book, it came from an article in *Nature* in 2003, in which Ernst Fehr and Urs Fischbacher of the University of Zurich explained the new experimental and theoretical results on altruism, including some I will describe here, which instantly seemed to me applicable to marine conservation, the realm in which cooperation might otherwise seem almost impossible. I got up before six o'clock in the morning in Anchorage to call Switzerland, where Fischbacher answered in his office in the late afternoon, and I groggily tried to connect his ideas with my hopes. Mostly, their research hadn't gotten to my questions yet. Likewise, I've met hardly anyone outside academia who knows that the ground has shifted. We know the truth about ourselves in our hearts, but we don't believe it as a society. Our political and economic system depends on relentless competition and coercion. We look only to those forces for solutions, trained in an intellectual tradition buttressed by two hundred years of law and practice.

In fact, the competitive system itself destroys the cooperative impulse.

Here's another example from the laboratory, this also from Zurich, but the settings look the same most places: fluorescent-lit rooms with rows of computers and undergraduates typing their responses to choices that come up on the screen. In this study, subjects played the roles of workers and firms, again with real money, bidding competitively on wages in each round of the game, making contracts, with the worker then contributing part of his or her earnings back for productivity to benefit the firms. Contrary to theory, firms paid workers well over the agreed bids, because well-paid workers voluntarily contributed more than necessary back to the firm, increasing profits. Now add another step: the firms were given the option of fining workers who didn't perform. Productivity went down. If they could be fined, workers wouldn't contribute beyond the minimum. Other kinds of experiments and real-life studies have confirmed that incentives—either fines or rewards—wash away the desire for voluntary cooperation. Once offered explicit incentives, people work only to the level required to get the money, even when willing to do much more as volunteers. Leaders lower us to their expectations.

The Enlightenment philosophers looked at the world around them to develop their theories of human nature—Hobbes wrote of man's anarchic tendencies during the English civil war. Now laboratory experiments tell us that some people almost always try to cooperate; some cooperate conditionally, when it seems to work; and some almost always try to take a free ride, or defect, unless disciplined by the rest of the group. Sometimes the defectors win the day and everyone competes with disordered brutality; and sometimes cooperators build a community in which members sacrifice themselves for the whole.

Imagine a community in which working together brings everyone a better outcome than competing alone—for example, a village of ancient Chugach Natives harvesting and butchering a whale. Together, everyone receives a share of whale meat larger than they're capable of taking individually. But a defector sneaks off from the cooperative work to fish, obtaining both halibut and his share of whale, more food than anyone else. After seeing the defector's payoff, everyone else goes fishing, too, and no one butchers the whale. Soon all are worse off, with fish but no whale. This outcome, familiar to anyone who has worked in a group, is theoretically inevitable in an anonymous Prisoner Dilemma, as the game analogous to the situation is called, even in communities in which many people want to cooperate and only a few prefer to defect. In real life, however, communities often do work together effectively, with social tools to keep the defectors in check: reputation and shame, communication, and shared norms of behavior. And even without any of those capabilities, cooperators can still beat defectors in experiments, thanks to the same internal quality that controls the Ultimatum Game—their willingness to sacrifice for fairness.

Everyone in a laboratory group receives an endowment of tokens at the beginning of each round of a game—tokens worth cash at the end—which each can either keep or invest in a pool. The pool pays back dividends equally to everyone based on the total invested. If you put in all you've got and the player on the computer next to you puts in nothing, you both get the same amount out of the pot, but since he didn't spend his original tokens on the investment, he ends up richer. Only a small number of defectors can quickly bring the investment rate to zero, and then everyone loses. But if players are allowed to buy punishments against each other—a costly act called altruistic punishment, the odd and wonderful name given by economists—the game usually stabilizes at high rates of cooperation and larger returns for everyone. Altruistic punishers must prefer fairness to money, since it costs them to zap the free-riders, but it takes only a few to support the whole community.

I imagine Will Langille in this role, and many other environmental activists. It only takes a few of them to stigmatize waste and shame polluters, producing improvements for society as a whole. But they receive no more benefit than anyone else. And the individual costs can be high. Langille's heroic work to conserve Alaskan forests made him a pariah in the communities he served, often threatened and verbally abused. Such a sacrifice without a special payoff doesn't make sense according to traditional economics, with its concept of selfish human motivations. It took controlled studies in computer labs to prove the point already made by history.

Playing these games comes naturally. (I participated in a test of one at Indiana University.) Describing the experiments is much more difficult, and the abstract scientific literature about their setup, strategy, and outcomes can be nearly impenetrable. But once you're in the game, the analogy to innumerable real-life situations quickly becomes clear. Emotions normally attached to financial transactions accumulate in the gut—the will to win, the fear of being a sucker, and, most of all, the sense of what the counterpart on another computer is feeling, an invisible person, without a name, gender, or race, who nonetheless comes to life in the theater of the mind. Watching an experimental test, I saw two women bubble with warmth and goodwill for each other, despite being anonymously linked by computer and unaware of each others' sex, age, or identity. A male graduate student glared at the screen and gritted his teeth as he and his partner mutually drove each other down. We're well made for these games. We can use the scantiest information as routes inside other people, mentally model what they may do, and similarly model our own potential actions. We choose what kind of people we want to be every time we interact. This is the capacity of theory of mind that—as I argued in the first part of this book—may or may not make a person, but certainly allows us to bestow personhood on one another.

How our complex capacity to cooperate evolved has been one of biology's most persistent puzzles. Charles Darwin struggled with this difficulty as he was devising the theory of evolution. Selfishness makes evolution work, and it's hard to see how selfishness can beget altruism. Organisms strive to pass on only their own genes to the next generation. Why? Because organisms without that driving purpose have left no descendants. The competition for energy and space and the development of defenses against predators and the environment all support the fitness to reproduce. Sacrificing for one's own offspring obeys evolution's rules, and so does giving to family, as long as those who benefit are related closely to those doing the sacrificing, carrying many of the same genes. Organisms that reproduce by dividing can work together, since they have identical genes. Colonies of bees and ants cooperate for reproduction because they share the genetic code of their common mother. Likewise, uncles can profitably invest in their nephews. Friendship can be explained as well, as long as its reciprocity remains roughly equal—essentially, a helping exchange. But it's not obvious how selfish evolution could produce people with a strong preference for fairness and cooperation with complete strangers.

Darwin thought he solved the problem with group selection, the idea that a tribe of cooperators, or moral men, as he put it, will prevail over a tribe that lacks loyalty, courage, and patriotism. Altruistic individuals within the group would gain along with their tribe and get to pass on their genes. But group selection fell into disrepute in the 1960s because the mathematics didn't work out: with just a little migration and interbreeding, a group would lose its uniqueness too quickly to develop a decisive advantage over competing groups. And there the problem lay, unsettled, when an anthropologist and an environmental scientist, Robert Boyd and Peter Richerson, offered a critical additional element: culture. Human groups can adapt quickly, within a generation, when members innovate and copy solutions to problems, quickly enough for one group to become dramatically stronger than another. A tribe that developed cooperation could conquer or assimilate groups of selfish defectors who couldn't work together. The winners would spread. Competition among cooperative groups would motivate tribes to bind themselves together in ever stronger affiliation.

When cultural evolution spawned stable groups of cooperators, genetic evolution could kick in to support that tendency. Natural selection would favor traits beneficial for succeeding within a cooperative group. Likewise, punishment of defectors would put them at a disadvantage in passing on their genes. If a village observed a cultural norm for helping butcher a whale, a selfish or lazy member who didn't pitch in might become an outcast and find it difficult to mate. People whose emotions helped them follow norms—with feelings such as shame and guilt— would gain a reproductive advantage. Social rewards for successful

warriors could drive genetic evolution for group pride, fraternity, and obedience. Of course, the old instincts of selfishness and family loyalty would
remain inside everyone, alongside the new instincts of social membership. Brothers might have to decide between family and nation, as in the
Civil War. Everyone would be born with a drive to share and a drive to
steal, to conserve and to consume.

This is a big idea, and one wonders how to decide if it's right. Fortunately, there is evidence. The coevolution of culture and genes certainly
happened in some more easily visualized cases. Culture seems to have to
come first in the development of human speech. Oral communication
would have begun before our physical vocal system adapted into a superb
instrument for it. The ability of adults to digest cow's milk is another example. Europeans and western Asians have it, while most of the world's
population does not. The genetic and physical map of where people can
tolerate lactose reflects the ancient map of the domestication of cattle—
biological evolution followed that cultural practice. Killer whales' ability to
adapt to the environment culturally may have reduced the need to evolve
genetically, an idea supported by a slow pace found in their cells' genetic
clock (called mitochondrial DNA).

The Ultimatum Game itself gives evidence of how cultures can evolve
cooperation to meet environmental challenges. It turns out that different
cultures play the game differently, despite all people being built essentially the same biologically. Joe Henrich, a former student of Boyd, went
into the jungle of Peru to try the game with the Machiguenga, an indigenous group subsisting by slash-and-burn horticulture in the tropical forest, without any apparent interest in conservation (although their small
numbers prevented them from doing much ecological harm). To his surprise, they played the game like the selfish rational actors of economic
theory. Classical economics predicts that the receiver will accept a mini-
mal offer in an anonymous, one-shot game rather than refuse it and get
nothing—and that logic seemed to work for the Machiguenga, unlike
American college students who would turn down an unfair split. At first
Henrich had difficulty getting his work published, as it contradicted what

had been found in so many university lab experiments with undergraduates (although, interestingly, business and economics students tend to behave more selfishly in the experiments as well). He and a team led by Boyd expanded the project with a diverse group of anthropologists and tested the Ultimatum Game in urban, rural, and preindustrial cultures all over the world—a challenging task in remote tribal villages, where simply explaining the game could take half an hour for a single subject. "People do think this is strange, but they are also keenly interested in it, because it gets you money," Henrich said. Besides, visiting anthropologists have many strange requests: "It just seems like another weird thing that we do with them."

The results showed an extraordinary range of cooperative behavior. In some groups, nearly every offer was fair, so few were rejected. In others, nearly every offer was unfair, but most were accepted nonetheless because fairness in such a transaction was not a cultural norm. In three societies, in New Guinea, China, and Russia, many offers were beyond fair—the giver tried to transfer more than half the money—and those excessive offers tended to be rejected like the low offers. In those cases, the potential recipient showed a willingness to sacrifice to maintain a status quo level of fairness, even when unfairness benefited him or her. Henrich's statistical analysis of the results showed that the identity of individuals didn't affect the results, but the characteristics of their cultures did, depending on their altruistic or cooperative practices and on how they lived—their need for cooperation in farming, hunting, or economic trading; their intimacy or anonymity in daily life; how closely their homes were grouped and their privacy; and the complexity of their communities. Everyone had the capacity for altruistic punishing, and all did cooperate to at least some degree, but each culture evolved differently in how it used those capacities to respond to its own environment.

Social norms of cooperation and exchange developed to match a community's ecological situation. Could cultural evolution also happen on a global scale, reprogramming humanity to fit our planetary niche? There is an example of this happening: declining birth rates. Human population was stable for most of our history. Before modern medicine and nutrition, lives were typically short. But cultural norms also held down population by limiting reproduction well below the maximum possible, with births averted by delayed marriage, extended breast feeding, taboos on remarriage of widows, religious celibacy, and other customs. Beginning with the Enlightenment, science and industrialization saved lives, and the number of people on Earth began increasing rapidly, from about 600 million in 1700 to 1 billion in 1800; doubling to 2 billion by 1927; and doubling again to 4 billion by 1974. Projections call for the number of humans to peak in 2050 at about 9 billion. That sounds like too many, but if birth rates had not declined radically starting in Europe in the nineteenth century, the outlook would be not for stabilization at 9 billion, but for geometric expansion and a rapid ecological catastrophe. In a purely gene-driven world, that might have happened.

Birth rates in Europe dropped beginning in France in the 1830s—not, as is often assumed, in lockstep with industrialization and economic development, and not even clearly correlated with declines in infant mortality. A twenty-year project has studied changes in birth rates by looking at historic records in six hundred individual European provinces. It found religion, language, ethnicity, and region were key factors. In Spain, fertility dropped in diverse urban and rural regions that shared an ethnic identity, while fertility remained higher in surrounding areas with similar economics but different cultural ties. France industrialized later than England and Germany, but its birth rate dropped fifty to eighty years earlier. Simply, an idea emerged and was passed along: that couples should stop having children well before menopause. Surely improvements in health and economics helped motivate the new norm, but once started it spread much faster than modernization—most of the change happened in just thirty years, from 1890 to 1920. "It would appear that new ways of living adopted by some change the landscape for all," concluded Susan Cotts Watkins, at the end of the fertility study.

Deciding to have a child feels so personal and special, but of course it's influenced by what's considered normal in the culture. Norms edit out many possible choices before we even consider them. (A woman is capable of having twenty children, and a man of having twenty wives, but even billionaires don't consider having families of four hundred.) From the perspective of living within the society, normal preferences seem fundamental and unchangeable—otherwise, we wouldn't be so obedient to them. Then one day, they've changed. They changed for family size, and for public conservation of resources, and now, perhaps, for the morality of emitting carbon to the atmosphere, or even for the idea of sufficiency—the belief that we don't need so much wealth and power, that we can stop accumulating wealth before we demolish ocean ecosystems. Ideas, led by a vanguard, spread invisibly, until one day society is simply different. How does that happen?

16.

Catalysts for Conservation

Something happened during the generation born after the Civil War to produce many Americans who wanted to preserve nature. While the West remained wild and immense—by our standards hardly scratched—members of a powerful social group began to appreciate that the environment was exhaustible and in peril and wanted to rearrange basic relationships between government, people, and the land to prevent its destruction. The advance from first awareness to preliminary action took about thirty years, a time span similar to the lag in our own day's response to climate change. In both crises, experts delineated the hazard early, but only as damage accumulated did a cultural process gradually unfold of understanding, acceptance, and the determination to act.

At the time, Americans were living through industrialization's biggest jolt—they were the first occupants of overscaled modern cities, the first to experience culture and politics through mass media, the first to bow down before machines that had grown beyond the limits of individual human control. They first experienced the reassuring but deadening capsule of technology, which today can be all-encompassing. Hospitals handle birth and death, we're largely unfamiliar with killing or gathering our food, and our waste flushes away, unseen. We're sealed off from our fundamental animal functions. Many Americans have never defecated outdoors and are afraid to do so. Americans of the 1890s were using toilets for the first time—they had lived in both worlds. Writers of the era expressed the weakening connection to nature, sometimes profoundly and sometimes with unsophisticated nostalgia barely acknowledging the real source of its sense of loss.

The early conservationists didn't have words for ecology, environment, or conservation, at least not with the needed meanings, but they managed to express the problem of keeping our species within our ecological niche. George Perkins Marsh, who started his career as a Vermont lawyer, explored the issue in an influential book first published in 1864, *The Earth as Modified by Human Action*. He explained the fragility of ecosystems, the special qualities of old-growth forest, the potential to permanently damage soil, wetlands, rivers, and ocean, and he argued that environmental abuse similar to that rampant in the United States had brought about the collapse of ancient civilizations around the Mediterranean Sea. Marsh used a legal term to define humanity's proper relationship to the earth: usufruct, which means a loan for use only, with the obligation to leave the borrowed item as it was found. Under the heading "Destructiveness of Man," Marsh wrote:

> Man has too long forgotten that the earth was given to him
> for usufruct alone, not for consumption, still less for profli-
> gate waste. Nature has . . . left it within the power of man
> irreparably to damage the combinations of inorganic matter
> and organic life, which through the night of eons she had
> been proportioning and balancing, to prepare the earth for
> his habitation, when in the fullness of time his Creator
> should call him forth to enter into its possession. . . . Man is
> everywhere a disturbing agent. Wherever he plants his foot,
> the harmonies of nature are turned to discords.

Gifford Pinchot read Marsh's work when he and the book were both twenty-one years old, as he was thinking about his course in life. He ulti-mately would dedicate himself to the principle of preventing waste in general, conserving resources for future generations, an abstract goal and one whose success could benefit him only intangibly at best. Although independently wealthy, he worked without respite for President Roosevelt, traveling constantly, using train rides back and forth across the country to catch up with his letters and other writing. When Roosevelt called various conservation conventions and national commissions that Congress wouldn't fund, Pinchot paid the bills. He would also write the speeches to be presented and draft the resolutions that the bodies approved. His unyielding conservationism exposed him to hatred and vitriol. He was fired and elevated as a martyr. He wrote in his autobiography, "The greatest of all luxuries is to work yourself to your very limit in a cause in which you believe with your whole soul." Pinchot's life doesn't make sense without the theory of altruism; moreover, his story is a psychological and political case study of how a driven individual can catalyze altruistic impulses into national policy.

The Pinchot family developed in step with America in the nineteenth century. Gifford's grandfather made a fortune helping to strip the trees from the rolling hills around Milford, Pennsylvania, and northward, floating them to market down the Delaware River to Trenton and Philadelphia. Gifford's father, James, increased the family's wealth as a merchant in New York City, supplying the rich and the rising middle class with wallpaper and other interior decoration at profit margins so large they embarrassed him. When the railroad bypassed Milford, killing its industrial trade, James remade the dusty town as a picturesque tourist destination, convenient to New York, at the head of the lovely Delaware Water Gap, and began replanting trees. He built a castle-like house there, Gray Towers, and hung environmentally conscious art by Hudson River School painters, including a haunting image of clear-cut woods by Sanford Robinson Gifford, after whom he named his first son. In full partnership with his equally indomitable wife, Mary, James trained up his oldest son like a topiary hedge to fulfill a very specific role in public life—to be the nation's forester.

The raw material couldn't have been stronger. Gifford grew tall and handsome, adept at manly pursuits in sports and the outdoors, a fast study, and, most important, energetic, charismatic, and painfully honest. His letters home from boarding school and college included news of his own minor misdeeds, inviting back letters loaded with guilt-inducing criticism, sometimes about infractions so small as to be incomprehensible to a modern reader, as when his father wrote, "Your mother has just received another letter *without a date.* Your carelessness and inattention to our wishes is painful." Gifford ultimately joined in the chorus, scolding himself without mercy in his private diaries for transgressions such as reading for pleasure, eating sweets, or spending a day at rest, while also admitting periods of depression, for which he also blamed himself. Pinchot seems to have accepted without irony the stated values of the age— for courage and strength in doing good, for fraternity and group loyalty, for purity and virtue—and he granted few exceptions to himself or to anyone else who fell short. Similarly, stopping the waste of the country's natural resources became a crusade of national self-improvement. Those around him were either invigorated by his righteous energy or irritated by his zealotry. The obese President Taft came in for Pinchot's disapproval as a backslider in the cause, a moral fault. Taft wrote to his wife, "His trouble is that no one opposes his methods without arousing in him a suspicion of that person's motives."

Adolescents tend to hold strong, two-dimensional opinions, to be indignant with insincerity and pursue their beliefs with the youthful spirit of certainty. In Gifford Pinchot, those qualities lasted longer than in most. His delayed adolescence continued until his parents died in his forties— after the most important period of his work. Until then, he poured his heart into a private space, an unconsummated love affair. After studying forestry for a year in Europe, Gifford took on the forest management of the Vanderbilt family's Biltmore estate in the mountains of North Carolina. There, at twenty-eight, he fell in love with Laura Houghteling, also twenty-eight, an attractive woman from a rich Chicago family who was visiting the mountains to recuperate from an advanced case of tuberculosis. They treasured passionate moments together, all charmingly innocent: the day he addressed her by her first name; a quick embrace while a friend fetched a tea tray. When Laura's illness restricted her to bed, her modesty forbade Gifford from the room, so he read to her from beyond the doorway. The ever-superior James and Mary Pinchot disapproved of the match, as they had of their daughter's to a British diplomat, but Gifford finally received their support to wed Laura, then bedridden, once she recovered. She instead died five weeks later.

The family had worried about Gifford's grief, but he impressed them with his stoicism. In fact, he simply didn't accept that Laura had died. Within a few weeks he began communicating with her in the beyond. Before her earthly demise, the couple had read books on spiritualism—common then, in an era of séances—and Gifford and Laura had agreed they would never truly be separated. At night Gifford would stand outside the empty house in Washington where she died, feeling her presence. He soon began recording interactions with her in his daily diary—he heard her words, he joyfully communed with her spirit, he even began reading spiritual books with her. He behaved as if Laura's parents were his in-laws, calling Mrs. Houghteling "Mother," and told his own family he believed himself married to Laura. For the next twenty years Pinchot remained faithful, each day recording his contact with "my lady" in his diary: "not a clear day" when she wasn't in touch, "a good day" or "not a blind day" when he heard from her, and "A clear and beautiful day. Ten years and four months today," in June 1904—while at the height of his political power.

Pinchot's life in politics began at a high level, in 1896, as a member of a National Academy of Sciences commission appointed by President Cleveland to study the West's controversial new forest reserves and to define their value and purpose (this was the same group that visited the Langilles at Cloud Cap Inn, although Pinchot didn't come on that part of the trip). Pinchot, then a forestry consultant, had encouraged creation of the commission through his family's political connections and contributed a clever idea to sidestep a lack of support for conservation in Congress by recruiting the Academy to lend its prestige. He became the commission's secretary, although manifestly underqualified and a generation younger than the other august members.

Pinchot spent much of the group's three-month tour of the West with a new friend, John Muir, who, although twenty-seven years his senior and already a famous naturalist and writer, was still eager to romp through wild places with Pinchot and sleep under the stars. But Pinchot made an enemy of the stuffy and self-important commission chairman, Charles Sargent, in a disagreement that finally split the commission. Sargent's side, the majority, wanted to emphasize protection of the reserves, with a recommendation to deploy the army to defend them and bar most development. Pinchot's minority advocated sustainable use managed by a new civil service. Muir supported Sargent's view. The two positions, later hardening into poles of rivalry, ultimately disrupted Muir and Pinchot's friendship and, according to Pinchot's critics, divided and weakened the conservation movement for a century. But the legend of the Pinchot-Muir dispute is another story; what's important here is that Pinchot's view ultimately prevailed in the forest reserves, fortunately, for they barely survived congressional attempts at elimination even with the political understanding that their resources would remain available for human use. Simply opposing waste and favoring government control was radical enough. The alternative would likely have been for the forests to be returned to the public domain with land-rush conversion to private ownership, not preservation.

President McKinley made Pinchot head of federal forestry in 1898, and he began bonding with Roosevelt on a visit to Albany in 1899, a month into Roosevelt's brief term as governor of New York. Pinchot was still only thirty-three years old. "We arrived just as the Executive Mansion was under ferocious attack from a band of invisible Indians, and the Governor of the Empire State was helping a houseful of children to escape by lowering them out of a second-story window on a rope," Pinchot recalled. "T.R. and I did a little wrestling, at which he beat me; and some boxing, during which I had the honor of knocking the future President of

the United States off his very solid pins." Pinchot had come to New York
to study the Adirondacks and to summit the state's highest mountain in
a severe winter storm, which he did dressed in a sweater and cap. Roosevelt, forty, had been charging up San Juan Hill in Puerto Rico a year earlier, and a year later he was elected vice president, and a year after that
McKinley's assassination made him the youngest president in history.
They were bound to be friends, similar not only in their vigor, idealism,
and youthful success, but also in the fortunate birth and upbringing that
put this success in their path, as if they were destined together for where
their careers took them. They talked about that once, in July 1903, during
a long afternoon outdoors at Roosevelt's home on Long Island, rowing
and picnicking. Pinchot insisted that God's providence had guided the
president's rise to power. But Roosevelt had seen more of life than Pinchot, and, unlike Pinchot, had done so on a path of his own choosing. He
had lost his parents and his first wife at a young age, and he had been
defeated in elections as well as winning them—he believed in luck.

In Roosevelt's administration, Pinchot remained two steps down in the
organizational chart, building his tiny staff in Agriculture's Division of
Forestry, an agency without direct control of any forests, as the Department of the Interior administered all federal lands. But over time his informal status as a member of the president's "Tennis Cabinet" gave him the
power of a conservation tsar, directing and coordinating land and water
policy between members of the real cabinet, endowed with implied authority that erased the barriers between departments and essentially placed him,
a subordinate, between the president and his secretaries. Officials outside
Roosevelt's chummy circle disliked the arrangement, including the future
president, Taft, who as secretary of war had to take Pinchot's orders, and
Richard Ballinger, commissioner of the General Land Office, who opposed
the creation of Chugach National Forest, among other initiatives. Pinchot
breezily brushed off Ballinger's concerns that withdrawal of lands for the
forest would hinder the new railroads in Prince William Sound. Ballinger
finally resigned.

Pinchot pushed forward relentlessly, working and traveling incessantly, organizing conferences and writing scores of letters in a day. He was a magnificent administrator, and the agency he created reflected his personality—eager and idealistic, full of men like the Langille brothers. Always an eager joiner and organizer of clubs and committees, Pinchot reveled in pulling together a loyal team of believers. In 1905 he won legislation creating the Forest Service in the Department of Agriculture and transferring the forest reserves to it from Interior, renaming them national forests. Pinchot soon began scheming to bring all national resources that could be conserved under the control of a single new agency, the Department of Natural Resources, although that never happened.

Pinchot and Roosevelt both believed their conservation work was changing the direction of civilization. Roosevelt, justified by his popularity and the good he believed he was doing, assigned himself powers no other president had possessed. He harnessed Pinchot's skill and drive to accomplish far more than most presidents, and used Pinchot's talent as a speechwriter and ghostwriter to enhance his own legacy as a thinker and wordsmith. Pinchot, for his part, gladly wielded Roosevelt's magic wand to remake the nation's relationship to its land and rivers. He enjoyed spending time at the White House and recorded each of the president's complimentary remarks carefully in his diary. Roosevelt's personal regard equated to his own power, so keeping track was natural, but one also hears in these entries the tone of a son too often denied praise by his real father—such as when Pinchot tells his diary in 1902 that the president is starting to use his first name more frequently, or in 1907, when he writes, after receiving a political attack, that "T.R. said to me, 'I have never known a more distinguished public servant than you are, or a more efficient one—there couldn't be.'" Like a parent reassuring a child, Roosevelt goes a little too far, and, like a child, Pinchot accepts the praise at face value.

Pinchot's real parents grew jealous of his relationship with Roosevelt. They called the president a vampire and complained, with justification, that Gifford's career had stalled and the overwork had damaged his health. For once, Gifford didn't listen. When Roosevelt repeatedly passed him over for promotions, he defended the president to his father and in his own diary—accepting, for example, that Roosevelt needed a Westerner for secretary of the interior rather than himself. The politics of each decision may have been defensible, but Roosevelt privately confided to a close friend that Pinchot's righteousness could be dangerous and he had to be closely managed, for otherwise, "his great energy would expend itself in fighting the men who seemed to him not to be going far enough forward." By granting power only by personal favor, Roosevelt kept Pinchot on a leash. When both were out of office, several years later, Pinchot drafted the conservation chapter for Roosevelt's autobiography, having received the promise that Roosevelt himself would add words crediting Pinchot for his part in the movement. Roosevelt broke his promise and minimized Pinchot's role. Yet the book as published gives Pinchot primary credit for the administration's conservation work—because Gifford's mother, Mary, demanded that credit in a blistering, shaming letter to Roosevelt, the sort Gifford himself used to receive, but which he would never issue.

If Pinchot did regard Roosevelt as a father, the president's handpicked successor, Taft, became the worst of stepfathers. Taft compared disadvantageously to Roosevelt in every way: absurdly fat, a weak campaigner, indecisive and easily swayed, and skeptical of the reach of his own power. In Roosevelt's last years in office Congress pushed back against his conservation legacy and he chose and campaigned for Taft as a bulwark to protect it. Taft's success as secretary of war and his warmth and skill at personal politics convinced the Tennis Cabinet he could do the job, and on the campaign trail he delivered a Pinchot-written speech extolling conservation policies. But once elected, Taft cut off the Roosevelt team and began working with new men who were their enemies. His public remarks also worried the true believers. To make sure Taft remained loyal, the outgoing president met alone with him and Pinchot over a long evening in his White House study. Roosevelt made Taft promise to continue his conservation policies, point by point, and to rely on Pinchot, who recorded, "T.R. asked Will to call me in if ever there was trouble about Conservation." Perhaps the meeting guaranteed that Pinchot could keep his job at the

Forest Service, despite Taft's private appraisal of him as a crank and a radical, but in the new administration Pinchot's role would go no further than his job description. Later, as if to show who was boss, Taft made a point of discarding a speech written for him by Pinchot and his friends when addressing a conservation conference Pinchot had set up. He joked that the audience had already heard enough of Pinchot's words delivered by the other speakers. Instead, Taft noted that conservationists tended to let their imaginations run away with them.

In his autobiography, Pinchot describes the dread of the last weeks of the Roosevelt administration as the circle of fortunate friends who had played and worked together in the glow of that unique moment in American history saw their special status ending and their achievements in danger of being reversed. Upon leaving office Roosevelt immediately departed for an African safari. That day broke Pinchot's heart. Within a month of Taft's inauguration he detected that all of Washington had fallen into listlessness—he wrote that "Hardly anybody seemed to care about anything anymore"—as if the entire city had lost its best friend, not just him.

But it wouldn't be fair to say those emotions alone drove Pinchot's resistance to the new administration. Taft removed his closest allies from their jobs and replaced them with business-oriented adversaries, including bringing back Richard Ballinger, now as secretary of interior, an appointment that placed the man Pinchot had previously defeated over creating the Chugach in a higher office than he had ever been able to attain himself. The smooth cooperation between the resource agencies became trench warfare. Ballinger and Taft began overturning important conservation decisions, seeking to privatize irrigation projects and pushing to grant corporations perpetual rights to sites for hydroelectric dams. Rather than defending against the rising opponents of conservation— such as the growing, organized movement in Alaska to do away with Chugach National Forest—Taft seemed ready to speed along the process of dissipating what had been saved by Roosevelt.

Pinchot wouldn't let that happen. Taft and Roosevelt were right—he was a dangerous zealot, at least from their perspective. Fired with moral outrage, he took on the role of an organizational whistle-blower. He hadn't compromised with death when it took his fiancée, and no more would he stand by for the loss of the work he believed in with his whole soul, even at the cost of his own political future, and that of his party and president.

17

Whistle-Blowers to Save Alaska

W histle-blowers are altruistic punishers run amok. Any enforcer of social norms pays some cost—even a raised eyebrow of disapproval can backfire—but over the long term, groups support those who uphold their standards. A whistle-blower, on the other hand, may call attention to the faults of a group itself, occasionally causing its demise, as in cases of corporate corruption that lead to business bankruptcy. From the point of view of the group, a whistle-blower violates norms of loyalty and obedience and deserves the full power of social retribution. Research shows that the more severe the organization's wrongdoing, and the more fundamental its bad practices are to its success in the wider world, the greater its brutality against a whistle-blower, which can include character assassination, financial ruin, and social obliteration, leading to family breakup and the loss of mental and physical health. A broad U.S. survey found that whistle-blowers reporting to outside authorities suffered life-changing negative consequences such as firing and career blacklisting nearly 80 percent of the time; fewer than 2 percent of cases had positive professional outcomes for the whistle-blower. And these punishments against whistle-blowers worked: organizational loyalty is high. A third of U.S. workers said they had seen illegal or unethical behavior on the job, but only a small percentage became whistle-blowers. Often it takes a special personality to blow the whistle, one bordering on sainthood—that is, it takes someone with qualities so out of the ordinary they're almost impossible for most people to comprehend.

Some researchers have tried to develop a profile of the typical whistle-blower to assist businesses to avoid hiring them. Others try to understand and empower whistle-blowers as reform agents for broad-scale social ideals such as conservation—ideals requiring allegiance to the whole world rather than any group within it. A Forest Service employee association encourages whistle-blowers with the goal of forging public values "based on a land ethic that ensures ecologically and economically sustainable resource management."

Outwardly, whistle-blowers don't share predictable characteristics—they don't match up well on income, social or professional rank, education, gender, or race. What most do have in common is a solid core of beliefs. Whistle-blower saints tend to hold a concept of justice like that of Cicero or Grotius: that law is natural and unchangeable. Often they see no real choice when faced by a wrong, because acquiescing would leave an intolerable sense of being personally corrupted. They tend to be hard workers strongly committed to the stated goals of their organizations, loyal and patriotic, and uncompromising to the point of rigidity. In other words, naïve. On psychological tests they often show a poor ability to perceive how others see them and ineptitude at fitting their behavior to varying social situations. Rather than heeding the cues of those around them, they respond to simpler, universal motivations. But despite being idealists, they're also likely to suffer from low self-esteem. If that aspect seems odd, consider Gifford Pinchot, whose extended adolescence so closely fits the typical whistle-blower profile: he didn't smoke or drink, he remained faithful to a dead fiancée, he worked to exhaustion, and spent his money in service of his cause, devoting his all to his organization—yet he avidly sought verbal approval and he privately chastised himself for meaningless vices. Perfectionists always fail on their own terms.

Whistle-blowers can also be misguided or wrong. The strictly moralistic view can be destructive. One person's sense of incorruptible right can differ from another's. Ethical rules, although seeming universal, can change like other norms. Madison Grant, a Boon and Crocket Club big-game hunting friend of Roosevelt (and the founder of the Bronx Zoo), campaigned to stop market hunting and to regulate killing of furbearers in Alaska, a clear good after 150 years of wanton extermination. In 1902, Roosevelt signed a bill Grant helped write to impose hunting seasons, bag limits, and prohibitions on selling wild meat in Alaska. Without notice, Alaska Native peoples lost their primary cash income; their subsistence food supply became illegal for most of the year, including the only times animals were present in many places. But Grant blamed the Natives for killing the game and preferred their demise to the death of animals. Grant believed nonwhites, including Native Americans, should be subjugated by the master race—by war, if necessary—and in any event stopped from re-producing. Eugenics held a central place in the progressive program, also endorsed by Pinchot and Roosevelt as a form of conservation—the con-servation of superior human heredity (as I'll document in Chapter 18). This actually makes sense, in a disturbing way. If these men's dedication to norms evolved from strong group affiliation, then a moralistic concern for conservation could grow along with a sense of superiority, loyalty, and purity.

Pinchot became a whistle-blower on the grandest scale. That story begins in 1904, on Will Langille's first visit to Prince William Sound, when he reported the massive spread of phony land claims in the rush for oil, coal, and minerals around Katalla, Controller Bay, and the Bering River. Pinchot forwarded Langille's report to the General Land Office in the Interior Department, which had authority, but noted in a memo to a staffer that he didn't expect anything to come of it. The land office was notoriously corrupt and incompetently managed, and it administered laws that didn't make much sense. In the American frontier tradition, settlers and prospectors could obtain land and its resources by staking a claim, working it, and proving up, which meant getting title by establishing they had met certain standards of development. The laws envisioned hardy pioneers spreading civilization and didn't fit the industrial scale of modern economics. All coal lands more than fifteen miles from a rail line, once properly claimed by a prospector, were for sale to the finder for $10 an acre, but an individual could obtain no more than 160 acres and a group was limited to 640 acres. Such a small parcel of property could not support the cost of building a remote coal mine, handling facilities, railroads, docks, and all the rest—this even Pinchot recognized. Speculators and developers got around the size limitations in the land laws by using dummies—men who put up their names only, for a small bribe—and then consolidating the claims later. They managed the laws' other requirements in various fraudulent ways.

The land office did send an agent to Alaska in 1905 to investigate the Bering River situation and found, among about nine hundred obviously bogus claims, thirty-three that might have merit, although even they were flawed. Clarence Cunningham had filed the claims, totaling 5,280 acres, using powers of attorney from various people with addresses in Washington, Idaho, and Ohio. They obviously had planned to consolidate the claims, but who could prove their original intent? The land lay more than twenty-five miles inland, up the Bering River from just east of Katalla, and then through the brush and spruce forest and along granite-bouldered Canyon Creek to a set of ridges, as high as three thousand feet, that separated the flow of the Bering, Kushtaka, and Martin River glaciers and their iceberg-strewn lakes. Getting there wasn't easy, and getting around required climbing through thick woods and undergrowth and up and down the creek valleys and draws that split the ridges. But it was a beautiful area, with coal evident in many seams reaching the surface.

In 1906, on Pinchot's advice, Roosevelt froze disposal of federal coal and oil to help push through a proposed leasing law that would compensate the government rather than giving the lands away. Valid claims already on the books could still be pursued, but those were few in Alaska, with the possible exception of the Cunningham claims. Consequently, these claims, if approved, would become an Alaskan monopoly—a valuable one, since coal coming into Alaska by ship from British Columbia or beyond cost several times the projected price for locally produced coal. The next year, 1907, Cunningham sold a half interest in the claims to the Alaska Syndicate led by the Guggenheim family and J. P. Morgan, which would use the coal to run a railroad from the sea to its copper mine at Kennicott. The nation's notoriously monopolistic captains of industry now controlled a potential monopoly on Alaskan coal.

As mayor of Seattle and a successful attorney there, Judge Richard Ballinger got to know the city's investors in Alaska long before going to Washington, D.C. Some of his friends had their names on the Cunningham claims. But despite his anticonservation connections, his reputation for honesty and ability got him a job in the Roosevelt administration to clean up the land office early in 1907. In short order he modernized its processes, fired old cronies—including a classmate of the president's—and brought in young, talented staff. The efficiency of the operation improved rapidly. But Ballinger chafed at Pinchot's informal power in Washington and the aggressiveness of the conservationists. He fought a series of battles over Pinchot's initiatives during his one year at the land office, including his opposition to creation of Chugach National Forest, which potentially affected some of his Seattle friends. Ballinger may have been as moral a man as Pinchot, but in another way: he believed in the chain of command and the limits of the law, the ethics of authority and order. And, unlike Pinchot, he had been reared on practicality, a self-made man who had earned respect in the freewheeling West.

Among Ballinger's bright young men was Louis Glavis, who took over the Portland division of the land office at age twenty-three, already with several years' experience as an investigator. Glavis's intensity and righteousness could have disappointed no one. He loved the thrill of a promising investigation, when the strands of a case lay ready to be picked up, but would come together only with perfect finesse and decisiveness, and with just one chance to do it right. He looked the part of an intrepid investigator, too, handsome, square-jawed, and tall, and he had a voice as powerful as a tugboat captain's, perhaps to compensate for a slight stutter.

On the job Glavis sniffed for corruption like a hound aching to run, and it didn't take long before he got a whiff. A fellow agent, Horace Jones, came to Glavis's hotel in Portland to tell him about the Cunningham claims. Jones had been ordered to launch an extensive investigation of the claims as a criminal conspiracy, but then Ballinger himself had said instead to make only a cursory examination and report as soon as possible. When Jones insisted on digging, he was transferred to the Salt Lake City office. Why wouldn't Ballinger want a thorough review? Glavis began inquiring, wrote to Washington, and traveled there at the end of the year to meet Ballinger, where he was assigned to the case before heading back to the Northwest.

Before Glavis could start investigating, however, a friend called on Ballinger at his D.C. office: a former Washington governor and holder of one of the Cunningham claims. He complained of the delay in their approval. Ballinger ordered many of the claims "clear-listed," meaning they could proceed directly to being patented, the final step in giving up government ownership. Glavis, back in Portland, was alarmed. He objected and won a delay, then dove into his investigation with vigor. Believing someone in his own agency was colluding with Cunningham's partners, he grabbed a night train from Portland to Seattle before they could be alerted of his intentions. He found Clarence Cunningham at breakfast the next morning, cheerful and compliant, willing to open his books. Flipping through a journal, Glavis stumbled on the key to the case, an agreement dated five years earlier in which the claimants committed to consolidating the coal lands once they received title. The timing of the agreement proved their fraudulent intent. Glavis, fighting to conceal his excitement, talked his way out the door with the journal in hand and had it copied. Glavis then led a team of investigators across the country to call on the claimants simultaneously so they couldn't coordinate their stories.

Over the next eighteen months, the investigation and countermoves webbed through deeper layers of detail, fractal in their complexity, producing enough confusion and contradictory evidence to fill books. Books were eventually written, some supporting each side. No one ever proved Ballinger committed a crime, but his actions for the Cunningham group were at least unethical. During his year running the land office he sent them the confidential communications of his investigators, interfered with the inquiries, and wrote and advocated for legislation to allow the claims to be consolidated, as well as policy changes that could have cleared their way. When he left office he took up legal work for the claimants as a lawyer in opposition to the government and lobbied his former subordinates on their behalf. When he returned to government as secretary of the interior a year later, Ballinger attempted to approve the claims using the new law he had supported that allowed consolidation.

Despite his gross personal conflict of interest, however, Ballinger's integrity remained intact in terms of policy: he believed in development in Alaska, which required coal, and the Cunningham claims were the only potential source of coal as long as Congress failed to enact a new mineral leasing law. Taft knew Ballinger's views when he selected him as secretary of the interior. So did Pinchot, who saw the appointment as evidence that Taft intended to betray his commitment to Roosevelt's conservation policies.

As Taft and Ballinger prepared to move into their new offices in spring 1909, Pinchot and the rest of Roosevelt's Tennis Cabinet rushed to nail down every possible natural resource for conservation before Inauguration Day. Nine days before leaving office Roosevelt signed a proclamation expanding Chugach National Forest to include most of the Cunningham Claims, giving Pinchot a hand there, although not control. Bureaucratic warfare between Pinchot's Forest Service and Ballinger's Interior Department erupted immediately upon Taft's taking office, with Ballinger winning several rounds administratively and Pinchot counterattacking through his political contacts and popularity with the press. By focusing his fire on Ballinger, not Taft, Pinchot retained his viability in the administration and his leverage on President Taft through his connection to Roosevelt. Praising Taft as a conservationist, he left open the possibility that the president could reform and fire Ballinger.

That July, Glavis turned to the Forest Service for help in investigating the Cunningham claims, which now lay within a national forest. Will Langille and a team of three others trekked for two weeks over the ridges between the glaciers, establishing that some of the claims had been chosen for their rich timber, not coal, another reason to invalidate them. For Glavis's superiors at the land office, already more than impatient with him, consorting with the enemy in the Forest Service was the last straw. Although they already had concluded privately that the Cunningham claims could not be approved, they took Glavis off the case, partly out of fear Pinchot would gain control of a potentially damaging political issue. Glavis, his worst suspicions confirmed by his removal, directly confronted his superiors with accusations of corruption. They began to wonder if he was mentally unstable—their private communications suggest they didn't think they'd done anything out of the ordinary. Meanwhile, Glavis used a friend to sneak the coal files out of the office to a public stenographer for copying, who then delivered the copies to the Forest Service.

The newspapers anticipated the "war between Pinchot and Ballinger" breaking into the open the next month, August, at the National Irrigation Congress in Spokane, where both were scheduled to speak. But the more important meeting happened the night before Pinchot's speech, when Glavis called on him in private and laid out the evidence of Ballinger's interference with his investigation. Pinchot listened, read the documents, and immediately made a critical decision: to set up a personal meeting between Glavis and the president. His stated motivation for this move befits a simple and honorable whistle-blower: Ballinger worked only for Taft, so only Taft could decide what to do. But Pinchot had a broader view than just Alaskan coal claims. He had already started blowing the whistle on the entire administration's backsliding on the conservation agenda. By putting Glavis's allegations in Taft's lap, Pinchot forced the issue. Either Taft would have to repudiate Ballinger, or he would have to accept ownership of a political time bomb.

Pinchot sent Glavis to Chicago to work with a Forest Service lawyer polishing his indictment of Ballinger, then on to Taft's summer home in Beverly, Massachusetts, a seaside rental north of Boston. Glavis and Taft sat alone in the president's study. After Glavis introduced the situation, Taft began to read the documents, and not only the summary, but fifty more pages Glavis had brought from the files. Hours passed. Taft sent his secretary to cancel an afternoon motoring outing with his wife. He asked Glavis a few questions, made comments about the unfairness of the press in reporting decisions that had gone against Pinchot, and finally ended the meeting.

Taft would have been wise to turn the matter over to a disinterested party at that point, but instead he called in Ballinger and his staff. Since Glavis's accusations had been written to paint the situation at its worst, Ballinger was free to add explanations and nuances that made that presentation appear unfair and dishonest. Taft looked no deeper and decided to exonerate Ballinger publicly and to fire Glavis, and moreover to take Ballinger's side in his conservation fights with Pinchot.

Taft didn't want to fire Pinchot and Pinchot wouldn't leave without being fired. The next time they met privately, in Salt Lake City, Pinchot told Taft he would keep fighting Ballinger regardless of the president's instructions, and that Taft might have to dismiss him. Taft, in response, said he wasn't going to worry about getting elected to a second term—perhaps a dare, or maybe declaring he was resigned to his fate only six months after taking office. In a follow-up letter, Taft ordered the fight to stop. But Pinchot believed conservation was a moral issue, and he could not compromise his morals. Like other whistle-blowers, he saw no choice—he had to stop Ballinger, and now Taft, by any means necessary. Thanks to Glavis, he had the means. "It never occurred to me that there was any other course I could follow," he later wrote.

Pinchot arranged for Glavis to publish his accusations in *Collier's,* one of the nation's most popular magazines, and assigned Forest Service staff to help him draft the explosive cover story. A stream of articles followed alleging Ballinger's corruption, also largely instigated by Pinchot's office. Congressional hearings were called. Ballinger's side sought to steer them against Pinchot for promoting accusations the president had already dismissed. Pinchot outflanked him—he relished being fired—writing a letter that was read on the Senate floor defending the publicity work as a patriotic act forced upon the Forest Service to protect the property of the United States. Taft took the statement as a direct rebuke, as Pinchot had intended it, and fired him, although Taft was tortured by the break with Roosevelt that the firing represented (Roosevelt was still hunting in Africa). As Pinchot anticipated, the political situation froze Taft and Ballinger's actions in reversing conservation decisions; with the hearings coming up and Pinchot becoming a martyr, they couldn't afford to make matters worse.

The theater of the hearings would be familiar to anyone who remembers Watergate or Iran-contra: the partisanship of the special joint House-Senate committee, its chairman trying to protect the president without seeming to; the ornate room with a massive chandelier; brilliant lawyers playing to the press; and the packed spectator gallery, its front row occupied by two irrepressible women, Gifford's mother, Mary, and Roosevelt's beautiful and outrageous eldest daughter, Alice. They carried on a witty commentary together, audibly deflating the overstuffed politicians.

One wonderful moment, remembered by Glavis years later, came during his testimony about a town-site fraud near Cordova—a town created to sell lots with no real possibility of development. Glavis avoided saying the name of the town, but finally was badgered to do so. It was named for Senator Nelson, he said, the same hostile, tobacco-chewing Republican chairman then presiding over the committee, "in appreciation of some legislative favors he extended to those people."

Nelson blanched and mumbled as laughter rippled across the room.

Alice Roosevelt said, "Look at the chairman. He's going to faint."

Mary Pinchot responded, "No, he's not. He's just swallowed his cud."

Pinchot's team understood their audience was the nation beyond Washington, which followed the hearings closely for four months. Gifford's brother Amos kept the newspapers updated, sending out handy, simplified summaries of the action. Glavis and Gifford testified unflappably, paragons of rectitude. Their side successfully linked conservation, Roosevelt, and anticorruption and set Ballinger up as the opponent of all three. Ballinger's blundering defense and shifty testimony played into the trap—he earned the nickname Slippery Dick—by attacking conservation and the accusations of corruption together, as if the arguments over policy and honesty were one and the same.

As so often happens, what finally caught Taft was the cover-up, not the crime. His attempt to discredit Glavis at his home in Beverly, briefly successful, was based on his ostensibly judicious care in reviewing the entire, enormous case file, and receiving a detailed report from the attorney general, before he reached his own conclusion clearing Ballinger. But that didn't ring true to the leader of the Pinchot-Glavis legal team, the legendary Louis Brandeis. He studied the context and details of the documents the administration submitted to the committee and came to believe Taft couldn't have examined everything or prepared his letter exonerating Ballinger in the time available. An amazing stroke of luck proved Brandeis's suspicions. As Glavis told the story, a freelance newspaperman in Pinchot's camp had lost most of his income due to administration pressure on his publishers, so his wife began working as a seamstress; one of her sewing clients turned out to be the wife of Ballinger's stenographer. This stenographer had witnessed a lawyer and Ballinger himself writing Taft's letter, and then saw them burn the drafts in a fireplace. Brandeis put the stenographer at the witness table. Soon after, Taft admitted he had exaggerated his review of the record, had exonerated Ballinger with a letter Ballinger himself wrote, and had submitted to Congress a backdated report by the attorney general that hadn't existed yet when he supposedly relied on it to make his decision.

Pinchot had kept up correspondence with Roosevelt, who received letters by runners in his camps in Africa, and when his testimony was complete Pinchot went to Europe under an assumed name to meet the former president. On their first day together, in a town on the Italian Riviera, they talked all day and until late at night. The emotional currents in that conversation must have run deep. Roosevelt, although cautious of Pinchot's extremism, admired him and was indebted to him, and must have felt a sinking sensation that Taft had betrayed his conservation legacy, having fired the friend left behind as its custodian. Pinchot, moving into his endgame, surely knew how to pull Roosevelt's strings, but he didn't have to lie to do so, for he believed Taft was killing the progressive cause. He must have ached for a return of the days of the Tennis Cabinet.

But more than emotions were at work. Taft had lost the country's support—only partly due to Pinchot's efforts—and Roosevelt's popularity was rising even as his policies were on the wane. He formed the belief he might be morally obliged to run for president again if Taft didn't reform. The Republican Party had controlled the White House for all but eight years since the Civil War. It represented both Wall Street conservatives and middle-class progressives supporting reforms in social and conservation policies. Democrats might be able to win if they could add those progressive voters to their own base of Southern voters, farmers, and the poor (and if they didn't lose too many urban workers to Eugene Debs's Socialist Party). Taft's orientation toward business interests opened the door for the Democrats, who were grooming candidates with strong progressive credentials. That fall of 1910 the Republicans lost heavily in the off-year elections. Soon Roosevelt returned to the stump, again speaking from texts written by Gifford Pinchot. Now, out of office and with the whistle-blower's sense of certainty, these speeches became far more radical in calling for progressive changes, all wrapped in moral language and the logic of conservation. The remake of government he invoked would have transformed the nation, and the related beliefs he held but kept quiet about on the campaign trail, would, in a second Roosevelt administration, have encouraged the eugenic engineering of America.

18

Conservation and Eugenics

In the summer of 1910, home from his safari, Roosevelt called for re-modeling the American governmental structure of decentralized state and local authority, known as federalism, which was established by the nation's founders. He had returned to acclaim, his popularity towering thanks to comparison with the inept Taft. The Pinchot-Ballinger affair remained unresolved, with Ballinger still in office and the Cunningham claims still pending. The drama of the congressional hearings was fresh in public awareness. Perhaps Taft's weakness also helped Roosevelt sharpen his own thinking. The moment of clarity came in Kansas in late August, when he gave a speech quoted across the country, among his most fa-mous, its text written by Pinchot and little altered by Roosevelt. It was known afterward as the New Nationalism speech.

Going well beyond the program he had pursued in office, Roosevelt posed increased national authority as the antidote to corporate power that had grown beyond the reach of any other force in society. The federal government would assure workers got their due, including a living wage, compensation for injuries, time off work for civic and family life, and protection of public health; it would see that companies were run hon-estly and openly; it would stop monopolies and regulate utilities and railroads, taking them over if necessary; it would rout out corporate in-fluence in government, with political contributions limited and publicly reported; and it would deliver more direct democracy to the hands of voters. Most radically, this vastly empowered national government would transform the U.S. economy to reward only merit, using graduated estate and income taxes to pull down the fortunes of the very rich.

The states that originally ratified the Constitution had faced none of these problems· and never consented to a national government strong enough to solve them, but once corporations could span the nation—and Roosevelt viewed corporate combination as an inevitable consequence of the industrial age—then only a central authority even mightier than they were could prevent a few rich men from controlling the country's laws, natural resources, and workers' lives. Corporations already did control much of that, and the workers weren't going to stand for it. Roosevelt's speech reads as liberal today, but at the time his way represented a white middle-class alternative to real class warfare fought in the streets by a rising socialist movement. New Nationalism offered federal power to manage the economy and tame the exploitation of people and resources, but it asked for labor peace in return. Instead of class conflict, all would join as equals in allegiance to a shared national identity stronger than the old links to community or state—a step in cultural evolution to broader group affiliation.

We live today within norms generally accepting of an intrusive federal presence in our lives, but it wasn't always so. Americans had to learn nationalism—flag worship and the pledge of allegiance were promulgated in that era, too. The federal government didn't seem equal to many tasks. When Gifford Pinchot was a young man, exploring the idea of going into forestry, a recently retired secretary of agriculture told him forest management would never work in the United States because the country lacked "a centralized monarchial authority."

Later that same year, Pinchot had attended the Paris International Exposition of 1889, the site of the brand-new Eiffel Tower, where he felt overwhelmed, deeply impressed, and driven forward by the immense forestry exhibit. The great world's fairs changed many lives: they were society's premier tool for acculturating its people to the new. They promoted more than amazing technology. They also demonstrated the new relationships among people that the machines brought with them, including affiliation to the symbols of national rather than community identity—monuments, mass communication and transportation, mass-produced goods, and celebrity. Contemporaries believed the fairs reduced class strife and political violence. In the United States, before electronic media, each fair attracted a substantial fraction of the entire population, and those who couldn't attend read saturation coverage in the press. In the world's fairs, civic leaders produced self-contained models of a hoped-for future in order to mold ideal citizens to live in it.

The utopia exhibited at American expositions included the richness of the country's natural resources and the superiority of its dominant race. The first U.S. fair, in Philadelphia in 1876, commemorating the nation's centennial, presented Native Americans as hideous brutes fit for extinction—a message validating that year's warfare against the indigenous people of the Great Plains, including the Battle of the Little Bighorn. At the New Orleans fair in 1885, comparative displays of skulls showed how Indians, Eskimos, and other lower races resembled criminals or animals. The enormous fair in Chicago in 1893 displayed living American Indians and other indigenous people on a honky-tonk midway where they were continuously jeered and ridiculed. Anthropologists arranged the races along the walk in a supposedly evolutionary progression from the lowest to the best, at which point viewers emerged from the noise and chaos of the carnival upon the quiet of a pristine new city, built for the purpose on an immense scale and painted pure white. The symbolism conveyed an idea of evolutionary ethics wherein white Americans could grow through racial purification from an animalistic, selfish nature to become higher, more cooperative beings.

These ideas developed at Ivy League and other universities and museums of natural history and anthropology in New York and Washington, in learned societies and scientific literature. When later fairs focused on the West, the link between natural resources, morality, and racism was taught ever more explicitly. The great Louisiana Purchase Exposition in

Saint Louis came in 1904, a critical time for the West as Roosevelt's conservation program hit its stride. Westerners sent pieces of the landscape
to demonstrate its value: for example, from California, part of the trunk of
a giant sequoia, and from Alaska, ancestral totem poles removed from
coastal villages. They sent live people, too. The mastermind of the fair's
anthropology department promised to "represent human progress from
the dark prime to the highest enlightenment, from savagery to civic organization, from egoism to altruism. . . . The method will be to use living
peoples in their accustomed avocations as our great object lesson."

Authorities shipped to St. Louis indigenous people from Alaska and the
Philippines, pygmies from Africa and giants from Patagonia, and many
famous Native Americans, as well—my grandmother, age seven, encountered Geronimo there, a pathetic figure in an Apache chief's regalia displayed on a platform. Given ten cents by her mother, she paid him for his
autograph, which he painfully scratched in block letters on an index card;
Geronimo then took her dime to another booth for a piece of apple pie.
Roosevelt and his daughter Alice (whom my grandmother also met at the
fair) toured approvingly, the president having sent word ahead via Taft,
then secretary of war, to have the Filipino savages dressed in properly mod-
est clothing (they wore bright silk trousers until the fair's Board of Lady
Managers certified loincloths as acceptable and more in keeping with the
exhibit's authenticity). Native people camped out for display according to a
plan designed to show the relationship of their racial types. Scientists ex-
tensively measured and tested these people while exhibiting them—their
physical size, senses, abilities, intelligence—all of which, apparently, proved
the superiority of whites. Some human specimens who died were sent for
dissection; the brains of three Filipinos were collected by the Smithsonian.

By the time of the San Francisco fair in 1915, the racists had moved
from justifying white conquest over other races to the conservationists'
goal of efficiently using the resources the dominant culture had thereby
obtained—from documenting the path of nature to intervening in its
progress. As gardeners and foresters would thin weak genetic strains and
nurture the strong, so eugenic campaigners called for planned racial

improvement through sterilization of people deemed inferior, beginning with anyone with a disability, and encouraged breeding by the racially superior. The U.S. Department of Agriculture helped foster an American Breeders Association that included research on humans, with funding and support from the Carnegie Institution, the Harriman railroad fortune, and the founders of the Kellogg cereal company, among others. The former president of Stanford University convened the Second National Conference on Race Betterment at the San Francisco fair, and the Race Betterment Foundation mounted an exhibit, with pictures of its illustrious supporters, including Harvard University president Charles Eliot and Gifford Pinchot.

Pinchot had entered the eugenics movement during the Roosevelt administration, joining several of the president's other friends. He solicited contributions from scientists and social activists advocating eugenics for a three-volume National Conservation Commission report to the president at the end of his term in 1909. Roosevelt transmitted the report to Congress with the statement that it was "one of the most fundamentally important documents ever laid before the American people."

The report's volume on "National Vitality, Its Waste and Conservation," by Pinchot's friend, Yale economist Irving Fisher, reads like a manifesto of the progressive political movement that Roosevelt sought to lead, and its words were echoed in the New Nationalism speech of the next year. Ten multifaceted recommendations called for a national administration of public health; ending air and water pollution; food and restaurant inspection; worker safety and child labor regulation; working-hour restrictions; health and safety inspection of prisons, asylums, factories, and schools; antidrug and antialcohol laws; safe drinking water; enforcement of antispitting laws; improved sewage and garbage removal; pest control; building safety inspection; school nurses and health instruction; universal athletic training; healthful changes in clothing, architecture, ventilation, food preparation, and sexual hygiene; and elimination of poverty, vice, and crime. And then, recommendation number ten: "eugenics, or hygiene for future generations," with forced sterilization or marriage prohibition for people with epilepsy or mental disabilities, or for criminals, the poor, and "degenerates generally." And creation of a new social norm benefiting eugenically favored marriages, making "degenerate" marriages as taboo as incest. The report concluded, "The problem of the conservation of our natural resources is therefore not a series of independent problems, but a coherent, all-embracing whole. If our nation cares to make any provision for its grandchildren and its grandchildren's grandchildren, this provision must include conservation in all its branches—but above all, the conservation of the racial stock itself."

More than a dozen legislatures passed eugenic laws over the next ten years, which, by 1970, had authorized forced sterilization of sixty-four thousand Americans with mental illnesses, epilepsy, disabilities, or criminal records, or who were simply poor. At least thirty states passed laws forbidding marriage of eugenically unfit men and women, and twenty-eight outlawed interracial marriages, including six that put antimiscegenation in their constitutions. Those laws stood until 1967, when a Virginia couple, Richard and Mildred Loving, validated their marriage before the U.S. Supreme Court—after a county sheriff had burst into their bedroom with a flashlight and arrested them, despite a District of Columbia marriage certificate hanging on the wall. Four states also prohibited sexual relations between Native Americans and whites. Progressives codified existing racial segregation in law as well. Roosevelt wrote that thinking men of both races wanted race purity.

Roosevelt worried about the loss of a special American quality of strength and ingenuity that supposedly had evolved among whites on the frontier. As eastern European and Jewish immigrants flooded into the country with their big families, and with the birth rates of white Protestant Americans declining, Roosevelt warned of impending "race suicide." He wrote, "I wish very much that the wrong people could be prevented entirely from breeding; and when the evil nature of these people is sufficiently flagrant, this should be done. Criminals should be sterilized, and feebleminded persons forbidden to leave offspring behind them. But as yet there is no way possible to devise which could prevent all undesirable people from breeding. The emphasis should be laid on getting desirable people to breed." His ideal American family lived on a farm with six white children—and lesser procreation represented failure of patriotism and a moral flaw, a rejection of the basic responsibilities installed in men and woman by nature. Roosevelt dispatched Pinchot to study the problem with the Country Life Commission. Continuing that work, the American Eugenics Society, one of various such organizations Pinchot belonged to, sponsored hundreds of Fitter Family contests at rural fairs, wherein couples would take intelligence and physical tests and submit to medical exams to become certified as worthy for breeding.

How do we make any sense of this behavior? How could progressives who worked for conservation, national health insurance, and the rights of workers adopt an ideology of hatred against the weak? In some ways, the inconsistencies reflect the diversity of a temporary political coalition. A lot of money and establishment power backed the eugenics supporters—a list that included John Kellogg, Henry Ford, John D. Rockefeller Jr., Andrew Carnegie, George Eastman, Woodrow Wilson, Herbert Hoover, George Bernard Shaw, H. G. Wells, Alexander Graham Bell, and many eminent anthropologists, psychologists, and biologists, including the founder of the movement, Francis Galton, who was Charles Darwin's cousin. Joining was smart politics. Roosevelt wanted women to stay home with large families; Margaret Sanger, the mother of Planned Parenthood, wanted smaller families and gender equality—but both were involved with the eugenics movement. A desire for power is hardly an excuse, however, especially for powerful opinion leaders such as Roosevelt and Pinchot, who constantly invoked moral authority for their policies.

Another excuse: Roosevelt and Pinchot believed in science and expertise, and eugenics seemed scientific. But even if accepting the faulty genetic reasoning, their impressive intelligence should have seen through the flimsy claims of racial differences; G. K. Chesterton, Clarence Darrow, H. L. Mencken, and other less famous writers grasped the errors and pointed them out. Pinchot's own conservation commission report, in the volume written by Fisher, contained statements an intelligent person should have seen through, such as when it blamed the demise of American Indians and Hawaiian Islanders on their own sexual immorality—rather than the government-sanctioned violence and theft of land justified by the scientific racists' own theories. Madison Grant's famous book, *The Passing of the Great Race,* would be laughable if it weren't so revolting: writing in pseudoscientific language, Grant denies the very right to life of members of other races, using as evidence nothing more than his own prejudiced stereotypes. In Grant's final analysis white Americans were not racist enough. "They lack the instinct of self-preservation in a racial sense. Unless such an instinct develops their race will perish, as do all organisms which disregard this primary law of nature."

This goal of creating a more racist society informed much of the cultural work of the institutions led by Roosevelt and Pinchot's peers—not only the world's fairs but the American Museum of Natural History, the Smithsonian, and others. Grant was an influential friend of Roosevelt and phrases and ideas from his writing crept into Roosevelt's. Oddly, the improvement of the dominant race meshed with the New Nationalism's utopia of a merit-based society. In theory, without money or class to distinguish them, the sexual attraction between men and women would be guided only by natural selection. Those unbiased choices would automatically sort mates by the proper eugenic criteria, matching the best to the best—white to white, intelligent to intelligent, and so on. The idea was absurd as science or social policy, but it does reveal the extent of the collectivism envisioned in Pinchot's brave new world.

The program Roosevelt advanced in his New Nationalism speech called for a stronger sense of national affiliation than ever before, a feeling of membership powerful enough to allow the federal government to regulate daily life, to deny the use of resources in favor of the future, and to redistribute income and inheritance to create economic equality. He asked for cooperation on a grand scale. Racism would glue together a national identity capable of incorporating the rest of progressivism. In theory, that idea makes sense. Today's economic experiments show that group cooperation quickly collapses when outsiders can profit from the contributions of insiders without fear of punishment. Moreover, altruistic punishers are more willing to act when a victim of unfairness is a member of their own group. That's common sense. To buttress a group's willingness to cooperate, enhance the members' sense of belonging and their hostility toward nonmembers—teach them that they're special, superior, and under threat. Bind together an American majority by equating its white racial dominance with Americanism.

I'm not saying Pinchot or Roosevelt schemed dishonestly to increase American racism. Roosevelt's philosophy could be inconsistent—he also spoke eloquently of the ability of nationalism to transcend race. The evidence suggests Pinchot and Roosevelt rode along with the eugenicists rather than leading their movement. But eugenic ideas slid frictionlessly into Pinchot's worldview, that rigidly moralistic construct of conservation, efficiency, and merit. Racist science emphasized that round heads and short bodies denoted an inferior type; in his autobiography, Pinchot described Richard Ballinger as "a stocky, square-headed little man." These issues mattered beyond personal prejudices. Eugenics thrived in the United States until it was discredited by the revelation of the Nazi death camps it had helped inspire. Grant's book particularly incited Hitler, who wrote him a fan letter, calling it "my Bible" before inscribing its hatred upon the flesh of millions of people. As witness to that horror, our country escaped the eugenic path, but the concepts are far from dead, as a quick Internet search will reveal.

There's a hangover for conservation, too. The American environmental movement remains predominantly white and middle class, detached from minorities, immigrants, and the poor along the same lines of class and color that existed a century ago. We're liberal and say the right things, but in the 1980s and 1990s, mainstream environmental organizations debated opposition to immigration using arguments differing in little but terminology from those eugenicists would have used (as I'll discuss more in Chapter 23). These connections don't by themselves undercut calls for conservation or implicate anyone as prejudiced simply for wanting to protect nature. But they do illuminate the ethical hazards that come with the kind of centralized power Roosevelt sought to accomplish his goals.

Alaska Native carver and counselor Jim Miller brought up the Nazis in one of our first conversations. He didn't distinguish between Nazi genocide and the genocide against Native Americans. In the eugenicists' world, Jews and Eskimos each were merely a lower rung. Writing in 1915, Henry Fairfield Osborn, an influential president of the American Museum of Natural History, used the supposed impossibility of educating Eskimos as a basis of his scientific argument that northern Europeans were a higher step in evolution. Jim deals with this ideology every day when he studies photographs of traditional forms for his carving. Anthropologists guided by racism stripped the region of the originals a century ago—to see them he would have to travel to big-city museums. And he also deals with it every day in his counseling practice, with men and women who internalized lessons of inferiority and carry on the oppression against themselves, through depression, self-destructive anger, and alcohol. Miller believes community healing depends on reclaiming personal value again.

The racists remain his adversary in every shadow, even in the shadowless village clinic where Jim works and where we talked about eugenics. "We think that's history," he said, "but what's the trickle-down? In this building there is very free and easy access to birth control. Any type of birth control you can imagine, and if you still find yourself pregnant, there is free abortion. There's no polite way to say it—to cut down on breeding. It's not just accessible, it's promoted. Kill your baby. And when you talk about values changing, when you no longer see your children as a blessing, that is some really bad stuff."

I felt uncomfortable. I support free birth control and legal abortion. I had to stop and think. It's true that the eugenicists debated how to promote family planning among the inferior but not the dominant races. It's true that free family planning services often focus on poor and minority communities. Historians have documented—as neither Jim nor I then knew—that some of today's major organizations for population control grew directly from the eugenics movement, like branches on a family tree. Good motives inspire this work, to save nature and improve human existence; but the eugenicists had precisely the same motives. I wouldn't charge family planning advocates with racism, but I'm not a victim of genocide. Victims shouldn't have to analyze the motives of their oppressors. Once our wise scientists and philanthropists unleashed this monstrous hatred, it lived and transmuted uncontrolled, deformed the society itself, and now, somehow, the descendants of slaves and displaced Indians are partly responsible for our redemption—by forgiving us and by loving themselves.

When we're ready to be a single human family we can use history to diagnose our unhealthy relationships with each other and with nature, to heal rather than to blame. Many Alaska Natives remain hostile to environmentalists, despite often sharing their goals. Some environmentalists' elitism, purism, and good-versus-evil worldviews still reflect the attitudes of their intellectual ancestors. Norms live in the culture like genes, manifesting themselves unexpectedly, the way a child's big ears appear from an ancestor of whom no picture or name remains. The Pinchot-Ballinger affair is generally forgotten, but we learned from it how to talk about natural resources. Although the antagonists' true disagreement concerned policy, their battle became moral, the pure knights of conservation against the corrupt barons of privilege. That conflict of good and evil continues, an archetype repeated every time virtuous environmentalists stand against greedy, self-serving developers. Like members of a dysfunctional family, those involved in these fights often seem activated more by their roles and learned anger than by the ecological or economic realities of their conflict.

The strategy of rallying a parochial group against a common enemy worked for progressives who combined racism, nationalism, and conservationism until their Nazi copiers discredited the formula. Roosevelt's New Nationalism speech stated that case with language that hasn't lost its inspirational ring.

Of all the questions which can come before this nation, short of the actual preservation of its existence in a great war, there is none which compares in importance with the great central task of leaving this land even a better land for our descendants than it is for us, and training them into a better race to inhabit the land and pass it on. Conservation is a great moral issue, for it involves the patriotic duty of insuring the safety and continuance of the nation.

But the politics of division can't help the earth now. Nature itself is endangered by threats that come from no specific villain or location. The oceans grow warmer and more acidic, marine mammals are contaminated, dead zones spread, plastic bottles flip from wave tops to beaches to the guts of birds. Who shall we fight? No one is innocent. Categories won't help us—nation, race, good and evil—for they have little to do with humanity's need to fit within a global ecological niche. If we're to cooperate on the scale that's needed, our goodwill must swamp those barriers and knock them down.

Surmounting the divisions between people for the benefit of the environment is not impossible. We are healing racism. Madison Grant's book sounds sickening to a contemporary reader. Our new way of hearing such language was made for us by good people working for change, who didn't give up even when the goal of racial tolerance seemed hopeless. We can unlearn Roosevelt's lesson of conflict. We must, if we're to work together to save the oceans.

19

Fantasy Meets Reality at Katalla

On a bright spring afternoon sunlight sparkles off Orca Inlet in front of the still snowy top of Mount Eyak and warms the mountain's green westward slope, terraced near the bottom by streets of moss-roofed clapboard houses and false-front businesses and the harbor in the lap of the town of Cordova. On a sunny spring day the retreating dampness of winter exhales rich, heady air. Birds are back on the sound, swarming the spawning herring. Billows of white sperm stream through dark green water. Such a day arrived on May 4, 1911, a Thursday, rich in possibilities, when A. J. Adams, the president of the Cordova Chamber of Commerce, organized a protest to bring attention to the still frozen Washington bureaucracy holding back development of Alaskan coal. Several hundred men gathered at the foot of Cordova's C Street with shovels before marching together to the ocean dock, where they began tossing a pile of British Columbia coal into the bay.

Richard Barry, agent for the Alaska Steamship Company, which owned the dock and coal, tried to reason with the protesters and then called the mayor, who came with the chief of police and demanded that the men disperse in the name of the United States. A few began to move off, but seeing others continuing to shovel they joined in again. The mayor and chief left and the afternoon continued without unpleasantness. Women standing by cheered the protestors on, one declaring, "The spirit of the

Revolution is not dead and Alaskans should show themselves to be men and not longer tolerate injustice"—or so reported the highly partial *Cordova Daily Alaskan* that same day, under the headline CORDOVA HAS BOSTON TEA PARTY.

The previous evening, in Katalla, a day's boat ride to the southeast, citizens had gathered for a driftwood bonfire on a sandy beach, celebrating the good weather and cheering when someone threw into the fire papers representing Roosevelt's conservation proclamations, which had stopped coal sales and created Chugach National Forest. As night advanced a group arrived carrying a dummy wearing a sign that said GIFFORD PINCHOT. 'CONSERVATIONIST' AND FRIEND (?) OF ALASKA. The Katallans hung the figure from a spruce tree, then tore it down and threw Pinchot in the fire, before the party wound down and people went home.

On May 5, after everything had returned to normal, the story broke across the country on the wire services. News flashed: "authorities are unable to cope with the situation and Mayor Lathrop has sent an urgent message to Governor Clark who is in Washington for troops" and "a telephone report from Katalla today said rioting continues there and that a mob had burned Gifford Pinchot in effigy." A sleepy little town can't sustain a riot—either editors didn't know that or they eagerly exaggerated the news for its potent political impact. The *Seattle Times* warned of mass revolt and of Alaska possibly seceding from the union. Letters and telegrams of support for Cordova's freedom fighters flooded into town. The secretary of the interior wired Governor Walter Clark, calling on him to suppress the lawlessness, but promising the coal would be opened as soon as possible. Pinchot's National Conservation Association issued a statement saying he always had wanted the coal developed.

Skeptics noted at the time of the coal party that the Guggenheim-Morgan Alaska Syndicate controlled the town of Cordova, owned the steamship line, the dock, and the coal, and that it had leapt to the defense of the demonstrators—the *Seattle Star* maintained the syndicate had organized the whole thing. No charges were filed by the district attorney, who said after a visit, "I have concluded my investigation and return to my official residence by the next boat, satisfied that law and order will be maintained in the future as in the past by the high-class citizenship of Cordova."

Almost a year had passed since the end of the Pinchot-Ballinger hearings without action on the Cunningham claims. A fight had broken out on the floor of the U.S. House of Representatives over a failed Alaska coal-leasing bill, pitting the territory's nonvoting delegate, James Wickersham, against a debate opponent; the sergeant-at-arms took up the House's enormous ceremonial mace to intervene, but the fight ended before he could decide what to do with it. Wickersham carried on a campaign to abolish Chugach National Forest that would last several years. Pinchot had written an open letter to Taft calling on him to step in personally and cancel the Cunningham claims, since neither Ballinger nor his lieutenants could be trusted. Ballinger never escaped the cloud from the scandal and finally quit, but still the coal claims lingered unresolved. Taft couldn't approve them, and doubtless knew that if he canceled them Pinchot would hold that as proof of having been right all along—as he did in fact do, with a press release, when the new secretary of the interior, Walter Fisher, invalidated the claims, a month after the Cordova Coal Party. With the presidential election campaign approaching and Roosevelt threatening to run against him for the Republican nomination, Taft likely hoped Fisher, a friend of Pinchot, would help restore his administration's conservation credentials. But the president's vacillating policies pleased neither conservation nor development advocates. He declared in a special message to Congress, "If the development of Alaska has been retarded, it is those scandalmongers, who raise the cry of 'fraud' and 'grab' whenever any action is taken by the administration affecting Alaska, who are alone responsible."

Both sides joined the fight on the basis that the coal promised immense riches—the typical Alaska illusion that Will Langille had warned of, dating back to the gold rush of 1898, and which frequently surrounds new energy and mineral prospects in Alaska to this day. Glavis's original 1909 article in *Colliers* stated that the Bering River coal claims would supply the entire country and "a monopoly of them would be a national menace." Pinchot, writing in the *Saturday Evening Post* in response to the Cordova Coal Party, asserted that the output of the Cunningham claims passing through Controller Bay could supply all the West Coast's energy needs for twenty years. Neither man had been to Alaska. The emphasis on the coal's value justified the fight to keep it out of the hands of the Alaska Syndicate and to obtain payment to the government for its removal. Pinchot's adversaries played the game, too, promising that if only the syndicate could control the coal—without paying a royalty—it would build a railroad to extract it, a smelter on Prince William Sound, and even a rail line to the Yukon River to open the entire territory.

Interior Secretary Fisher traveled to Alaska that August of 1911 to see the Bering coal fields for himself, and there, as if caught in the vortex at the boundary between fantasy and reality, he barely survived an early fall storm in the windy, shallow, and hazardous harbor that had been the subject of contention. The coal lands, supposedly the focus of so much investment and railroad building, still could be reached only by canoeing upriver and then tramping through tangled undergrowth on the sides of steep mountains. Fisher badly sprained his ankle during his three days in the bush. On returning to the sea, strong winds and rising waves overwhelmed the launch carrying the secretary, the governor, and their entourage of VIPs and journalists as it attempted passage to a government cutter anchored outside Controller Bay. They saved themselves from swamping and capsizing by beaching the boat (the cutter went out to sea for safety). As night fell, miles of rough terrain, thick brush, and flooded streams separated the party from shelter in Katalla. With the help of an Indian canoe discovered along the way, Fisher limped into town on his hurt leg at 1:00 A.M., drenched and exhausted. He stated for the press that "Controller Bay does not possess the natural advantages that have been claimed for it."

This truth had already been evident for at least four years, when Katalla's brief heyday suddenly passed. The Alaska Syndicate's original purpose in investing in the region, to get to the copper at Kennicott, did require a railroad, and the company had covered its bets by starting to build lines from several ports. In 1907 money poured into Katalla for the syndicate's construction work and for a competing line, and from other developers, both sincere and obviously fraudulent. The tracks crossed and rival construction crews attacked one another's work with dynamite, hundreds fighting battles with pick handles. Syndicate workers dipped theirs in pitch so they could tell who to hit. The town blossomed along a grid of lettered and named streets, with running water, telephones, electric light, a school, hotel, restaurants, stores, bakeries, a laundry, a machine shop, a plumber, a drugstore and confectionary, a women's clothing store, lawyers, stenographers, a doctor, a dentist and a barber, and more than a dozen saloons, dance halls, and casinos (but never a single church).

The *Katalla Herald* published its first issue on August 10, 1907, extolling the railroad and marine improvements, headlined: AN ASSURED SUCCESS; NO DOUBT ABOUT THE KATALLA BREAKWATER NOW UNDER CONSTRUCTION. PLANS OF PROJECTORS WISELY MADE. Shortly, the fall storms wiped out the breakwater and docks, a loss of millions. In November a *Herald* headline read, hopefully, STILL ON THE MAP. But the paper didn't last long, as the disaster persuaded the Alaska Syndicate to shift its work fifty miles west to the new town of Cordova, and smart businesspeople followed, draining Katalla of its economic vitality. The Copper River and Northwestern Railroad, completed in 1911, ran from Cordova to the copper mines at Kennicott, its locomotives powered by oil rather than coal, making prodigious profits for the Alaska Syndicate, which imported the labor it needed and exported the ore and profits. The population of Cordova crashed with the completion of railroad construction; the conclusion of the gold rush drained Alaska of people generally that same year.

In August 1911, Pinchot began talking about seeing Alaska coal for himself, consulting advisers who included Roosevelt, and finally set off with a senator from Washington state, with a couple of reporters in tow. With the presidential nominating process approaching, Pinchot attacked Taft in another article in the *Saturday Evening Post,* calling him untrustworthy, partly on the basis of a decision to remove Controller Bay from the national forest (by this time, apparently, Americans were so familiar with the issue they needed no further geographic explanations). He argued that renominating Taft would push progressives to the Democratic Party, leaving the Republicans as the United States' minority conservative party. Keeping the Alaska coal issue working for his side would help Roosevelt win the nomination instead.

As Pinchot's ship approached Cordova a telegram went ahead from a critic in Seattle—and was reported nationally in the press—encouraging townspeople to physically prevent him from landing. No welcoming committee met the boat, but Pinchot's group got a ride in the town's only car to the Windsor Hotel, and he found the people who met him there friendly enough. In Seward, the coastal town to the west, with its own short railroad, Secretary Fisher had been greeted a month earlier by a sign that said,

> You must not mine your own coal nor cut your own wood.
> All reserved for future generations.
> (Signed) G. Pinchot, Pinhead.

But Fisher explained to gathered townspeople that coal would not be given away, and moreover he developed a realistic assessment of the limited value of the Bering River coal field. When Pinchot arrived in Seward he was able to hold a civil public meeting there to explain his views. In Katalla, the many boarded-up saloons, dance halls, and other buildings were posted with signs announcing, "Closed, the Result of Conservation." After dinner in a gambling hall, Pinchot held a public meeting, noting afterward, "The feeling, while personally very pleasant, is evidently bitter against the Conservation movement, which they believe has ruined the development of the town." These were people who had held on to investments in Katalla when the railroad left for Cordova.

Pinchot's party went up the Bering River in a motorized canoe, a dugout made of a single spruce log thirty feet long and seven feet wide, owned by a Native from Chilkat. The beauty of the river reminded Pinchot, in spots, of the Delaware Water Gap near his home. He tramped happily over the mountains for a few days in the September rain, climbing the stream gullies and ridges to see the coal—observing that it was not of the quantity or quality promised—and, in vivid notes, described the views when the clouds lifted from the mountaintops, revealing vast glaciers and peaks receding to the distance, berg-dotted glacial lakes, and, nearer, the autumn red, russet, and yellow of dwarf willow and cranberry carpeting the mountainsides and valleys. On their way back to Katalla the brutal fall wind and rain pinned down Pinchot's group near Controller Bay, as it had so many others, and drove ashore and holed a Forest Service launch sent for him, and the group had to strike out for seven miles to Katalla, through brush and water and oil spilling into the slough from the wells. When the storm broke he was glad to finally leave the town, "the end of creation," as it seemed to him.

On the steamer back south, in mid-October, Pinchot thought, as he always seemed to, that his skill as a speaker and the rightness of his point of view had won over a great number of those he had met. Some did credit his courage in coming to Alaska, but many Alaskans' hatred for Pinchot and what he stood for burned brightly for more than fifty years more. His follow-up article about the trip in the *Saturday Evening Post* made many perceptive points about Alaska resources—including the colonial destruction of its fisheries—and noted that "an Anglo-Saxon community of so high a type is entitled to . . . a degree of home rule." Then he turned his attention to getting Roosevelt elected.

How distant the presidential election must have seemed for those left behind in Katalla, where no one had a right to vote and news came mainly when a break in the weather allowed a passing steamer into the dangerous harbor: news of Roosevelt winning the primaries only to lose the Republican nomination to Taft at a corrupt Chicago convention; Woodrow Wilson, a progressive, nominated by the Democrats after forty-six convention ballots; Roosevelt forming his own progressive Bull Moose Party, holding his own convention in Chicago (Doug Langille was present)—but many progressives moving instead to the Democrats.

Wilson won the three-way race, Roosevelt second, Taft third, and we're left with the politics we have today—Republicans conservative, Democrats progressive.

A fight over Alaska conservation helped cause these events—or, at least, a fight over a fantasy of potential riches from Chugach National Forest, Controller Bay, and the orbit of Katalla—and yet neither the town nor even the coal really mattered to the politicians in the end. Bering River coal ultimately failed a quality test by the navy for use on its ships and was never mined. Alaska coal remains a marginal industry today owing to the cost of production and distance to markets. Indeed, Alaska stagnated economically, even with a government-built railroad, until federal spending on World War II—and conservation was not to blame. Katalla itself, irrelevant and lost to collective memory, faded away quickly in the decade after Pinchot's visit. The last businesses shut their doors, and the streets of abandoned buildings lay empty for use by just a few families who remained to fish and work at the oil wells and refinery.

An ornate dark-wood bar from one of the Katalla saloons ended up in the Alaskan Bar on Front Street in Cordova, but mostly the empty houses and shops waited quietly for people to come back. Mae Lange, born in Katalla in 1920, explored the old buildings with her brothers and sisters, taking the big mirrors from the saloons to make a playhouse, which they built all of mirrors, some whole and some broken to fit. One big general store was locked tight, but finally the children found their way in through a hatch behind the counter, discovering jewelry and sundries still on display in showcases and, upstairs, apartments with clothing still in drawers. One had belonged to Katalla's only famous resident—Barrett Willoughby, a best-selling novelist who had left years earlier, but whose dress stand and neatly made bed were still waiting for her return. "They'd leave and they always thought they'd come back," Lange said. One year when the Lange family ran out of candles for the Christmas tree, Mae's brother scavenged through the abandoned houses and came up with enough flashlights to decorate the tree with their battery-powered bulbs.

In 2007 I spent a rainy evening in Cordova with Mae, her daughter

Sylvia, and Sylvia's daughter, Melina, going through cookie tins stuffed full of old photographs, hearing stories told and traded about the family members in images taken at places all over the sound. As the child of an Alaska Native mother, Mae wouldn't normally have been allowed to live in Katalla—Natives had to stay in a squalid separate village of the same name—but Mae's father was Danish, having arrived to unload materials for the town in 1903. Sylvia grew up in Cordova, but used to spend sum-mers in her childhood at a traditional fish-gathering camp at Chilkat, a dying Native village on the Bering River, where she knew Sewak, Chilkat's last resident, a century-old Tlingit whose eyes had turned blue with cata-racts. He told her of the night his best friend in Katalla's Native village got drunk and went to shoot all the white people, and Sewak had to shoot him in the back to stop him. One member of Mae's family tried to pass for white with powder and white gloves; another looked pale enough to pass without trying. In the movie theater in Cordova the usher didn't know where to seat them, in the Native section or among the whites. Sylvia's fa-ther had an Aleut mother and a white father, and was born on a fox farm on Peak Island; that grandfather worked at the mine at Ellamar, but he couldn't get a good job because, with a Native wife, he was a "squaw man."

Mae survived racism without bitterness. Neither did she resent the conservationists, although she remembers her father shaking with anger over their decisions, which she didn't understand at the time. In the end, the protection of Prince William Sound in Chugach National Forest saved the land for the Natives' eventual recovery and ownership—unlike the national parks or land that became private. As I'll explain in the next section, Alaska Natives became the largest owners of land and businesses in the sound and across Alaska. Sylvia has helped lead those companies.

Gifford Pinchot's life followed a new path after Roosevelt lost. His fa-ther died in 1908 and his mother, in her old age, dominated him less ef-fectively before her death in 1914. During the 1911 trip to Alaska, the dead fiancée, Laura Houghteling, rarely made contact. Almost every diary en-try says, "Not a clear day," or he wrote nothing about her at all; in 1913, it was never "clear"; and the next year he stopped mentioning Laura. During the Bull Moose campaign he had met a real woman, Cornelia Bryce, his intellectual equal and a stronger personality, for she had rebelled against her own rich and controlling parents, and their whole controlling social system, refusing to enter society as a debutante and instead taking up radical causes such as women's suffrage. She ran for Congress three times. Pinchot had remained faithful to Laura and had never shown interest in another woman, despite the assets of his wealth, power, and good looks—according to the evidence available, he was a virgin on his wedding night with Cornelia, at age forty-nine, in 1914.

Cornelia's influence changed Gifford. In the later part of his life his ideological rigidity and racism disappeared: he matured. In politics he focused more on humanitarian causes, advancing environmental issues such as stopping unhealthy industrial pollution and promoting clean water and sanitation for the poor. He was ahead of his time, as most conservationists preferred to concentrate only on natural landscapes. As governor of Pennsylvania, Pinchot prioritized helping ordinary people affected by the Depression, including seeking economic aid for unemployed factory workers and advancing road construction projects for the rural poor in previously neglected areas. He revised his racial views in 1929 on a nine-month family cruise to the South Pacific. Prepared to find dirty, murderous savages, he instead met people with a lot to teach the industrialized world about living in harmony with the environment and each other. His influential book about the trip, *To the South Seas,* helped start a literary reappraisal of indigenous cultures as models to be emulated rather than assimilated. He also came out early against European anti-Semitism. Cornelia remodeled Gray Towers, the family castle in Milford, Pennsylvania, as effectively as she had remodeled Gifford, opening it to the outdoors and adding her sense of humor, including a patio table that was an elevated pool—guests sat around the edge and the food floated between them on specially made little boats. Gifford died in 1946, active in environmental causes to the end.

Mae Lange's family stayed in Katalla long after the other people were gone. When she was eleven her mother fell in love with the Cape St. Elias lighthouse keeper and moved to Juneau; her father raised six children alone in Katalla, making money from intermittent jobs, including trail clearing for the Forest Service, but mostly living off the land. One of the boys got up early each morning to light the fire in the schoolhouse, where the teacher taught Mae and other children through the eighth grade. When she reached sixteen, Mae married a prize fighter and moved away to Cordova (Sylvia is from her second marriage). Mae's father stayed in Katalla even after the refinery burned down and the town breathed its last sigh. He rowed to the Bering River to gillnet salmon by hand until, in his late seventies, he moved to Cordova, in 1945. A few squatters stayed into the 1960s, including Mae's brother, who moved from building to building, dismantling each to burn as firewood. When Mae and Sylvia visited together in the late 1960s, the family's old house had shifted so much that the window seemed to be a door—Sylvia stepped through and fell to the floor. But as much of a wreck as the place was, Mae's body memory guided her hand into a gap behind a cabinet, where she found her late father's glass tobacco cuspidor still hidden in its secret spot.

In Alaska's coastal rainforest, where everything seems alive, time has a transformative will. Torrents of rain bring up hedges of alders on the long gravel prisms of unfinished railroad lines. A locomotive rusts, still waiting for steam miles into the wilderness, on the flats in front of Bering Glacier. Milled wood softens, sprouts fungus, and sags; a roof falls under a load of snow with a whump, but no one is present to hear. A great earthquake raises the land, leaving the Katalla waterfront six hundred feet away from the river, the timbers of wharves becoming intertwined with bushes and saplings, which grow half a century into substantial trees, their trunks lining up next to pilings where ships once tied. In midriver lies the steamer *Portland*, the ship that carried a ton of Klondike gold to Seattle in 1897, starting the rush that brought Will Langille and everyone else. It hit a rock off Katalla in a snowstorm in November 1910 and scuttled here, and then the earthquake lifted it, and its boiler and rusted ribs stick up from little more than a creek. Buildings decay until there are no two boards together, just rows of divots, filled-in cellars, indicating where a street once ran and another crossed, thick forest now, the sky closed off again by hanging spruce bows. A fishing guide who comes here during the salmon run in the Katalla River pokes around to find a pile of empty bottles from one of the saloons, but turns up little of significance: "The wreckage of humanity. If you rip the moss up you find lanterns, bed frames, billiard balls, and clothing. Nothing of value, just the things of the time."

I met this guide, Dave Salmon, on a trip to Katalla with Rick Steiner, a university extension agent who became, following the *Exxon Valdez* oil spill, an international environmental activist. Rick was trying to conserve the Bering River coal fields. After Alaska Natives received their lands from Congress in 1971, the Native corporation for the region, Chugach Alaska, selected the coal fields from within Chugach National Forest as its own. In 1987, the coal lands passed to a Korean businessman, who said his new studies showed enormous reserves of exceptional quality, and, as if possessed by the ghosts of Katalla, proposed to build a railroad to Controller Bay and ship from there. Chugach Alaska, meanwhile, wanted to build a road through the forest to cut timber from the area. Steiner worked to raise money to buy out the coal reserves and block the road. When the deal appeared to be coming together he wrote to supporters, "Teddy Roosevelt and Gifford Pinchot must be smiling from wherever they are in the cosmos!" But instead the Korean company turned down offers as high as $7 million. Now ten years have passed with no resolution.

Our four-seat plane from Cordova landed on balloon tires on the vast sandy beach of Softuk Bar, at the mouth of the Copper River, where we slept in a Forest Service public-use cabin. In the morning we planned to walk the six miles along the shore to Katalla, but during the night a ferocious storm erupted from the east. A dim September storm, as heavy as wet flannel, with rain and sand that turned the wind into a biting thing. The breaking surf couldn't complete its curl before being blasted into frothy wind debris. By 2:30 P.M. we decided the wind had dropped enough to hike into—or, at least, we had convinced each other we were both the sort of hikers who do these kinds of things that most people hate and avoid. We felt sand between our teeth. At a long stretch of wet, slippery boulders, Rick, a very tall man, fell down several times, and we probably should have turned back, because it was getting dark and on the return we would have to climb back along the wave-battered bedrock cliff at Point Martin. Besides, there were huge brown bears in the twilight, some near enough for us to startle them from their feeding. We needed at least some daylight for all this. Yet we kept on. When we reached Katalla the only goal was to find Dave Salmon's camp and get advice on finding a quicker return route. We did find him, but in the failing light we had no time to stay. Dave sketched a map of a pond, a beaver dam, an old rail line, an abandoned house—one of those pencil maps that could mean nearly anything—and we plunged into the woods where the town once had been.

The storm's wan light gave up long before penetrating to the forest floor and the faint suggestion of a path we were trying to follow, hiking as fast as we could. Bare branches like thin bones reached across us, their fine twigs sleeved with moss like tattered clothing, the dangling strands tipped by glistening drops of rain. I stepped up to an ancient iron wheel, part of an old washing machine, and just beyond it was a house—an attractive, Craftsman-style house at one time, with shingled gables and narrow double-hung windows—but half the roof was gone and a young tree had grown up in one of the rooms. Near here was the fallen granite tombstone for two young men. Mae remembered when four boys drowned in the river while making a drunken run for booze, in 1929; these two, the White boys, were the sons of the town's founder, the man who drilled the oil. He died here, too.

We hiked into a swamp, following a line of mossy stepping-stones between the broad, wet leaves of skunk cabbage—too regular to be stones or logs, they turned out to be railroad ties. Then a hill as straight as a road embankment, an old rail line. On among great, dark trees, towering and groaning in the wind, the sound of surf never absent, past clapboard railroad sheds that were still dry inside. Finally, how would we get back to the shore for the rest of the beach walk to the cabin? Rick turned one way and I turned the other. We argued briefly, then broke through the brush, back to the flying sand. Walked past the great bulk of one watching bear, then another, into darkness. Back to the present, the cabin, the light of a propane lantern. The past receded like a menacing dream.

We had copies of Gifford Pinchot's diaries with us. I read aloud passages about his visit to Katalla in 1911, the fight over the coal and the campaign to abolish the Chugach by delegate Wickersham—the rhetoric, the corruption, the exaggeration and bitterness, all so familiar from our present-day Alaska politics. Rick said, "That's the fascinating thing to me, that we're in almost the identical political situation today with these forests as we were one hundred years ago. Some things just never change." That seemed true to me, and not only because of the ceaseless repetition of the same fight. The plague of fantasy also continued, extremes generated from the ideologies of people struggling at a distance over a place they didn't really understand. Pinchot and Roosevelt thought they could make all the decisions themselves from the far remove of Washington. A lasting solution to fitting humanity into our ecological niche in Alaska, or anywhere else, would instead require social change involving local people who are truly qualified to make choices where they live.

Part IV

PROPERTY RIGHTS AND
COMMUNITY RIGHTS

20.

Chenega Destroyed

Chenega's little bay faced south, partly enclosed by tiny rock and spruce islands and backed by the steep, wooded slope of Chenega Island. For hundreds of years it was a perfect spot for a village, until the earthquake in 1964. A waterfall splashed down bedrock to the crescent-shaped beach, and at its far end an island hugged the shore, creating a pond-smooth passage just big enough for paddling. In that southwestern part of Prince William Sound mountain peaks stand as if freshly extruded from the earth, solid and uncompromising. In some places billows of stone remain barren, immune to redemption by the rainforest. Few valleys divide the mountains; instead, the ocean steals into the narrowest cracks between their walls, branching and connecting beyond all probability into bays and channels, some of them miles long but only as wide as a river, navigable corridors into the heart of the forest. Seals and humpbacks and killer whales traveled these water labyrinths, as did Chenegans, in their handmade skiffs and kayaks. The people were poor, but they were survivors. After invaders left behind salteries empty of herring and gold and copper mines slowly rotting, after so many towns and villages withered, becoming merely stains on the map, this ancient community of Chugach Natives spoke its traditional dialect of Alutiiq and slowly repopulated. They had the southwestern sound to themselves.

During summers before the earthquake, families rode north on fishing boats all named for the cannery at Port Nellie Juan to fish salmon and buy the year's supplies on credit from the company store. After commercial fishing, they put up their own fish for the winter in riverside camps. They'd go to Cordova, too, on the eastern end of the sound, to dig razor clams for the industry there, competing to see who could gather the most. In spring the men hunted seal together. They caught herring with dipnets to eat pan-fried and families gathered kelp covered with glistening berries of herring roe. They would picnic on sunny spring beaches and collect goose eggs from the winter-dried grass behind the storm berm, children cooking their own seal meat morsels on the ends of sharpened sticks over the campfire. The mail boat occasionally brought news from the outside world and a few cash-bought foods, mainly flour, sugar, tea and coffee, and candy for the children.

As in old times, entertainment for boys and girls mainly came from adventures on the beach, throwing rocks and catching animals; for adults, stories and jokes, visiting. They called on neighbors and family in their little houses—weathered shacks, tin-roofed, connected by boardwalks at the top of the beach like something deposited by the tide, each house facing a different direction, a collection as beautifully disordered as any nature could arrange, fewer than twenty buildings in all for around seventy-five people. Andy Selanoff recalled his childhood there, his grandparents welcoming guests for breakfast, each bringing his or her own teacup, and seating them around a bleached flour sack on the floor to share bread and smoked salmon. The teapot and sugar bowl were empty fruit cans. When supplies ran short mothers dispatched their sons and daughters to borrow from a neighbor, but in those days the word "borrow" did not imply repayment. "I recall the years we lived together as a village were full of joy," Andy wrote recently. "People lived close to nature, they traveled, hunted, and camped together, and no one was left out in village life."

He went on:

When I was younger, elders respected everything around them and what they believed in. Nothing was taken from the land without genuine respect, because it meant survival. Back then a hunter would never butcher a seal in the same place it was shot. It would be taken to a different place and butchered. This was done so the next seal that came along could use the same spot because it was clean. The same method was done in the bear den, so that another bear could use it to hibernate. These locations were marked in memory, year after year and time after time. The need to survive formed a family, and a family formed a community, and a village was born.

A human community can fit into its place in the world. Chenega did. But communities all differ. Like other living things, communities take forms shaped by their environments and the ways their parts connect. The social DNA of a community is written in its laws and political traditions— how people relate to each other and what they believe they own or share. As I'll explore in the chapters ahead, these characteristics can predict whether a community harmonizes with the land and sea or consumes them like a cancer.

As Easter approached in Chenega, the men hauled loads of clean pebbles from the beach to spread around the Russian Orthodox church. Women decorated inside, making bright crepe paper flowers to surround the gilded icons. The church had no priest, but each religious holiday was carefully observed. On Good Friday all the village dogs were gathered and taken by skiff to a little island in front of the village to wait until Easter festivities were over so they wouldn't disrupt the observances.

The sun shone warmly on a Friday evening in late March 1964, shrinking deep banks of crusty snow that remained on the steep slope dividing the boardwalks from the school, high above the village. (That night was Good Friday elsewhere, but still Lent in the village, as Russian Orthodox Easter came weeks after Western Easter that year.) Children were chasing birds and throwing rocks on the beach, waiting for a movie to start in the school, a corny old Vincent Price feature called *House on Haunted Hill*. Families were eating dinner or finishing steam baths.

Little Carol Ann Kompkoff was walking out on the dock to the out-house with her older sisters.

Avis Kompkoff had finished her bath and was dressing herself and her baby.

Margaret Borodkin, thirty-five at the time, was at her mother's house, having just finished dinner. The house began shaking violently, walls cracking, furniture crashing down, and something fell from the ceiling and pinned her to the floor, hurting her leg and hip. The earthquake lasted more than four minutes, long enough that many people believed it never would stop.

The dock where Carol Ann stood waved like a ribbon. Boulders on the beach bounced like rubber balls and boys jumped to stay atop them.

Avis left the house in her slippers carrying her baby. The bay seemed to boil.

Margaret saw her mother, Anna Vlasoff, outside the door, frantically running around and calling for someone to help release her daughter from the fallen debris. There was a roaring that seemed like jet planes and the house exploded with the impact of water and Margaret swirled, rolled—and passed out.

Avis heard someone yell, "Run." She carried the baby up the hill, sinking into the deep snow, losing her slippers. Steve Eleshansky carried his baby daughter right behind her. The wave roared in. When Avis looked back, Steve and his daughter were gone.

Carol Ann, age three, and her two sisters, Julia and Norma, were still on the dock when the bay emptied of water. Their father, Nicholas Kompkoff, came for them. He carried the two younger girls and told Julia to run with him, but the wave came like a fast-rising tide and knocked Julia down, pushing her ahead and then pulling her back toward the bay. Nicholas reached out to catch her as she passed but instead lost Norma into the water as well. The ocean sucked both girls away from him.

Some of the children playing on the beach froze in terror during the shaking and didn't run for safety when adults shouted to them. Men in the village ran toward the danger, some of them never to return. The fog of fear cleared from Kenny Selanoff, twelve, only when he saw a respected man running away like a rabbit—if that man was running, Kenny would run, too, holding on to his brother, George. They saw their aunt Dora Jackson rush back into her house with her daughter, Arvella, and then saw the wave take the house away.

The second wave broke with crushing force on the village, not like the first that had risen like a tide. Kenny and George held tight to a clothesline pole while water washed over them. Kenny saw Daria and Willie Kompkoff and Phillip Totemoff caught running for safety when the wave hit. It smashed them against their house with a horrendous impact and demolished the house; he saw Willie and Sally Evanoff and their granddaughter swept away. He saw Richard Kompkoff give his life trying to save Anna Vlasoff, who wouldn't leave her daughter, Margaret Borodkin, trapped under the debris in her house. Kenny never was able to forget what he saw.

The wave lifted a father carrying two sons up the hill and set him down, standing, on a ridge, where the water receded and left them safe. It caught a woman and rolled and battered her before leaving her ashore, unharmed but stripped naked except for one anklet.

The wave tossed Carol Ann and her father, Nicholas, who had already lost two daughters, across a creek and into a snowbank. A light pole fell on Nicholas and he dropped Carol Ann in the water. An uncle managed to grab her by the hood of her jacket, which had a stuck zipper; the zipper held fast, Carol Ann stayed in the jacket, and survived.

The second wave had destroyed everything but the school on the hill, washing high enough to flood its cellar, seventy vertical feet above the bay. The third wave cleaned up what was left. The bay filled with debris, lumber from sheds and houses, pieces of the church, furniture—everything that had made the village—floating calmly in the dimming evening. Dazed survivors gathered at the school, then an aftershock alarmed them, pushing them to move even higher, into the woods. Men built a bonfire to warm wet, ill-dressed people.

In the morning those sheltered around the fire on the hill looked down and saw nothing of their village—even the water of the bay was glassy smooth and empty. The wreckage and bodies had disappeared, drifting away with the tide. A Coast Guard plane flew over and they waved for help, but the village had been wiped away so cleanly that the pilot didn't realize anything was wrong: the place looked normal, with friendly people waving. But the pilot of the regular mail plane came out to check on Chenega and gave the world the news that it no longer existed. Twenty-six of its members were dead—more than a third.

Margaret Borodkin survived. The earthquake had trapped her and swept away her home and family, but somehow, when she regained consciousness, she was floating on boards in front of where the village used to be. Those on the hill that night recognized her voice crying from somewhere out in the bay. She heard their cries, too, children and parents looking for one another, families absorbing their losses, the injured calling for help. No one could help her. She tried to paddle with one hand while holding on to her floating island with the other, but it was no use. She grew so cold she thought she wouldn't make it. Finally, a boat of hunters returning to the village from the sound approached her, its bow nosing through the debris. On board, wrapped in three sleeping bags, she faded in and out of consciousness, but ultimately recovered.

But Margaret could not tell the story of what had happened. Many could not; some still haven't. Only when her people's leaders decided to record the history in a book, a solemn memorial, did she set down her memories on paper. Even then, in her midseventies, she couldn't escape the vision of her mother crying out for someone to help her, refusing to leave while the wave approached. She wrote, "It's been forty years since that tragic night, and in all that time, I've never gone back to Chenega. I just don't think I could stand to see where my home used to be, where my mother last stood, where my entire village was swept off the face of the earth."

I've been there with my family on a sunny June day on one of our Prince William Sound camping trips. We've felt the cool waterfall, still splashing down the bedrock, climbed up to the white schoolhouse, which is collapsing now, and found bits of broken plates amid the gravel on the beach. The old women from Chenega can still recognize those china fragments and know to whom each belonged.

Like any living thing, even a healthy community can be killed. The earthquake broke Chenega's link to its place, and the village barely survived as a cohesive cultural group. In some ways, the crisis of the earthquake has never ended for them. Restoring a connection between a people and the land and ocean is hard, uncertain work. But the Chenegans tried, heroically.

21

Ownership in a Liquid World

In the cold three hundred fathoms below Knight Island Passage, next to Chenega Island, rockfish with long spines and enormous eyes and mouths devoured shrimp and other alien-looking creatures, which themselves foraged on whatever dead organic matter sank beyond the deepest shafts of daylight into the zone of eternal darkness. Rockfish grow and reproduce slowly, rarely straying from a familiar patch of seafloor, but they persist: some live over two hundred years, centuries spent without seeing the sun in unchanging deep canyon waters. They encounter the light of our world only when pulled to the surface, and then they swiftly die. The rockfish's vulnerable swim bladder allows it to adjust its depth like a submarine, slowly inflating and deflating with a gland able to emit and absorb buoyant gas. When the earthquake suddenly shifted Chenega Island southward fifty-three feet and its steep underwater slopes gave way in huge landslides, the pressure shock blasted the fishes' delicate mechanism and they bobbed to the surface, their skin turning bright red like buoys and their swim bladders exploding up their throats like pink balloons. Red dots covered the passage—thousands of ancient fish—and in part of Port Valdez, too, where a wave threw sand 220 feet high on the shore and severed two-foot-thick tree trunks one hundred feet above the water. Who knows where else rockfish floated up: ten thousand miles of shoreline were affected by the 1964 earthquake, mostly unobserved.

A wave just as big as Chenega's hit the next cove over with enough force to break boulders off a cliff. Waves swept land all over Prince William Sound—and Seward, the Kodiak archipelago, and Kayak Island—as a piece of the earth's crust 600 miles long and 250 miles wide tipped and slid. In King's Bay waves cleared timber 110 feet above the water and stripped vegetation and soil down to bedrock. Water erased abandoned mines, canneries, and ghost towns as well as living communities. In marshes and muddy clam beds, waves ripped away soft sediments and flung the animals and mud onto beaches and woods, or dragged them back into deeper water, smothering seafloor life. Fishermen found their crab pots buried as deep as twelve feet under sediment. Waves poisoned lakes with salt water and dammed and diverted streams.

A salmon biologist returned to a field camp on Olsen Bay a week later to find it buried in foot-thick ice chunks that had been thrown ashore, with rocks still falling from the cliffs. Crabs, flatfish, and *Fucus* seaweed (popweed) hung from the trees. Heaps of dead clams lay in long rows. The earthquake had raised the camp six feet. The strip of rock coated with barnacles and mussels had become dry land. Intertidal streambeds where pink salmon had laid their eggs arose from the sea and the water carved new channels. At the south end of Montague Island, the land emerged by as much as thirty-eight feet. Scientists in skiffs with outboard motors checked the uplift at hundreds of spots in and out of the coves and passages of Prince William Sound, using a surveyor's level and staff to measure the distance between the stranded line of barnacles or popweed and the new water level, adjusting for the stage of the tide. By midsummer, fresh barnacles and popweed grew at the new tide line and land plants had begun to colonize the newly uncovered seafloor, making the amount of uplift even easier to measure—simply the distance from the dead to the living growth of each species. In the southern sound, where land rose more than the twelve-foot range between low and high tide, seafloor met air for the first time. Every shoreline animal died. Red and green algae and plantlike animals called bryozoans turned white in the sun and rain, leaving rocky shores as bright as the snowy mountains above them. Sea stars and small fish died in piles, and drifts of small shells could be dug up by the shovelful. Huge strands of kelp hung from a former reef like stingy hair from a mostly bald head. Streams cut deep gullies in the mud of the seafloor, meandering and changing course as they searched for new beds.

The sea set to work at once reshaping beaches. The earth's crust had pivoted like a seesaw on a line running along the western edge of the sound; east of the line the land rose, but to the west it sank, including the Kenai Peninsula, Cook Inlet, and Kachemak Bay. Water entered forests and, briefly, algae, barnacles, and periwinkles attached themselves to living trees; then the trees died and the salt water they sucked up preserved them like standing driftwood—many stood more than forty years later. I've watched them all my life, wearing away gradually like standing white stones. At sunken shores the old beaches disappeared under the tide and new beaches built themselves of material carried in currents from fresh sites of erosion, where the sea dissolved land it had never touched before. Where the crust rose, the earthquake stranded the old beach far above the waves. The long lines of gravel storm berms eventually dirtied enough to support moss, grass, and saplings, finally hiding with the earth's other geologic bones under its skin of soil and roots. A special smell of growth gathers in these shadowy places of thick young trees.

At both kinds of shores, raised or sunken, the sea's mysterious intelligence remembered the beach's shape from before the earthquake. The slope-break between the steep upper and gradual lower beach moved and reset, a new storm berm built at the top, and even the subtle ridges and steps on the beach face, which seem random and hardly noticeable, slowly faded back into existence at their former relative locations. The ridges, bars, and runnels of broad silty shorelines disappeared when the sea level changed, down at the lower part of the shore that is usually covered by the tide. But through a complex, many-step process, they gradually refocused back into the old patterns, displaced but accurately recalled, needing only time, the cycles of wave and wave, low water and high, the orbit of the earth around the sun and the moon around the earth, and the flush of rivers bringing new silt and sand, until each grain had been placed accurately in the geometry that had existed before.

The earthquake made suddenly clear that rock is no harder than water; or, to be more precise, that everything is liquid. The earth is a bubbling molten ball with a skim of green on its surface as relatively thin as the mildew that grows between bathtub tiles. Wave patterns on the ocean differ from the waves written in the sand and buried in the rock only by the tempo of their changes. What appears permanent, bedrock itself, seems so only because of the short span of our lives and our attention.

On the floor of a shallow Perry Island lagoon the tide arranged pebbles into a paisley pattern of curling channels upon which danced complex overlays of bright lines—sunlight projected through ripples pushed by an erratic breeze—and into the clear water I launched my body and felt the warmer layer at the surface and the colder water underneath, and felt how a kick of my foot or a paddle with my hand could mix the layers, which shined ghostly swirls onto the bottom. To live fully in this liquid world is to swim.

Some communities, like people, seem able to swim in the dynamic cauldron of earth and water. Old Chenega did. But now legal institutions instead relate us to land and water with square-corner rigidity. The power of these laws on our thinking becomes clear once you draw the contrast between property and land. The one, property, is a concept as abstract as plane geometry, and the other, land, is a physical reality with which you can connect and grow.

Occasionally, the liquidness of the land itself forces the law to acknowledge that property is a mental construct rather than a reality. Where property lines are set by natural boundaries, such as a river or beach, the legal implication of physical change depends on the ability of people to perceive the change. For example, a big U-shaped bow in the Arkansas River in eastern Oklahoma marked the legal boundary between lands of the Cherokee and Choctaw; much later (with different owners) the river changed course, cutting through the bow. Did the property line change with the

river, or did the boundary remain stationary, in line with the river's former course? Common law said the property line would move along with
the river if its bed rerouted slowly, by erosion and its opposite, accretion,
but if the flow cut across all of a sudden, then the property line would stay
where it had been in physical space, a process called "avulsion" (the word
is borrowed from medicine, and means to rip off part of the body). The
Oklahoma Supreme Court decided in 1938 that the difference between
erosion and avulsion depended on whether or not a person watching the
river could see the change happening in real time. The illusion of perma-
nence mattered most. Only a river cut-through or an earthquake could
shatter the illusion, forcing the law to admit that property is an overlay on
the physical world, not a part of it.

In Alaska, the state government owns coastal waters. The line between
that ocean property and private property on land follows the high-tide line.
When the earthquake ripped the two lines apart, some people ended up
owning seafloor and others saw the waterfront recede away from their
land. With the entire region moving sixty feet horizontally, figuring out the
exact location of any property became complicated. Land represents durable wealth, our "real" property—longer-lasting than buildings, businesses,
or books, or so it seems—but a property right in fact is only the social
meaning of papers filed at the recorder's office or the courthouse. Maps
seek to tame the land, but the land changes. Records connect names to the
maps, but the people die or wander away. In Alaska, property lines had
never mattered as much as resources—the fish, wildlife, timber, and minerals that powered conquest and motivated ownership. In the decades after
the Alaska purchase of 1867, before Congress set up a civil government,
prospectors in the area of a local mineral discovery would elect a recorder
of gold claims from among themselves, with voting power coming from
the ad hoc group of whoever happened to have wandered to that particular
wilderness beach or riverbank. Arrivals simply took land; when gold didn't
pan out and they left, property rights that had been sanctioned only by
their transient communities disappeared as well. Files of deeds matter
when people covet the land and lose meaning when they abandon it.

So it happened at Katalla, and almost at Cordova. The closing of the copper mine at Kennicott let the air out of Cordova. Dick Tapley remembered—he was eighty when I interviewed him twenty years ago, and blind, but still gillnetting with his seventy-year-old wife, Tina, on the Copper River flats. "At that time I was fishing in Bristol Bay, back in 1938, back when the railroad closed and the mine shut down. . . . A lot of them had the idea that they were just shutting down so they could cut the wages, and that they'd reopen. Until they started shipping out the locomotives. Then they knew it was over. The ones that sold out, they practically give their places away. But the fishermen lined up to buy them. . . . And I'd had a big season. And what I mean by a big season is that I made twenty-eight hundred dollars. Then I came back to Cordova and all the mine people and railroad people were leaving and trying to get rid of their houses. And I bought my house furnished for one thousand dollars. And that was with a piano."

Cordova remade itself as the "Razor Clam Capital of the World" by digging in the vast mud flats in Orca Inlet and at the mouth of the Copper River. But the 1964 earthquake lifted up the flats, drying out the clams. By summertime, the stench of rotting clams hung over the town. The clam industry never returned. The bottoms of bays had become the lower part of the tidal shoreline, where clams normally would live, but bottom mud wasn't suitable for clams—too gooey, black, and sulfurous. Moving water eventually could rehabilitate the clam beds by cleansing the mud and mixing it with gravel and sand, as the upper beach had resorted itself, but more than forty years later many of the raised bay bottoms still have not recovered. Beach scientist Jacqui Michel looked at such places in protected waters like Herring Bay and Bay of Isles, on Knight Island—bays without waves, as they are surrounded by mountains. Sediments had hardly begun to blend into a normal lower-beach mixture. Since the earthquake, the progress amounted to a thin layer of coarse pebbles on top of the old bottom mud. No one knows how long a shore like this will need to stabilize. "I would say it will take at least a thousand years," Michel said.

The state government sent crews to fix the salmon streams in the late

1960s. They succeeded in at least 30 of the 180 spawning streams affected, removing barriers and carving channels to stabilize wandering water courses. Pink salmon were spawning in the wrong places—the gravel in which they needed to place their eggs had been scoured away by waves, submerged by sinking land, or carried off by faster currents on raised shorelines. In some places, the streams simply became too steep for salmon. Six years after the earthquake, salmon returns remained poor. On Montague Island, where the land had risen the most, only twenty thousand fish returned, compared to seven hundred thousand before the earthquake, but the changes were too large to repair.

Tatitlek villagers noted a scarcity of clams, crab, cockles, and seals—the entire ecosystem had suffered—but they just worked harder or traveled farther to hunt and gather. For the commercial salmon industry, however, economic stress that had been worsening for twenty years reached a crisis. Federally permitted overfishing had driven down salmon catches across the territory for decades and through the 1950s. Local control of fisheries became a key rallying cry for those favoring Alaska statehood. But when statehood arrived in 1959, the problem persisted. The fish didn't come back even under the scientific management of state biologists. In 1967, the Alaska-run salmon fishery disastrously failed.

A solution came over the next decade—spectacularly so—with roots in a new system of ownership. The recorded papers with which law assigned land to individuals now would encompass the sea as well. The right to fish, like a deed to a parcel of land, became a thing that could be bought and sold. Once those rights were set, fishermen could form organizations to build hatcheries, producing more salmon for themselves to catch. They could transform the sound ecosystem to serve their needs, and they did.

22

Clem Tillion and the Ownership of the Sea

The World War II generation transformed Alaska from free-for-all public domain into property owned by individuals, corporations, and governments—both the land and the fish in the sea. Few had as large a part in the division of the ocean as Clem Tillion. An admirer of Theodore Roosevelt, Clem was born, like Roosevelt, to a patrician New York family—colonial-era landowners from Long Island—but he rebelled and left in search of adventure before finishing high school. He lied about his age to join the military and fought the Japanese in the Solomon Islands. A foxhole buddy told him about free land in Alaska. He arrived on a steamer in 1947, kicked around, hopped a freight train into the mountains of the Kenai Peninsula, hiked and floated a route similar to Will Langille's down the Kenai River on a raft, and eventually emerged on Kachemak Bay to explore by dory. Other veterans arrived at the same time, fresh blood for a withered frontier. They bought large parcels of land cheaply on the rocky and easily navigable south side of the bay, land which Langille had always excluded from the Chugach National Forest and which briefly had been populated before the herring boom destroyed itself in waste in the 1920s. For $1,400, Tillion acquired most of Ismailof Island, more than seventy acres in the ghost town of Halibut Cove, by then just falling-down salteries and a few alcoholic hermits and bachelor fishermen. Clem knew how to longline for cod. For the money to buy the land he went to an Anchorage bank with a letter of introduction from a Swedish seafood processor and fellow veteran he'd worked for, Squeeky Anderson, which read, "He's a crazy kid but catches fish." With that and a handshake he got his loan.

Clem created a memorable impression with his red hair, wiry body, and a grinning directness that would have appealed to Roosevelt, whom he often quoted along with his own aphorisms of pioneer wisdom, to wit: "A man's got a right to take a leak off his front porch, but only until it runs into the neighbor's well." (I've decided his accent is Burgess Meredith playing a pirate.) Clem arrived with a Roosevelt-style ego, too, and no difficulty imagining a community based on the bay's resources growing up around the family he would start with his new wife, Diana, a miner's daughter and visual artist who had grown up in Alaska. They built their home on the passage between Ismailof and the mainland, a glassy smooth natural harbor that unfolds from a narrow, inconspicuous entrance into a winding country lane of water between crenulated rocky shores—a community without roads, only skiffs and rowboats to shuttle between houses. The Tillions invited other families, offering free land to couples with at least four children who would settle and help fill a school. Some accepted and the town slowly grew. Clem homesteaded a water source on the mainland and built a distribution system. For cash he fished and shipped freight from the regional commercial center of Seldovia. The garden provided potatoes and carrots, the sea fish, the woods and duck flats meat—and the cash bought sacks of flour, coffee, and such. In the fall the root cellar would be packed with home-canned food and the loft hung with circular hardtack loaves on broom handles. The family lived in one heated room and an unheated bedroom. "The good old days weren't worth a shit," Clem said, with his sharp, piratical growl. "They were just the price you paid to get what you wanted."

I loaded my own family in the boat on a sunny day last July for the familiar half-hour voyage to Halibut Cove to talk to Clem. We observed the speed limit in the channel, admiring the rows of summer mansions that cling to cliffs on each side, the floating espresso stand with skiffs around it and parasols on the deck, a new floathouse under construction and the sound of hammering on more piers and boardwalks—some of them purely fanciful, like a walkway that encircles a small rocky island topped by a large vacation house belonging to Clem and Diana's physician daughter. Another daughter runs a stylish seafood restaurant on pilings. She decorated the outdoor dining area with life-sized ceramic sculptures and a tide pool touch tank. Guests cross from Homer on the restaurant's classic wooden passenger ferry to walk the boardwalks of the private island, see the floating post office, the beautiful boats and houses, and shop in the gallery and artists' studios. Clem and Diana's house stands back, as grand and solid as a manor, befitting the patriarch and matriarch of the community, built of large stones and timbers, its cellar stocked with fine wine as well as a year's groceries, which a dumbwaiter brings to an upstairs filled with rare books and original art. We sometimes bring summer visitors, who are always delighted, and in the last few years Clem has stopped calling me by my father's name.

Tillion entered the Alaska Legislature in 1962, where his friend, Jay Hammond, already served—two conservationist Republicans in a prodevelopment, Democratic state. Both had opposed statehood, preferring commonwealth status, like Puerto Rico—Tillion at one point favored national independence for Alaska—but those positions tell more about the pleasure they took in unconventional ideas than any radical ideology. Long evenings by lantern light produce good talkers, and Clem and Jay loved discussing history and home-grown philosophy. They knew how to prove a proposition by coining an ingenious, earthy proverb. Hammond's complex imagination sometimes produced policy ideas that outstripped voters' comprehension, but both men were trusted for the intellectual integrity of their quirky politics. Partly on that basis, they improbably came to dominate state government together in the late 1970s, as Governor Hammond and Senate President Tillion.

Prior to Hammond and Tillion's rise both political parties favored resource development with gold rush fervor, but many migrants of the era were instead attracted by the wilderness itself. Scary, out-of-control growth driven by big oil in the 1970s also brought supporters to the conservation side. The flood of money and people that followed oil brought a crime wave, rampant inflation, a housing shortage, pollution, and traffic jams—a wild, chaotic boom similar to the Klondike gold rush. Hammond, like Tillion, had fought in the South Pacific before traveling north to settle a bush paradise, in his case on Lake Clark, on the west side of Cook Inlet, where he supported himself as a fisherman, guide, and pilot. As residents of their ideal rural worlds, Tillion and Hammond's political popularity rested on resistance to unnecessary change. Hammond ran for governor in 1974 calling for a buyback of oil leases in Kachemak Bay that had been sold by the Democratic incumbent, Bill Egan. (Hammond won, but the legislature balked at condemning the leases until a marvelous fiasco in 1976, in which a portable drilling rig got stuck in sixty feet of Kachemak Bay mud and had to be removed with explosives.) Being Republican and battling the oil industry did not create a philosophical conflict. Hammond and Tillion equated conservatism with conservation—both simple principles of caution.

"I have grandchildren," Tillion told me, "And they shall have some fish, too. I never thought of myself as anything but a temporary organism that was going to be here for a few years and pass on, and the only thing that counted were my children, my grandchildren, and my great-grandchildren. And they might not understand what you were doing. They don't have to. But how can you be a conservative, and stingy with money, if you aren't stingy with trees and fish and everything else that is the natural resource of your country?"

Statehood advocates had argued that Alaskans could bring back their salmon by removing ineffective federal fishery managers, but after the new state took over, runs remained low through the 1960s, even as many more fishermen joined with new boats and nets to divide the catch into ever smaller slices. To allow enough fish to escape the nets and spawn, state biologists had to cut fishing periods and sometimes close fishing altogether, resisting pressure from fishermen who faced ruin as their increasing investments in equipment sat idle. Unable to exclude each other, they tried to reduce competition by killing seals and sea lions, and they even shot whales. The waste and inefficiency of this competitive system, called, in academia, open access to a common pool resource, seemed to validate the pessimistic predictions of economists. Calculations based on the assumption of selfish human behavior showed that any such resource and industry would inevitably spiral into disaster.

In 1968, ecologist Garrett Hardin succinctly stated the problem in an immensely influential paper in *Science* titled "The Tragedy of the Commons," in which he imagined a shared village pasture to which herdsmen were free to add more livestock until it was destroyed by overuse:

> Each man is locked into a system that compels him to in-
> crease his herd without limit—in a world that is limited. Ruin
> is the destination toward which all men rush, each pursuing
> his own best interest in a society that believes in the freedom
> of the commons. Freedom in a commons brings ruin to all.

Tillion was well aware of these ideas as he worked to limit the number of fishermen who could join Alaska's salmon industry. Hardin's solution followed Hobbes: only the power of the government could save selfish people from themselves. The true focus of Hardin's essay, however, is often forgotten: he called for government regulation of human reproduction to prevent overpopulation of the earth, reaching the same conclusions as the conservation-oriented eugenicists of fifty years earlier, although for reasons of environmental sustainability rather than for racial purity. Hardin's pessimism about human nature led him to advocate a state so powerful it could control citizens' most intimate decisions. His belief that humankind was incapable of large-scale cooperation left no room for the kind of self-organized community ownership of resources that Chugach Natives had lived for centuries.

Tillion believed private ownership offered the only solution to the tragedy of the commons in Alaska's fisheries, because no other plan would give individual fishermen a financial stake in conservation. Hammond initially opposed limiting salmon fishing licenses, fearing outside

fishermen might win rights over locals—including Natives. The economic survival of many villages depended on commercial fishing. When
a program finally won approval from the legislature and the voters in
1973—after several tries over a series of years—it awarded permanent
limited-entry permits to fishermen, which they could use or sell, allocating the permits with a scoring system that awarded almost half to rural
people fishing in their own areas.

Despite the intentions of the program's designers, however, private
ownership itself proved the undoing of many fishing villages. Some Native
fishermen in Prince William Sound, from Tatitlek or the destroyed village
of Chenega, missed out on getting permits because their catch records
didn't match up with their actual fishing—in their informal, cooperative
fleet the person who caught the fish wasn't always the one who sold it to
the processor. Those who did get permits in villages had trouble keeping
them. Transferable permits rose quickly in market value, often higher than
$100,000; for a family in a cash-poor village, a permit might be the only
large financial asset. In a crisis, it might be sold, or during a downturn lost
as collateral on a loan or seized for unpaid taxes. The need for investment
in new boats and equipment to stay competitive with other fishermen
sometimes made it hard to hold on to a permit. A successful fisherman
who got hurt or retired might sell a permit to the highest bidder, likely
someone from outside the village. But once a permit left a village, it was
unlikely ever to come back, as few village Natives had the money to buy
one. Since the program started in 1975, 27 percent of rural Alaska permits
have left. Tatitlek fishermen had thirteen permits in 1978, but only three
in 2006.

Fishermen have grown older on average, too, as the young can't afford
to buy permits. A strong young man like Clem Tillion in 1947 would
have to struggle and save much longer today—the average age for acquir-
ing a permit is around forty. Tillion was willing to accept that outcome
for Alaska because he cared more about the fish than the fishermen. Villages losing fishing permits reminded him of North Dakota towns drying
up because of the demise of the family farm. He said, "You got cheap

bread because of that. Is it sad? Yes. The market might not be the fairest, but it serves the public best. . . . To me, all things are economic, so my position has been: I want to protect the resource, and then I want to sell it so it will bring in the maximum amount of money from Outside. It's not very complicated. It's Adam Smith, *The Wealth of Nations*."

Clem's belief that tradable permits would give fishermen a financial stake in ecosystems quickly proved correct. In Prince William Sound, where salmon returns had gone from bad to worse in the early 1970s, permit holders banded together to build salmon hatcheries as soon as the limited-permit system went into effect. Armin Koernig helped lead the movement—he had escaped East Germany in 1948, eventually to own one of the big boats in Cordova called seiners for their long, curtainlike nets, but he nearly lost it when seining was closed due to failed salmon runs in 1972 and 1974. Fishermen contributed labor and a share of their catches to the new Prince William Sound Aquaculture Corporation, which also took out large government loans. It built a hatchery in an old cannery in broad, placid Sawmill Bay, in the southwestern sound, where fresh water for the egg trays flowed down from a lake in the mountains of Evans Island.

The hatchery seemed to be a free food machine: pink salmon the size of minnows swam from pens by the tens of millions in the spring, fattened up for two years at sea, and returned to the same spot, where relatively few were needed to provide eggs and milt for the next generation. Fishermen on contract caught enough fish to pay hatchery expenses before permit holders rounded up the rest for their year's income. As the corporation grew, fishermen voted in a mandatory 2 percent tax on their catches and built a second large hatchery on Esther Island. The corporation took over management of three more state hatcheries as well. It became the largest operation of its kind in North America. Cordova fishermen proudly wore aquaculture corporation T-shirts proclaiming, "We hatch 'em, you catch 'em." By the 1980s, salmon catches in the sound were hitting all-time highs. Salmon fishing all over Alaska entered a golden period of huge harvests and high profits for fishermen.

The hatcheries seemed at first to be a complete success and vindication for privatizing the salmon permits, but with time the story became more complex. The Armin F. Koernig pink salmon hatchery in Sawmill Bay certainly produced a lot of fish—I watched the release of 180 million tiny salmon from net pens one spring day. Their schools looked like billowing black clouds in the water. Days later and miles away I saw them in thick groups whenever I peered overboard. But more fish doesn't necessarily mean more food. Some biologists believe the fishermen's hatcheries, built to manage their ecosystem and bring stability to their lives, instead simply displaced wild spawning fish, distorting the natural food web. Studies show that when larger groups of salmon go to sea they eat through more of the ecosystem's total production of energy, potentially producing a larger harvest for fishermen but less for other animals. Besides, in 1978, the climate shifted in the North Pacific, as it does now and then, and altered the entire ecosystem in favor of salmon; the Koernig hatchery opened in 1977. The fishery probably would have rebounded anyway, although it's impossible to know by how much.

Limiting salmon permits did not produce more food in the ocean. It allowed one group of people to get together so they could mold the ocean system to their needs. And, without perfect understanding of the system or foreknowledge of their own needs, not all the results suited the fishermen, either. Some years in the 1990s, when salmon prices crashed, far too many fish returned to the hatcheries and salmon were stripped of roe and ground up or hauled offshore and disposed of whole. Even today Native people who live near the hatchery complain of the death and stench from piles of hatchery fish trying to return to their natal waters to spawn: since human-hatched fish have no spawning streams, the salmon flop and die by the hundreds and thousands in tiny rivulets and other inappropriate spots on the shoreline. I'm not against the fishermen; my point is that assigning ownership of a part of an ecosystem to a particular group has consequences, including potential diminishment of other parts of the ecosystem that remain unowned.

Returning to Clem Tillion—he became more involved in fish politics than fishing. He negotiated for a fish treaty with Japan on behalf of the

U.S. State Department, helped craft a law expelling foreign fishing fleets from within two hundred miles of the U.S. coast, and served on the federal council that managed those waters. He sold his own salmon and crab permits and most of his gear in 1976 to avoid being accused of a conflict of interest, even though he didn't have to—he sold the permits cheap to keep them in Halibut Cove, where they have remained. In 1990, the Republican governor, Walter Hickel, made Tillion his "fish czar," and he forced through federal privatization of halibut and cod, a program even more radical than the salmon permit system. The salmon fishermen still had to compete among themselves for a catch; the new cod and halibut system instead assigned fixed ownership shares of the annual harvest to fishermen, so they knew at the beginning of every season exactly how many pounds of fish each was entitled to remove.

At the time, halibut seasons had shrunk to one-day derbies; anyone with a skiff and a longline bucket would catch whatever they could during that wild rush. The new system would allow fishermen to extract their quotas all year, whenever market conditions and weather were prime, saving lives, improving quality, and raising the prices they received. But fishing communities revolted over losing one of the last open-to-all fisheries, and at the government giveaway of fish worth millions to big, outside operators who wouldn't even pay a royalty—the quotas went free to whoever had fished in the previous years, including large ship owners. (Roosevelt or Pinchot would have wanted a royalty, too, but Clem didn't have the power to require it.) Tillion took the full blast of fishermen's anger. Twenty communities and a unanimous State House passed resolutions in opposition. Even Hickel lost heart. When Clem traveled to coastal towns he sometimes needed state trooper protection. In the end, he won the fight, but he was kicked off the fisheries council. On the day of final approval in 1993 he told a newspaper: "It's over now; it's the law of the land. Is it a monumental change? Yes. It's as much a change as the fencing of the Plains."

Around that time his wife, Diana, said, "I get mad. I get mad six months at a time. Clem, he hasn't gained from serving. It costs us in money and it costs us in time. I'm a martyr to these fisheries, and it makes me mad because the people who gained—these stupid jerks—they gained from what he has done and he hasn't. It makes me furious."

Considering his belief in the necessity of self-interest to motivate people, it's interesting how much Clem Tillion gave for the cause of fisheries management—certainly more than he or his offspring would directly benefit from. But life in Halibut Cove could be much worse. He drives the mail boat, chops wood, charms the tourists, receives tributes and awards for his accomplishments. Some fisheries experts cite the system he helped create in Alaska as the best in the world; others point to damage suffered by coastal communities when the new legal system alienated them from their resources, with a loss of freedom and economic autonomy and the creation of permanent classes of haves and have-nots. Clem doesn't apologize for his narrow focus on efficiency in producing valuable fish rather than for social goals or the broader ecosystem. He acknowledges that Prince William Sound hatcheries release hundreds of millions of predatory mouths that may consume more than their share of total biological productivity—but Tillion believes intensive management of the sea is the price of feeding a growing world population.

People come from far away to see the cove, where Tillion rules as a sort of benevolent feudal lord (the family trust still owns most of the island). In the summer, water taxis run on regular schedules to the trailheads in Kachemak Bay State Park. It's common to meet tourists there wearing white tennis shoes and cotton sweaters tied around their shoulders, hiking on paths formerly trod only by fishy rubber boots. In 1971, Clem helped create the park when logging threatened the dark spruce backdrop of mountains on the south side of the bay (he excluded the best coastal lands where people could live). A couple of decades later, as the rest of the world arrived, a team of park rangers swooped down on his daughter's boat with a search warrant and charged her with dropping off visitors on park beaches without a permit. The Tillions also lost a cabin in the park and Diana gave up raising mussels there partly because of red tape.

In the old days, no one owned much of anything but it was as if everyone owned the whole bay. Now a lot more people are around, everything is owned, and the bay and the world are smaller. When Clem started the Halibut Cove community, he foresaw it being based on the area's resources. There are a few wealthy fishermen living there now, but mostly affluent Anchorage people use their houses for brief summer visits. Real estate agents sell expensive lots touting the cove's prestige. "It hasn't come out exactly as I wished," Clem said. "But with the sea otter moving in and wiping out all the crab, the resource base changed from one that you harvested and made a living, to one that's sacred under United States law and you can't harvest it."

It took me a moment to realize what Clem meant: that the rebound of sea otters from Russian fur hunting had crashed the Kachemak Bay crab fishery, and that the blame belonged with federal policy for protecting overpopulated otters from hunting. Otters do fill the bay—sometimes a boat ride back and forth to Homer is a slalom to avoid them—and perhaps they were partly responsible for holding down crab numbers since the crash in the late 1980s. But, if so, that's because otter numbers have returned closer to what they were before 1741, when the Russians arrived; it may be that decades of abundant crab were an anomaly of an ecosystem depleted of otters. Or maybe not. Biologists don't understand the disappearance of crab, and a list of suspects would have to put otters below overfishing and the climate and ecosystem shift of 1978.

But I did appreciate Clem's regret for the bay's lost resource economy. I remember camping on the beach on Homer Spit as a teenager among a temporary community of college kids and drifters who were hoping for cannery work or a fisherman's crew share. Spit rats they were called, and their plastic sheeting flapped over driftwood tent poles and their tables were made of wooden pallets or cable spools. People without money or transportation dedicated their days to constructing elaborate homes of beach debris while waiting for the fish to come in. The spit is a gravel bar reaching five miles into the bay, like a natural pier for the town, which otherwise has only shallow, mudflat access to the sea. Camping there is the closest thing to going out in a boat while staying on land. At night smoke from salty wood and beach coal drifted above the pebbles, a sharp, vivid odor that mixed distinctly with the muddier scent of marijuana. Under a glowing midnight sky a bottle of sweet, potent Yukon Jack passed around the fire, eyes glistening in the flickering light as young people talked of maybe hitting a big payday and told stories of somebody who once did. I remember waking one morning to find a ship beached next to the tent—it had stranded during the night's high tide for a low-tide inspection of the hull—and having someone walking by offer to sell me a doughnut.

The town outlawed the spit rats in the 1980s and the cannery burned down in 1998. The company didn't rebuild because the salmon industry was on the ropes and no one had been able to fish crab or shrimp in the bay for years. By then, Homer Spit wasn't mainly a fishing port anymore—it was a tourist trap with picturesque fishing boats to look at. Hundreds of huge RVs parked on gravel pads; one RV park was wired for Internet and cable TV and had a putting green. The road near the harbor was lined on each side by cheaply made booths and offices for tour boats and fishing charters, choked with traffic and double-parked cars and people window-shopping, like the New Jersey shore, but 30 degrees cooler. Even the garbage had changed. A volunteer coast walk each fall documented the evolution of washed-up refuse over the span of twenty years, a gradual transition from lost commercial fishing gear to fast-food wrappers, water bottles, and other urban junk.

I don't think an excess of sea otters caused this transformation. An excess of wealth, consumption, and waste spilled out here, and many other places, overrunning local economies that had been connected to the rise and fall of the sea. Clem wanted the fish for his grandchildren, but he also wanted his community to stay connected to the fish, and private ownership didn't accomplish that. The world's best fisheries regime also proved inadequate for managing the ecosystem. In Kachemak Bay, crab isn't alone in its decline—also down are shrimp and clams, seals and sea lions, and even the otters recently have been affected by some kind of epidemic. Sportfishing charter boats drive hours, well beyond the bay, to haul in big halibut;

once, you could catch a one-hundred-pounder from the city dock. The huge herring runs have not returned over the span of a human lifetime.

Many people sense that the ecosystem has diminished, but we can't pick out a cause. Cause and effect don't link in a way that we can comprehend and express—too slippery, too many invisible connections, and too much hidden in dark, wild waters as inaccessible as outer space. What does seem clear is that establishing private ownership of the money fish wasn't enough. Garrett Hardin's metaphor, the tragedy of the commons, was far too simplistic. New laws controlled access to certain species as securely as the grid of lot lines on land, as if the net of the law could strap down everything worth owning. But it turned out that the money wasn't the important thing. Something else slipped through the net that was as good, or even better, and that couldn't be privately owned.

23

An Alternative to Tragedy

Students in a computer lab at Indiana University in Bloomington earned money by moving a Pac-Man-like avatar across their screens to eat up diamond-shaped tokens, each worth a penny. Ideally they could go home with $22 each, but only if they allowed the tokens, like fish, to reproduce over the course of four-minute sessions rather than consuming the breeding stock all at once. But with five players competing in an open-access system like an unmanaged fishery, the tragedy of the commons played out as predicted: all the tokens were gone in little over a minute, leaving students three minutes to look at an empty screen when they could have been harvesting sustainably for the whole time. They earned only a third of the maximum. In the next version of the experiment, an all-powerful state took control—in this case, experimenters Elinor Ostrom and Marco Janssen—and drew property lines on the screen to keep each student in his or her own area. As Hardin or Tillion would predict, the students conserved more of their tokens and earned more money, although they still did not earn the maximum possible as a group, because some overharvested or were inefficient.

So far the experiment confirmed Hardin's gloomy outlook, but Ostrom and Janssen tried another variation. They put students in the open-access environment again, tokens free for all, but with one change: they allowed players to talk to one another between rounds. In that experiment, the students created their own cooperative systems that worked even better than the private-property version of the game. After an initial round in which they learned by experience the consequences of unrestrained competition, players devised ingenious ways of restraining themselves. Some divided the space into areas like private property; others divided the harvest by time. One group agreed that when tokens neared depletion they would wait together at the bottom of the screen for regrowth to occur. A group of fifteen players found a way to enforce their agreement. They agreed to harvest for periods of ten seconds followed by breaks of twenty seconds. Everyone remained quiet during the twenty-second break times so that if a clicking keyboard were heard—indicating someone was illicitly harvesting—everyone could resume together, destroying the advantage of the defector.

Lin Ostrom could hardly have been surprised. She began her academic career before the invention of experimental economics by studying a self-organized community of groundwater users in southern California almost half a century ago. She studied police departments and school boards. She gathered hundreds of case studies from obscure journals from all over the world that documented the familiar human experience of working together to share and solve problems. Lin said, "I was looking for the optimal set of rules. But what I found was this chaotic set of rules—just many, many, many. I thought, 'I'm seeing an immense amount of cooperation. Hardin isn't wrong in all settings, but he's wrong in a lot.' . . . You just do this one change—allow humans to reach out to other humans—without structure—and they say, 'Hey, this is kind of stupid. If we just work together, we can do a lot better.'"

In her part of the academic world, Ostrom won the debate over Hardin years ago. Citations of her landmark 1990 book, *Governing the Commons*, which refuted the tragedy theory, nearly equal his 1968 original. The National Academies of Sciences published a volume on the field of commons studies she helped create, and *Science* published her review of changes since Hardin's paper on its thirtieth anniversary. In October 2009, she became the first woman to win the Nobel Prize for economics.

But outside academia, the tragedy idea lives on, an eloquent metaphor for a commonsense view of human nature that happens to be wrong. Policymakers solve problems by privatizing resources, centralizing power, incentivizing desired behavior, and harnessing competition—and in the process crowding out and demoralizing cooperative impulses that still emerge when people have the chance to use them in the ever-narrower community realm unoccupied by government or corporations.

The economic and political systems built on Enlightenment thought have become their own reality. Competitive and coercive systems destroyed the opportunity for cooperation, and now it's hard to believe in the possibility of cooperation anymore. Or so a naïve member of our society would think before picking up Ostrom's books and articles, and those of her followers, and finding example after example from the real world, examples by the scores and the hundreds, by the thousands, proving the contrary. As she wrote when finally demolishing Hardin's claim of the inevitability of the tragedy of the commons in *Science:* "It is obvious that for thousands of years people have self-organized to manage common-pool resources, and users often do devise long-term, sustainable institutions for governing these resources."

Lin and Vincent Ostrom met in the early 1960s when she was his political science graduate student at University of California, Los Angeles— he was the seminar instructor who assigned her the study of groundwater sharing. Both were from poor families; both cared more about ideas than money; both were already married. They started a new life in Bloomington on a big patch of woodlands. The house they built—all low angles and natural wood—overlooked their own shadowy canyon of trees from big, rectangular windows. When it was done, they asked a local craftsman to make their furniture. He instead drew Lin and Vincent into his workshop to do the work themselves, and years later they had made their own sleek cherrywood chairs and cabinetry, and a spectacular dining room table with an edge that exposes the gnarled surface of the tree. It alone took a year and a half to construct. The experience of working collaboratively with their hands became an inspiration. In 1973, they established their own workshop on the Indiana University campus, the Workshop in Political Theory and Policy Analysis.

Vincent, fourteen years older than Lin, had already created a prototype of a political workshop at the Alaska Constitutional Convention in 1955 and 1956, which he attended in Fairbanks as a consultant from Stanford University. The Natural Resources Committee asked him to write a draft of their article of the constitution, but he refused. Instead he offered a tutorial on the law and history of Western resources over dinner at the home of one of the committee members, and the next morning facilitated a brainstorming session in front of a chalkboard in a classroom at the University of Alaska's School of Mines. The words copied from the board at the end of that day's session remain at the heart of Alaska's constitution. They enshrine the best ideals of fairness, sustainability, and common ownership held by the era's white Alaskans about the land and sea. (But the resources article made no recognition of Alaska Native land or subsistence rights—it represented the point of view of its makers.) Vincent's own convictions were enacted through the process itself, by organically involving the intelligence of many people, collaborating, and putting decisions in the hands of those directly affected. As a student of Native American cultures and of the political debates of the nation's founding fathers, he believed the real world's complexity and uncertainty could best be addressed by those closest to them. Half a century later, he was most proud of the Alaska constitution's flexibility. Rather than making detailed law, as many state constitutions do, it created an adaptable system for the Alaskans of the future to solve problems their own way, in their own towns, according to the values of their own times.

The workshop in Bloomington likewise built its products using the combined materials of theory and reality. Furniture makers' ideal forms must conform to function and to the peculiarities of their materials; likewise, scholars at the workshop, calling themselves artisans, developed a craft of weaving political philosophy with detailed empirical studies drawing on anthropology, economics, psychology, law, or any other tool at hand. The scholar-artisans, mostly PhDs, came from all over the world, especially southern countries, and carried their work home for practical use. A critical early project collected a vast but obscure literature on local management of common resources. Workshop scholars created a database of case studies that had been recorded by many different kinds of scientists to compare using a single analytical system. Collaborators applied mathematical and economic game theory to decipher how the groups self-organized. Field researchers took the games back to communities of resource users, recruiting as players, for example, indigenous South Americans on the banks of the same rivers that the games had been intended to model. People there recognized their own lives in the games, and the simulations gave them ideas for how to better share and conserve their real-life resources. At the workshop, ideas became a kind of common property as well, contributed and tested in rigorous but egalitarian open sessions that involved everyone. Big groups of authors contributed paragraphs and members picked over each other's sentences. The workshop itself became the subject of a workshop held every few years.

Lin said the university didn't know what to make of the workshop at first. Support was thin and budgeting a struggle. She and Vincent created a fund to pay the expenses of visiting scholars, who often came from poor countries. When Vincent got a job offer somewhere else, Indiana deposited a grant into the fund to retain him; honoraria and consulting money went in, too, and the Ostroms both contributed half of their own salaries over the years. Eventually they had amassed $2 million as an endowment. The workshop took over a closed frat house on a tree-lined street a couple of blocks off campus. It's a worn-out building, old-fashioned and grungy in spots, but remarkably homey. Former bedrooms lie at the end of narrow corridors, rooms open to rooms, and behind creaking doors I found visiting scholars, at laptops on metal desks, always cheerful to be interrupted for a conversation, although many were important people in their own countries—a reformist president of Liberia studied here while in exile. Lin bustles through the building, enthusing about their work like a proud mother—or grandmother: she is now in her midseventies and has long gray hair pinned back.

She stepped into a large classroom edged with furniture that looked like it could have been left over from the building's frat house days and sat down at a table in front of a two-way television, and a moment later began lecturing to a hall of more than 150 upturned faces in southeastern Brazil. Lin was always so sweet that her intellectual prowess kept surprising me. For the Brazilians she summarized key points about self-organized systems in the commons. Mathematical game theory predicted that altruistic punishing shouldn't happen, but after seeing it in a cooperative Nepalese irrigation system, Lin had modeled that same situation in the lab; she found that most people spontaneously set up their own sanctioning systems and those who didn't performed poorly as groups. To make sense of these results would require an improved model of the individual. This real-world individual could not be a rational automaton. He or she would exist within an ecological and social context, could deal with imperfect information, and could innovate, learn, and apply rules. The individual would understand when and how norms of trust and reciprocity would apply in the various concentric circles of human relationships—among family, work partners, community, or state.

With this complex individual in mind, Lin explained the attributes of resource commons where self-organized systems work: the resource must be viable (not already overfished, for example); there should be a way of monitoring the resource and its use; it must have some predictability; and its boundaries must match the size of the group sharing it. Practical rules, boiling down to a match between the ecosystem and the people using it. And the attributes of the users themselves: caring about the future, part of a community in which trust and reciprocity were possible, and experienced in the use of the resource. It struck me that the traditional Chugach people had all these qualities, and the Enlightenment invaders from Russia and the American gold rush had none of them. The barrier to conservation, therefore, lay in the people, not the place.

I wondered if the U.S. history of ecological conquest explained why Lin's work was accepted so eagerly in southern nations but so little here. In less developed parts of the world, informal relationships to nature persist in many places. The workshop's visiting scholars report examples all the time—sustainable fisheries managed by village agreement, community grazing of herds that protect the range, forests where traditional logging respects the ecosystem. No two systems are the same. People figure out their own arrangements and punish those who don't cooperate. But when top-down control enters the picture or outside competitors arrive who aren't part of the community, self-organized cooperation collapses. To work properly, the mechanisms of trust and reciprocity need both the responsibility of collective ownership and ongoing relationships to enforce cooperative norms. An economy and political system based on competition, private property, social anonymity, and centralized power—the United States—is not a fertile bed for cooperation to grow.

And yet it did at Indiana University. And the commons idea has spread into discussion of other shared realms—the Antarctic and Arctic commons, the atmosphere, Earth's orbit, the electromagnetic spectrum, the Internet, the commons of information, knowledge, education, civic discourse, and, of course, the open ocean. Any good that can be diminished by use but from which it is difficult to exclude users could be a common pool resource amenable to some sort of collective management. But Lin distrusts many new applications of the concept as merely airy metaphors. She doesn't want her well-bounded and well-tended definition of commons dissipated by outsiders' indiscriminate use (that's my metaphor, supposing that language is a commons as well, protected by altruistic punishers such as grammarians who correct other people's usage errors).

More fundamentally, she resists easy answers because the real world is complicated: sometimes self-organized communities work best, and sometimes private property and free markets make more sense, or even the authority of a distant government. Ideas about commons don't always fit. But there's a hunger for cooperative answers. So said Charlotte Hess, the workshop's librarian, who monitors and collects the rapidly expanding commons literature and published a book on the knowledge commons. She collaborates with Lin and understands her concern for rigor and precision, but added in a quiet voice that the concept of commons had already slipped beyond academic control into a space where the imagination dwells: "It has an attraction and strength at this time, because people feel so helpless, and feel that it's at the end of days, and that we're just going to watch it all run down. And yet there are these systems and communities that are resilient and refuse to give up."

Mead Treadwell studied the Ostroms' commons theories and passed them on to some of his powerful Republican associates in Alaska; Charlotte's husband died of brain cancer at the same time as Mead's wife, Carol, and they share that link as well. After his contact with the workshop, Mead became deeply involved in planning Arctic research on climate change as chairman of a federal commission, doing things like calling for construction of more icebreaking ships for scientists. The atmospheric and oceanic commons of carbon dioxide disposal worry a lot of people at the workshop—there's concern their work won't scale up to that global level, which may turn out to be the only one that matters. A visiting lecturer talked about selling carbon pollution credits and then distributing the profits to the public as a sort of dividend for the sacrifice of the atmosphere. Somehow, that didn't register emotionally in the same way as hand-net fishermen managing an ecosystem on a tropical beach. And emotions matter. The will to cooperate depends upon them.

The oceans are unpredictable, unbounded, and larger than the mind can grasp. They do not fulfill Lin's attributes of a resource commons amenable to management by a self-organized community. But such communities are the starting points. The problem is unimaginable in scale, nonlinear in shape, and infinite in complexity, and so may be the solution—in the interlocking relationships of human societies. And the solution may also be small enough for a single person to choose, which is important, since individual people alone are capable of making choices.

To unravel these seeming paradoxes, think of a lone woman picking up a plastic water bottle from a beach. Something clicks in her heart, like a tumbler opening a lock, and she feels differently about the ocean. She passes on her revelation, by example, to the circle of people who respect her. Small communities affect one another. The same people are members of many communities and larger institutions, and the networks intertwine. As new norms spread through the overlapping circles of affiliations, the social reality itself transforms. The new order may emerge rapidly, even if the process of change itself has been invisibly gradual. Or it may develop as imperceptibly as a shifting ecological baseline, so only historians or elders realize a change has occurred.

Scholars at the workshop have watched how cooperative norms develop. Xavier Basurto dove for scallops with Seri Indians on the Gulf of California. They walk on the ocean floor in rubber boots, wet suits, and weights, breathing through garden hoses attached to compressor apparatuses built of modified paint sprayers and beer kegs. After near extermination by Spanish and Mexicans invaders, a few hundred surviving Seri received collective ownership of remote land and water around Tiburon Island in the 1970s, and beginning in the 1980s they developed traditions of collective management of the scallop beds there. For example, at certain especially low tides the community gathers for celebratory scallop harvesting on certain sandbars; if they find scallops growing scarce, outside fishermen are expelled from Seri waters. Sanctions within the culture are nonmonetary. Xavier said the Seri love sexual humor and the women use it to humiliate their husbands when they transgress fishing rules: "My husband, who has a small penis, is fishing." After three decades of the scallop fishery, beds are depleted outside the Seris' area but remain productive within. Outside fishermen and buyers are allowed into the community only on its own terms, and for its profit.

Led by Juan-Camilo Cardenas, who pioneered taking experiments into the field, Maria Claudia Lopez participated in large-scale studies of ten villages in Colombia that shared use of fisheries, watersheds, or forests. Some villages had no roads. Maria Claudia described children gathering to watch the visitors eat, as they'd never seen a white person before. To travel safely, researchers had to make their purposes known—to avoid conflict with drug dealers—and keep their educational status unknown—to avoid kidnapping. It might take five hours to explain how a game was supposed to work, but the games did successfully model the resources and the relations of the people sharing them. Afterward, out of courtesy, researchers spent time explaining their findings to the villagers, who then used what they had learned to improve how they managed their commons. Leaders from other villages asked to participate, too. "It's an environmental tool to learn how to use nature," Maria Claudia said.

Together in one of the workshop's well-worn meeting rooms, Xavier, Maria Claudia, and I talked about the social power of Lin Ostrom's work, which has created connections all over the world. Maria Claudia pointed out that the workshop itself is a family, cooperative and egalitarian in every respect. I experienced that: the week I spent in Bloomington felt like Cultural Heritage Week in Tatitlek, full of shared food, ideas, and stories. Xavier agreed the workshop's ideas were creating a social movement both within its walls and in communities around the world, but insisted that was not Lin's goal. She wanted to advance knowledge, and knew a social agenda would affect her academic credibility. Maria Claudia, a PhD economist, expressed no such fear. She said, "I do this because I care about people."

The people the workshop helps belong largely to the same groups the scientific racists and their ilk wanted to stop from reproducing. The workshop's ideas also are more sophisticated than those of their ideological adversaries. Garrett Hardin's tragedy of the commons was inspired by doomsday predictions of population growth, which have proven overblown and simplistic, and by the false and untested assumption that human beings are always competitive. He opposed humanitarian aid to poor countries, hoping to stop their population increases through starvation and disease, and opposed immigration to preserve the United States as an island of wealth and environmental quality. Hardin believed compassion was a weakness, bound to be eliminated by natural selection, an idea closely analogous to the concern expressed by Roosevelt's friend Madison Grant about the suicide of the master race. In the 1980s, Hardin's writings helped found an anti-immigration branch of the environmental movement, many of whose members also supported English-only legislation. Intellectually, Hardin had simply updated Hobbes: Americans' power gave them the right to consume the world's resources unsustainably; therefore, ethical protection of the environment consisted of stopping more people from becoming Americans, not reducing the unsustainable consumption itself.

Enough environmentalists shared this point of view to bring about a national vote by members of the Sierra Club to oppose immigration in 1998. The proposal failed, but the existence of the debate suggests the durability of the equation of racism, nationalism, and conservation from a century ago. This kind of prejudice concerned Lin Ostrom. Intergroup hostility arises between the kinds of strong, exclusive communities that most successfully manage common resources. Scholars find the tendency out in the field—Basurto's Seri fishermen cheated outsiders without compunction—as well as in the laboratory with college students playing economic games on computers. This worry helps explain the philosophy of government Lin and Vincent propound: many levels of nested and overlapping affiliations based on different systems of human relationships and different kinds of ownership. Each of us switches from cooperation to competition and back many times a day. In theory, by living in numerous circles of self-government at once—a system the Ostroms call "polycentricity"—our parochialism could be conditional as well.

But how do you start to live that way, especially when you have to start from scratch? The Chenegans had been a self-organized community in harmony with their ecosystem, but after the village was destroyed, an outside legal and political system carved up the land and resources of Prince William Sound into various new forms of ownership. Rebuilding the community proved more complicated than constructing new houses—people had to restore a way of life with a new set of rules.

24.

Chenega Reborn

Larry Evanoff's mother was one of those who would decorate the Chenega church with crepe paper flowers for Easter weekend. When the earthquake started she and his father ran toward the church for shelter. Larry, fourteen, was seven hundred miles away, attending eighth grade at the Wrangell Institute boarding school for Alaska Natives in southeast Alaska. He knew nothing of the earthquake. On the Monday after the Friday it happened, he received a package of home-baked bread and treats from his mother. Later that day, he was pulled from class and told a tsunami had destroyed his village and killed his parents. He didn't believe it—he'd never heard of a tsunami, and he had the care package as proof that they were okay. Later that afternoon, Larry heard the news again through a black telephone from his aunt and older sister in Cordova. They said not only his parents were gone, but also an aunt, an uncle, his grandmother, and many others. He wanted to go home, but they insisted he stay at school. They were afraid he would drop out and never return. Larry spent three days in the nurse's office crying and sleeping before the principal told him to snap out of it—he had grieved enough. And Larry tried, but he couldn't stop thinking about his loss. "I thought at that point I had nowhere to go, nothing to do. I felt so damn alone."

Over the next years many men from Chenega died, taking their own lives or succumbing to alcohol and its hazards outdoors in Alaska. Larry couldn't see much of a future for himself, either. After a summer commercial fishing with his uncles, he resisted going back to school in Wrangell. But he was a promising student, a whiz at algebra, and finally his aunt and sister prevailed. He believes that decision probably saved his life, because at school that year he fell in love with his future wife, Gail, an Iñupiat from the Nome region, and she became his support. Gail was pretty, bright, and articulate. Larry said, "I could do math, but I couldn't do English. And she hated math. It was funny. In between classes we'd stop, we had five minutes, and we'd meet each other. 'And here are the papers for math.' 'And here are the papers for English.' And we had five minutes to explain. We got each other through."

After high school Larry and Gail planned to attend the University of Alaska in Fairbanks together, but Larry received a draft notice just before enrolling. After army basic training, they married in Fairbanks, and Gail got pregnant with their first child, William. The boy was eight months old when Larry first saw him, when he returned from Vietnam with four bullet wounds. They settled in Anchorage, had a daughter, Wannah, and Larry began working as an air-traffic controller. But he knew he had to get back to Prince William Sound. He said, "That wasn't a question. One way or another I was going to get down here."

Chenega's villagers had dispersed. The government moved them to Tatitlek the summer of the earthquake, first to live in tents and then in newly built houses that were much more modern than any they had owned before. But the two communities didn't mix. Despite many family ties with Tatitlek, Chenegans continued to relate as a separate village. They didn't fit in with the land, either. Their traditional waters for fishing and hunting were much too far away to use, and they didn't know Tatitlek's area. The sea looks different there, with broader spaces and bigger sky than among the steep, narrow channels near Chenega. Besides, food was scarce in the sound's earthquake-shaken ecosystem. Chenega people drifted away from Tatitlek to Cordova, Anchorage, and beyond. The old village remained abandoned and the land it sat on was absorbed into Chugach National Forest.

Nick Kompkoff Sr., like Moses, held his people together in diaspora. During the tsunami he had saved one daughter, Carol Ann, while two more slipped away from his grasp; he had three sons who survived as well. The family settled in Tatitlek. After training in Sitka, Kompkoff became lay reader in the church. When they moved to Anchorage he got his high school diploma, helped start a parish, and, in 1971, he became a priest. Carol Ann remembers him as a powerful, assertive presence. Kompkoff involved himself in Native politics, keeping up the village's status as a federal tribe, and he gathered the survivors to meet annually, when they always talked about their desire to return and how they missed the old village. Carol Ann said, "I seen him break down for his daughters, and all the things we lost, but I've also seen him stand up and fight for the people. If it wasn't for him, there wouldn't be a community."

As time passed, the job of reviving Chenega as a real village became more difficult. While Chenegans' emotional recovery slowly progressed, Alaska was being split up into state and private property. Although every adult in the village had fished commercially before the earthquake, the disaster destroyed that entire system a decade before limited-entry fishing permits were finally distributed. Nick Kompkoff Sr. himself fished for a few years while serving as a priest in Cordova, but gave up during the bad harvests of the 1970s. And the land was at risk as well as the fish. The 1958 act of Congress that made Alaska a state threatened all Alaska Native rights to their land by granting the state 103 million acres of federal land as its own—about a third of the Alaska landmass—with twenty-five years to choose whatever was most valuable. (Alaska had remained relatively free of property lines except for the national forests and national parks.) While the Statehood Act vaguely directed the state to respect Native lands, it made no provision for what that meant. Since Natives wielded little power in state politics and commonly faced racial discrimination outside their own communities, scant hope existed for a fair settlement.

Driven by this danger, Natives created associations in various regions—a Chugach Native Association in Cordova formed in 1965—and began filing claims with the U.S. Department of the Interior over vast swaths of Alaska. President Johnson's administration acknowledged the claims by freezing disposal of federal lands to the state or anyone else until the issue could be resolved. But legal experts considered the Natives' position weak: as we've seen, the legal system itself was founded on a philosophy that did not recognize the rights of indigenous people. Meanwhile, the freeze created additional racial animosity among prodevelopment whites against newly activist Natives, including men like Nick Kompkoff Sr. A Chugach leader, Cecil Barnes, wrote to the *Cordova Times* defending the claims and challenging centuries of mistreatment, including the loss of the Eyak village of Aleganik, which had been wiped off the map to make room for the Alaska Syndicate's railroad from Cordova to the Kennicott copper mines.

> Native rights have always taken second place to the rights of God-fearing white Americans and large companies of one sort or another. Perhaps there have been reasons for this. Perhaps the reasons were bound up in the white man's culture and superior training which taught him how to protect himself and how to achieve his ends in a society which is essentially his.

In 1968, with the discovery of oil on Alaska's Arctic coast, Natives again found themselves in the way of a resource project of national interest, but this time the land freeze made them more difficult to brush aside.

The blundering of the oil industry ultimately promoted settlement of Native land claims. Some Native groups, including the Chugach association, initially bargained away their claims to the pipeline's route in exchange for concessions on jobs, but pipeline builders broke these promises so swiftly and cavalierly that pro bono attorneys for the Natives easily won an injunction stopping the project, despite the weakness of their underlying case for controlling the land itself. The companies also bulled ahead without regard to the environment, advancing the absurd plan of simply burying the hot pipeline in Alaska's permanently frozen ground, which liquefies when heated. Environmentalists sued to force a full environmental review. Faced with the prospect of long delays, Alaska's political leaders pursued new acts of Congress to give the Natives their lands and exempt the pipeline from environmental laws, decisions that proved among the most important in the state's history.

The Alaska Native Claims Settlement Act passed at the end of 1971, allocating 44 million acres and $962.5 million to Native peoples' cultural regions and individual villages. At the time, Chenega was no longer a village and received nothing. Nick Kompkoff Sr. led the community in demanding Chenega be recognized as a village corporation and allowed to select its lands. Gail Evanoff helped in many meetings. She represented Larry, who was busy with work and didn't have her verbal talent or college training. The Native world had changed rapidly, as a new generation of college-educated young people took over leadership from elders who had traditionally guided village affairs. Only an agile and well-educated mind had any hope of comprehending the multidimensional snarl of law and regulation affecting Alaska Native land, tribes, and corporations. For Chenega, recognition required developing allies and making technical arguments. Elders helped prove their case by guiding tours of the lands they once used, but the fight had to be won in conference rooms full of business suits.

Chenega Corporation received 76,000 acres in southwestern Prince William Sound from within Chugach National Forest in the mid-1970s. Carol Ann Kompkoff remembered her pride when she realized her people had transformed from an excluded racial minority to powerful landowners. "They no longer looked at me as, 'Look at that brown-skinned girl,'" she said. The change struck her particularly when she met a banker who in the old days had refused loans to her father and other Native fishermen. "There we were, the next generation. I was on the Chenega Corporation board of directors. And who was there to pick us up from the hotel? The banker. He was driving the van and catering to every little whim that we had. And I turned to the others in the van. I said, 'Look who's driving us around. The one who said no to our parents.'"

But owning the land was not enough. Larry Evanoff became impatient to live on it. It seemed to him that the annual meetings of the sixty-nine Chenega shareholders always brought talk about reviving the village, but nothing happened. Once, Chenegans would simply have gone out and built their own houses. In the old village, men hauled wood and made simple structures with their hands and the help of neighbors; when a family left, someone else would move into the house. Villagers carried water from the creek on foot—there were no roads or vehicles other than the boats down on the beach and at the dock.

Native expectations for standards of living changed after the earthquake, but the bigger change came from outside. Federal agencies invaded villages, imposed their values, and tangled up civic life with their acronyms and petty rivalries. Earthquake reconstruction officials sometimes ignored Natives' wishes when they arrived to rebuild, treating them as dependents or serfs, demanding they work as subordinates without pay, and deciding for them where to put houses and how to arrange and allocate land—sometimes deciding foolishly, to the lasting regret of village residents, according to Nancy Yaw Davis, an anthropologist who studied North Pacific communities during those years. But to get government money for earthquake reconstruction and other social programs, village people had to shut up and do things the way agency employees directed. And after the agencies came they never went away again—outside government bureaucracies became as much a part of village life as outboard motors.

Chenega Corporation needed money. Its funds from the land claim settlement were nowhere near enough to rebuild the village. Most Chenegans were poor, and moving from the city to the wilderness even in a primitive way can be expensive. With the help of the Army Corps of Engineers, villagers chose a new site for Chenega. Elders couldn't stand to go back to the old cove—one visit was enough to show them that. It would remain a sacred grave site. They looked instead for a place that would be safe from tsunamis, had a deep port, and fresh water. A site on Evans Island suited them, eighteen miles from the old village over the water, on a bay within Sawmill Bay near the Armin F. Koernig Hatchery. An abandoned saltery still stood above an old dock. The Evanoffs lived in the saltery for a time, amid the spiders and rot, trying out the new area.

Back in Anchorage, Gail lobbied bureaucrats in state and federal agencies responsible for housing, community development, Indian affairs, and sewer and water systems, among others, a long and complicated task. One reason they resisted was because no one lived at the place where the Chenegans wanted to build. Larry and Gail solved that themselves in 1981, when President Reagan fired Larry and eleven thousand other air-traffic controllers who had gone on strike. Larry said, "I didn't have a job. I had a few dollars in the bank. And Gail said, 'Let's not talk about it. Let's do it. No more talking.' And Gail and I sold everything. I mean everything we had. And we took bags of clothes and we moved down there. We put up a cabin. And then, 'Hey, there's people here.'"

Larry ferried supplies in his twenty-two-foot wooden dory—seventeen barrels of fuel, on one run—over many slow trips, seventy-five miles from Whittier, crawling along with a sixty-five-horsepower outboard and praying for good weather to hold. The family was happy. Wannah, at nine, ventured alone through the woods and over the beaches of Evans Island, picking berries, fishing, playing with Barbies on the beach with clam shells and driftwood—she roamed free, drawn home only by dinnertime or the need to help with chores, carrying wood or water. Larry took her with him in the boat and taught her how to stay quiet while hunting seal. At night she listened to the grown-ups talk by lantern light, poking her head over the edge of the loft in their one-room cabin. Larry and Gail's partnership, built in adversity, came into its own. "This was my home," Larry said. "I grew up here. Gail didn't, but she was behind me one hundred percent. I lost both of my parents. This is my destiny. This is what I'm supposed to do. Gail was the brains behind most of it. She was able to get a lot of funds. And I was there to make sure the job got done."

Larry built a tent camp for construction workers who would put in roads. The contractor insisted on hot showers and a flush toilet, so Larry laid a PVC pipe up the mountain to a stream, running it to a shed with a generator that could heat the water, which then flowed through the shower and toilet into the ground. Next came the twenty-one houses, fabricated by HUD in Oregon and shipped in halves on three huge barges, the first of which went to the wrong village—the old village site—before Larry figured out the mix-up over the radio and brought it to the beach below new Chenega. Huge cranes came along with the houses to offload them. Larry supervised construction and made sure the job was done right. In 1984, families arrived to move in. Gail said, "They were literally coming off the boat and we were handing them the keys to the house. It was fabulous. Those who had not chosen where their house was going to be, it was—'Go over there, there are three houses, take your pick.'"

The first year, families got reacquainted. They already knew each other, but living in a village of less than a hundred residents was different than occasionally meeting in the city. Over the first winter on that quiet cove, where the tide rises gently and the stars blaze brightly at night, people learned to go visiting and tell old stories once again, to pitch in when they saw a neighbor chopping wood, to share a big catch of fish. Most of the settlers were young families like the Evanoffs. Older people tended to be too established where they were to move back, or they needed medical help the village couldn't provide. Nick Kompkoff Sr. moved to the village, but he was fighting cancer. As the priest, he oversaw the construction of the church, which the men made of pieces from the old saltery. He died in 1987.

Twenty years after the earthquake, most adult residents had never lived in a village. Carol Ann didn't know how to smoke fish, cut game, or prepare Native foods. She would plan dinners two weeks ahead using food from Anchorage. She had never learned to sing or dance, either; she hadn't asked her grandmother about the old dances, and then it was too late. Larry said the men all knew how to fish and hunt, but not the old techniques the elders had used. Besides, the places where village memory recorded the seals' favorite rocks and the fishes' gathering places were too far away to use. The young men sometimes harvested as much as they could, wasting fish and meat rather than leaving some for next time. "It wasn't like the old days where elders were there to coach you," he said. "It would have made a whole heck of a lot of difference."

Larry Evanoff's generation tried to hand off to their children what they did not receive from their parents. Larry has gray hair now. He and Gail project warmth, but also unassailable dignity. Wannah, now in her thirties, never asks him about the earthquake, the boarding schools, or Vietnam—she feels those subjects are off limits—but her parents live just steps away from her house and are a constant presence in teaching her children. When her oldest son took his first deer, Larry presented him with a gift, a skinning knife, as tradition dictates for a Chugach grandfather.

A new generation had the opportunity to grow with the village in their bones. Wannah and her husband, Dennis, with five children, were getting ready to add onto the little house Larry had supervised building twenty-five years earlier. Wannah lives within a stone's throw of the water and still encounters the hermit crabs and sea urchins that low tide leaves there for children to play with. Her children roam as she did, checking in once an hour. In the village quiet she can hear them off in the woods. At sunset the quiet is total—a single ripple would stand out—as if all living things were poised, waiting to release a breath. The mountain holds the village in an open palm, cradling it in front of smooth water. The sky is so dark and deep that the stars seem to float in it, the bright ones most buoyant, mixed with the planktonic cloud of the Milky Way. The old Chenegans must have felt as Wannah and Dennis do in this generous universe, watching the seasons bring good food—goose eggs and herring spawn in spring, salmon in summer, berries and venison in fall—as their children rapidly grow. And watching the winter come, when the seventy villagers mount a craft bazaar for each other at the school. That is when the humpback whales gather in the bay to feed on overwintering herring and the deep rumbling of their underwater song shakes the houses, surprising villagers from their TV watching or dishwashing.

"This whole waterfront is all singing with dozens of whales for a couple of months," Wannah said. "You can see the splashing and hear the singing. That vibration sound they do, you can feel it all the way up here. It's amazing. I don't know how you can feel it or hear it, those big notes, with that deep, loud sound. They make the season Christmasy. To me, it's not Christmas without the whales."

25.

Native People Become
Native Corporations

Back when my oldest son was a baby, I sat in the storefront of Steen's gift and furniture store with seventy-three-year-old Gail Steen and her school friend Tina Tapley and listened to their stories of how Cordova had changed since their girlhood, when the railroad was still running and they could take a hand-pumped car called a speeder out to the delta to pick blueberries. The night watchman would stop traffic—a vehicle or two over an evening—so the children could sled down the steep main street. Dances were held every Saturday. At night Tina saw lights out on the flats—people collecting cockles—and the fishermen chased herring schools in the dark by following the phosphorescence they stirred up in the water. It could take a week to get from Seward to Cordova on a steamer because the boat stopped at every mining camp and saltery on the way. They recalled hearing radio for the first time, music coming from the air. As the elderly often do, they bemoaned the changes in their town—although the changes seemed gentle enough to an outsider. The population had doubled in seventy years, from one thousand to two thousand. A burly, black-bearded fisherman walked by on the sidewalk. Gail said, "You see that big old fellow there, that big gorilla? He was a preemie. He was three pounds."

Since that conversation my baby son has grown over six feet tall and is getting ready for college. Gail and Tina are no longer living, but a modernized version of the store continues the family business started by

Gail's father when he moved to Cordova from Nome in 1909. The same old grocery store, drugstore, hardware store, gas station, and bookstore do business on Front Street, run by their owners. Still, there's no road link to the outside world. Metal roofs rust and paint chips away from window frames; after a rain slugs safely cross the street.

People have rooted here as firmly as the damp, drowsy ferns and moss at the foot of the tall rainforest spruces. They're a part of the sound, too. Even newcomers settle immovably. A friend who became a vice mayor and environmental leader arrived as an itinerant fisherman in 1974, just back from Vietnam. Riding in from the airport in a taxi he said, "I don't think I'm ever going to leave here. I love this place." He has finally finished building his house so he can retire in it. Another friend came around the same time as a volunteer recreation counselor and has been through every civic role since, including town historian and science center director. Craig Matkin, the killer whale researcher we met in Chapter 2, came to the sound to count fish in 1975 with a bachelor's degree. He floated for hours in a kayak next to resting killer whales, recognizing individuals by their tail and fin markings. To keep up his whale studies he worked as a fishing deckhand. After a good fishing season, Craig bought his own boat, but he continued his research, earning a master's degree, scratching for elusive and inconstant grant funding. A fishing mentor told him to keep working with his own money, saying, "To hell with these guys, they'll always keep you on the line and cut you off." Matkin followed that advice. What mattered was the place and its animals. His whales became the sound's most carefully and consistently studied creatures and his work gained scientific recognition. But Craig wouldn't leave long enough to earn a PhD. "This is where I want to live," he said. "It was all about being out here. Being *here*. *This* geography."

Oil tankers began crossing the sound in 1977, ships so large they scarcely looked like ships. Their flat black sides stood like steel cliffs in the water. In the corridor of Valdez Narrows boats appear the size of toys, but tankers fit more like sofas being maneuvered through an apartment's hallway—stiffly, gingerly avoiding the corners. Eventually, you're going to nick a wall. Any mariner with experience in these waters could see the risk, and Cordova fishermen allied with environmentalists to fight against the plan to use the Valdez port as the terminus of the pipeline when Congress debated legislation exempting the entire Alaska oil shipping system from environmental laws in the early 1970s. Tina Tapley remembered Cordova fishermen filling hearing rooms. "Some of those boys made awful good speeches," she said.

The oil industry, as an institution, has repeatedly demonstrated sloppiness and dishonesty in its time in Alaska: corruption at the Land Office dating to the 1950s; the errors at the very start of pipeline planning I mentioned in the last chapter, including the idea of burying the hot pipeline in frozen ground and breaking promises to hire Alaska Natives; the colossal waste and mismanagement of pipeline construction, mob-infiltrated and astronomically overbudget; and more recently scandals about illegal spying and dirty tricks against critics, criminal neglect of pipeline maintenance, ignoring safety rules, repeatedly punishing whistle-blowers, bribing politicians, cheating on state taxes in the billions, and using political muscle with the state government to impose taxes so low cheating became unnecessary. The list seems endless, even leaving aside an undermanned supertanker under charge of an unqualified mate and relapsed alcoholic captain. Yet it's hardly necessary to mention that fishermen lost the debate over tankers in Prince William Sound, even with the insuperable argument that there was bound to be an accident over decades of moving billions of barrels of oil through rocks, ice, and mountain corridors. As typical in Congress, the pipeline law passed with a compromise—all the environmentalists could get, but not enough to achieve their actual purpose. The U.S. Coast Guard would establish a vessel traffic system to guide tankers through Valdez Narrows and tanker operators would be prepared to clean up a spill. In less than twelve years complacency nullified the effectiveness of both those safeguards.

During the congressional debate the Nixon administration also promised a more meaningful protection for the sound, double bottoms on tankers (on a conventional tanker only a single sheet of steel lies between

the oil and the sea). But Congress didn't write that into law as a require-
ment and the Coast Guard, with direct authority, reneged on the commitment soon after the bill passed. Double hulls would have cost the industry
an additional 3 percent in ship construction while containing, by the government's estimate, more than 90 percent of oil spilled in U.S. tanker
accidents. It's common—indeed, usual—for industries to gain controlling
influence over their regulatory agencies, called "industry capture." The
Coast Guard appointed an oil industry trade group to write ship-building
regulations for the Alaska tanker fleet. The states of Washington and
Alaska briefly resisted with their own laws for double hulls, but were over-
ruled in court because the federal government had preempted their authority.

Another Cordova friend, Kelly Weaverling, who owns the bookstore,
explored Prince William Sound by kayak during those years, drawing seabirds in pen and ink, until he knew not only where the animals lived, but
also knew individual animals, the seal of a particular rock or the oyster-
catcher that defended a certain stretch of beach. Rick Steiner worked for
the University of Alaska in 1983 when he arrived in Cordova as an extension agent for fishermen and a fisherman himself. He settled in an office on
pilings in the boat harbor that shook whenever a boat pulled up to the
dock, and where an otter often swam below his window. Rick fell in love
with a New Age masseuse, floated the rivers and hiked over the mountains,
wore flowing clothes and sandals on his loose-jointed body, knew everyone
in town, played music with his friends. He performed in the Pursuit of
Happiness Puppet Theater. He had found paradise. Once, he said, all the
world was Eden; here, the virgin forest still stood.

But in 1988, the forest began to fall for the first time. And those cutting
the trees were hired by the Native corporations. The corporations owned
the land, at least in the legal sense of deeds and papers, but there was
another community here now with another kind of ownership—the kind
connecting the heart to a place, somewhat like the traditional link of
Chugach people to the sound. This is rude to say, because the first people
deserve respect, especially survivors of genocide. But why shouldn't this

new group of people have a claim at least something like that of Makari and the celebrants of the Feast of the Dead? Their Cordova community also lived from the sea. Their physical and spiritual connections to the ecosystem also provided meaning to their lives. Rick Steiner and his friends loved the sound as deeply as family. Gail Steen and Tina Tapley had never known another place in the world. They looked out on Orca Inlet their whole lives, season after season, for almost a century. They were disgusted by clear-cuts that grew as they neared the end of their lives. "All the Natives cried about how we need our land, we love our land," said Tina, who was a quarter Tlingit. "They are selling all their trees, and the checks they are getting are pitifully small."

Two kinds of ownership—one based on love and the other on paperwork—each projecting from different eyes that see a different place. One sees spirits, people in the trees and animals, and sustenance, the relationship of the body to the natural world, in which a clean, healthy, and abundant system of life enters partly into the flesh of the owner. That kind of owner cares about the flow of energy from the sun through the plankton and into the minds of thinking animals. The other kind of ownership, which arrived with Captain Cook, the Russians, and the gold rush, conceived of a place as a thing interchangeable with other things—measurable, divisible, tradable. Lines on maps; which is to say, metaphors that sever land from its function and natural continuity. I think of the masks laboriously carved for the Feast of the Dead and then thrown away, because their worth lay in the connections they represented, not their material substance. Natural history museums took them, stored them, owned them, but really owned nothing, because the meaning was gone. Millions of museum visitors cast uncomprehending glances into glass cases. Jim Miller carved in Port Graham, his most valuable pieces made from found driftwood and given away, his least valuable pieces made of bought material and then sold for money. The things we value most cannot be bought or sold.

If this idea of community ownership seems fuzzy or radical, step back for a moment and think of it in terms of rights. Property rights represent a useful system. The papers at the recorder's office maintain order between people. But the land doesn't know who owns it—the land moves, people die, empires rise and fall, and with those changes the meaning of legal documents is lost. Private-property rights last only as long as everyone respects the writing on the papers. But human rights last as long people eat, drink, breathe, and love. We recognize the deeper rights of communities even in the Western system of laws, which allow the state to override individual property rights, as when zoning laws tell owners how they can use their land or condemnation takes land for public needs. Towns protect historic buildings and preserve wetlands and open space they don't own. Nations override property rights to defend or reclaim cultural objects. No one can buy the world's great museums or libraries. When it comes to the important things, the sources of life and meaning, the oceans and the sky, individual ownership becomes irrelevant.

With the Settlement Act of 1971, Alaska Natives had received ownership of their land based on pieces of paper, shares of stock issued to each Native alive as of December 18 of that year. It was better than no ownership at all. Native and political leaders had reasons for choosing the corporate model. No one knew how else to transfer the land. Outside Alaska, sovereign Indian nations lived on reservations that were pits of poverty and hopelessness. To create more of those seemed unwise. Yet policymakers' minds dwelled so deeply in their own paradigm they seem never to have considered forms of ownership other than sovereignty or shares in corporations.

Perhaps community-based nonprofit associations could have owned the land, but profit was among the primary goals of the land claims legislation. Native leaders believed for-profit corporations created the possibility of economic development in cash-poor villages. White leaders, including Senator Ted Stevens, also saw Native claims as a gambit against the hated federal conservationists. In the case of the Chugach people, the land would come from the heart of the area set aside by Gifford Pinchot, even finally privatizing the Bering coal fields whose ownership had helped bring down the Taft administration. Statehood had provisionally wrested away 103 million acres of Alaska from federal control. Creation of Native corporations would put another 44 million in private hands. Historian Donald Mitchell said the debate among white Alaskan leaders reflected their perception of human nature and racial differences—some assumed Natives would inherently prefer traditional uses of the land rather than the hoped-for industrial development. But Stevens believed Natives, placed in a corporate system, would behave like any other corporate owner. Motivated to profit on their holdings, they would log, drill, and mine their land.

So many ironies. As we've seen, Roosevelt turned to patriotism and the unifying bonds of racial superiority to counter the new power of corporations and conserve. Seventy years later, Natives, oppressed by the racists, adopted the tool of corporations to defeat Roosevelt's conservationism. But the corporate model quickly proved unsuitable in much of Alaska. Normally, corporations are born from business opportunities, but here thirteen regional corporations and more than two hundred village corporations popped into existence with the expectation they would create businesses from thin air. For small corporations, settlement money barely covered the cost of lawyers to issue stock and file reports. Few of the new owners knew how to run a corporation; no one knows how to run a business in a place without a significant cash economy or a practical source of revenue. After their first decade most had done poorly, by every measure, and many were on the verge of failure.

Stevens addressed much of his forty-year career in the Senate to these problems. In 1986, he sought to bail out the Native corporations with a provision in the tax code that allowed them to sell their net operating losses, known as NOLs, to other, more profitable companies. Firms such as Marriott or Pillsbury would then write off the losses as their own tax deductions. After two years Congress stopped the provision as too costly to the Treasury, as it allowed major corporations to avoid hundreds of millions of dollars in taxes, but not before Native corporations across Alaska liquidated assets at fire-sale prices, booking huge paper losses compared to the appraised value of their land as of the date they had received it from the government. The need to generate losses drove the corporations to strip enormous tracts of timber in southeast Alaska. Chugach Alaska

Corporation made $48 million on loss sales, divesting itself of the Bering coal field to a Korean businessman and selling a forest on Montague Island, where Will Langille had tramped and enthused about emerald waters, to loggers.

And why not? Native culture was always based on using the land, not preserving it like a park. Sheri Buretta, chairman of the board of Chugach Alaska, interpreted the rise of Native corporate power as the redemption of her people's spirit of self-worth and assertiveness. I met Sheri at Heritage Week in Tatitlek, where she humbly taught classes of children to sew skins while listening to the late 1980s heavy-metal band Guns N' Roses, the music of her youth. Sheri grew up in Anchorage and admitted she didn't think she could live in a village, but she certainly seemed to belong. She was pretty and approachable (cute is the only word that fits) while projecting competence and strength.

Under Buretta's leadership, Chugach Alaska emerged from bankruptcy to become the second largest of all Alaska-owned companies, with more than six thousand employees worldwide fulfilling government contracts for diverse services, such as operating a job-training center in Roswell, New Mexico, and a missile testing base on Kwajalein Atoll in the Marshall Islands. The opportunities came from a law passed by Stevens that expanded federal purchasing laws for minority businesses to allow agencies to conveniently award no-bid contracts of any size to Alaska Native corporations (for other businesses owned by minorities or women the program has limits on contract size). Despite the success, Buretta was ousted from power in a secret board coup in 2004, only to scratch back into control after seven months of bitter, litigious maneuvering between two factions.

Brutal public fights for corporate control have become a familiar feature of Alaska Native life. This kind of conflict entered the culture with elections and the formal government required by outsiders. Before the Russians, anthropologists believe Chugach people organized themselves by listening to elders and successful individuals who had earned respect through their skills or judgment; during the Russian era, church lay readers became chiefs. The agencies that arrived around the time of the earth-quake wanted something more—interaction with elected institutions that would hold meetings, keep minutes, approve resolutions, and act as conduits of money and power. Elinor Ostrom found in many cases around the world that this kind of top-down reorganization could destroy self-organized communities. Elections are no panacea. Voting introduces competition, factions, and deal making. If individuals in a small com-munity are cooperative, they don't need to elect a leader to tell them what to do; if they're selfish, then they will seek advantages through elections, and so will those they elect.

By the end of the 1980s, Chugach villages had church leaders, chiefs, traditional councils, for-profit village and regional corporations, nonprofit cultural, housing, and health corporations, besides local government, school boards, and legislators—most of these entities requiring elections, often among a few dozen people. "It's a popularity contest to be able to run for these," Buretta said. "And it's so out of the ordinary for our people to run and ask for people to vote for them, and then you get into the fights that hap-pen, and discrediting somebody else to look better. It's such a destructive system. . . . These people are related to each other. They might be brothers, they might be cousins, they might have grown up together. Suddenly to have them pitted against each other, and to have this desire to have power and position—and what happens to the one that doesn't win? They might go on to the nonprofit [corporation board], and then you're really in trouble, be-cause they're going to fight from those positions."

But Buretta believed the power of for-profit corporations—their ability to compete and win—helped compensate a generation of Chugach people who endured the role of passive clients imposed on them by social service agencies. "Truly, that is not how those people are," she said. "They're not passive in their ability to make decisions about their communities, but it's almost like they're forced into that role in order to get the benefits that these people hold for them. We want our people to be strong.

"You know, the term 'dumb Native,' I heard that so much when I was growing up. I think there was a first generation that has gone to school just in spite of that, just to prove that that's not the case." Now that generation is running Chugach Alaska. "You cannot really quantify the sense of pride and self-esteem," Sheri said.

In 1988, that success still lay well in the future. With its money from NOL sales Chugach Alaska launched an ill-advised expansion that led to bankruptcy, building a huge computerized sawmill in Seward to process timber that it and village corporations would cut from their land. Together Chugach Natives owned hundreds of thousands of acres of the most beautiful and productive forest on islands and peninsulas across the sound and inland from Cordova, up the Bering River and on the Kenai Peninsula. From Cordova, Rick Steiner lobbied congressional staff to stop the net operating loss tax breaks in order to save the trees. He and his Cordova friends talked to organizations that might be willing to buy timberlands and let the trees stand. Sheri Buretta still bristles at the mention of Rick's name—she called him a zealot and the sound his playground. "Who is Rick Steiner to judge what a private landowner wants to do with their land?" she said. "The bottom line is logging is a renewable resource. It's like a farm." Indeed, Gifford Pinchot himself is quoted on the corporate website comparing forestry to harvesting crops.

Nineteen years after Tatitlek Corporation began logging its lands to feed the Chugach Alaska mill (now long gone), I went to see how farmlike the forestry had been. A clear-cut ran mile after mile along the steep south side of Port Fidalgo, beginning at Snug Corner Cove, where Captain Cook anchored and repaired his hull and Chugach Natives tried to capture his boats in 1778. Bob Sanford, a respected skipper and guide who married into Tatitlek, said the village corporation's contractor logged sloppily, leaving downed debris and crossed logs that rendered much of the land impassible, and made no effort at reforestation. "They raped and pillaged," he said.

After two decades the mess was still there and tiny trees had barely begun to grow back. Geometric clearings were bounded by mature trees like partially shaved hair. The silver stubble of weathered stumps and wasted logs contrasted with green brush and grass and little Christmas trees poking up here and there. A real forest would be decades in the future on the best of these clear-cuts. But on the worst, on steeper slopes, the damage of the logging had not yet stabilized. Roads carved into the mountainside had channeled water, causing washouts and landslides that were still stripping away soil. Some slides looked like claw marks, where vegetation and organics had been peeled back from the ridgetop to the sea level, leaving rock and gravel barren of life. Soil takes ages to accumulate here, especially on draws where snowmelt and rain continue to cleanse the rock. It takes practically forever.

For a human lifetime, logging has ruined this land. Of course people who felt they owned the sound with their hearts fought to stop it, regardless of the historical identity of those who owned it on paper—the shareholders of a corporation that pursued this kind of harvest for short-term jobs and a better balance sheet. Alaska Natives vanquished by Enlightenment law and violence had obtained the tools of the conquerors and emerged victorious in the invaders' own arena. And now they stripped hillsides of trees the way the Russians took otter pelts and the Americans took salmon. Captain Cook had stood on deck in that very cove considering the possibility of taking possession from Natives who didn't know what his concept of possession meant; they knew now. What right had Rick Steiner to judge a private landowner? A right based on a community's connection to the life of a place, like that of the people who first watched the alien system arrive here and carve the ecosystem into dead, saleable chunks.

I met no bad people on either side of the forest conservation issue in Prince William Sound. Individuals didn't create the conflict: it arose from incompatible systems of relating to the natural world. I suppose the same could be said of those who ordered oil tankers back and forth through the rocks without taking the care expected of a weekend fisherman. That's the puzzle that I've thought about over these decades, as the sea sickened— about the political and economic system that we made, and that makes us destroy our own beloved home.

26.

The Broken Covenant

Imagine a society that would take advantage of the efficiency of free-market competition when appropriate while also protecting space for communities cooperating in balance with nature. Is that possible? If not, then the fate of the oceans is already sealed. Personal feelings about nature won't matter if every inch of the world is controlled by political and economic institutions transcending human scale and lifespan that prioritize power and material wealth above all else. Individuals and communities can choose connectedness and meaning, but large corporations and governments cannot; they aren't alive, and for that reason aren't capable of the intimacy and emotion of belonging. These big institutions do have uses. We need one to counteract another, corporation against corporation, government against corporation, or government against government. We need them to manage large problems beyond the capabilities of any human community. But competitive organizations cannot create social norms for sufficiency. For that we need wise people with control over their own actions and environment. If communities don't have that freedom, then norms fulfilling our humanity will have no chance to evolve.

Perhaps such a free space is no longer possible. Geerat Vermeij taught that competitive entities are by nature insatiable for power. Even tiny

Prince William Sound villages weren't safe—corporate and allied government authorities spoiled their waters and split their communities. A society that can keep those powers at bay may no longer be possible.

When this question of balancing the powerful and the personal was new and clearer in people's minds, during the Enlightenment, Americans debated vigorously how to stay free. They tried to create a single nation containing multiple forms of political relations. Nested circles of autonomy would allow people to solve the problems of living together in their own ways in their own places. Vincent and Lin Ostrom called this kind of system polycentricity. Vincent believed such a system was implicit in the United States Constitution. He interpreted that document as a compromise between contradictory forms of human order that could allow both to exist. Since the Constitution represents the original and most powerful implementation of the Enlightenment's competitive paradigm, it's worth understanding how it came to exist—and evolve.

The constitutional covenant addressed the needs of two factions: the side of property and the side of community. The farmers of postcolonial America knew all about managing the commons as self-governing communities. Families, neighbors, and churches took care of almost all the services we think of today as the domain of government, if those services existed at all. They had fought against King George to rid themselves of interference from central authority, and they carefully guarded their shared spaces from outside control in the years after the Revolution. The democratic ideal required economic equality, which allowed citizens to work together cooperatively. And it abhorred the two-headed monster of wealth and power, which weakened community and, once unleashed, could intrude limitlessly into every area of life. These ideas had been developed before the Revolution and were deeply engrained in the generation that fought the war. Among the books that influenced the patriots most, *Cato's Letters*, published in London in 1724, had called for English law to redistribute property in order to create the conditions of equality necessary for democracy.

These considerations guided the writing of the Articles of Confedera-
tion, which preceded the Constitution by a decade. The Articles joined the
former colonies, but hardly created a nation. Each state contributed only
voluntarily, like members of the United Nations. The impotent Congress
mattered so little that states didn't always bother to send representatives.
Sovereignty resided with state legislatures. They, in turn, were kept on a
short leash by towns, the locus of communities. Legislators usually held office for only a year between elections—just six months in Connecticut and
Rhode Island—and strict term limits sent them home before they could
become ensconced. They represented small areas, sometimes one legislator
per village. In Pennsylvania, the entire public acted as legislators: to make
law, elected representatives would propose a bill, but it had to be broadly
publicized and debated in local conventions before it could come to a final
vote. Americans kept power in their neighborhoods; they could take care
of themselves. They believed that a democracy had to be small, because it
depended on the personal knowledge and relationships of those making
the decisions. Some thought the states were already too large and needed to
be split into smaller units, not combined into one nation.

But this system was far too democratic for James Madison and Alexander Hamilton. State legislatures composed of rural rabble alarmed and
disgusted Madison; they forgave private debts and redistributed property
for economic equality. A Virginia landowner, Madison sought a powerful
national government to protect private-property rights. Speculators in
western lands needed a strong federal army to get rid of the Indians so
paper rights to frontier property could be traded more profitably. Since Native and African Americans wouldn't be party to the constitutional agreement, it would grant them no protections. Hamilton saw further ahead. He
wanted unified nationhood to promote capitalism, believing large-scale
commerce could eventually make the United States strong enough to
emerge as an imperial power, rivaling the great European nations with its
military might. In his vision, national government ruled by lifetime presidents and senators would subsume the states and promote a market
economy spanning the continent. In 1786, Hamilton and Madison joined

at a conference of five states concerning a tariff dispute between Virginia and Maryland, calling for a national convention to discuss provisions to make the federal government stronger. Preparing for the Philadelphia meeting, Madison read deeply of the Enlightenment philosophers—Thomas Jefferson sent crates of books from Paris—and concluded that the nation needed a central government as strong as the monarchy of King George had been.

Madison believed the worst of human nature; like Hobbes, he equated majority rule with anarchy. His constitutional architecture of separated powers would protect Americans from themselves, using conflicting self-interest to prevent any single faction from becoming too powerful. As he wrote in *The Federalist Papers*:

> Ambition must be made to counteract ambition. The interest of the man must be connected with the constitutional rights of the place. It may be a reflection on human nature that such devices should be necessary to control the abuses of government. But what is government itself but the greatest of all reflections on human nature? If men were angels, no government would be necessary.

As we've seen, government often isn't necessary, at least in groups that can function as self-organized communities. But rather than attempting to unite people in common cause to solve shared problems, Madison conceived a mechanism to sum and cancel out their self-interest. Strong competitors and reciprocal deal makers succeed in such systems—cooperators and altruists lose the game to aggressive and crafty politicians and are rapidly flushed out. The Constitution would bring back the rule of the strong. Its leaders would be insulated from citizens by distance, numbers, and indirect electoral systems. And this new national government would be empowered to negate acts of the more democratic state legislatures.

Many Americans of the time, perhaps the majority, rejected the new Constitution, its centralization of power and antidemocratic tendencies, especially people in rural communities already functioning well informally or through direct democracy. Called antifederalists by their opponents, these revolutionaries had no motivation to cede control to an unaccountable, faraway government run by conflicting strangers. City dwellers, property owners and the mercantile class, on the other hand, tended to favor a stronger federal government that would provide more order for the economy, including protection of wealth. Madison and Hamilton got less than they wanted at the convention in Philadelphia, and even then the constitution won ratification only narrowly, slipping through thanks to the antifederalists' lack of organization and less-moneyed and less-educated advocates. The document gave the federal government only a short list of essentially national powers, with the states and citizens keeping everything else. To make sure, the antifederalists insisted on a Bill of Rights, added after ratification as the first ten amendments. Besides listing the rights of individuals and groups, it reserved all powers not specifically granted to the federal government to the states and the people. The framers settled for a compromise.

Vincent Ostrom believes the Constitution with the Bill of Rights represented a covenant between the two sides. It protected different forms of order within one nation. Self-organized communities survived in many places. He first studied one on common grazing lands among cattlemen in Wyoming in the 1940s. When he helped write the Alaska Constitution as a young professor in the 1950s, he already was thinking about how words could enable a political entity to respect different forms of property— private, governmental, and common—without assuming that ownership of resources is only for money or consumption. He believed the Alaska Constitution would restrict private-property rights to permit collective action. The experience put him in the role of one of the federal framers and he began a long career in constitutional interpretation. He's an old man now, and rarely at the workshop he started with Lin at Indiana University, spending most days among his books and Native American art in the house overlooking the wooded canyon outside Bloomington. But he still speaks clearly in his past writings, as when he conceived of how a collection of diverse communities could scale up into a nation:

If we view a federal society as a covenanting society capable of generating rich assemblages of associations, we would expect to see social units of one sort or another, formally independent but choosing to take each other into account, functioning in mutually accommodating ways to achieve many different patterns of order.

How would this work in real life? As Alexis de Tocqueville observed when he studied the country in the 1830s, Americans practiced democracy through their voluntary associations. "Not only do they have commercial and industrial associations to which they all belong but also a thousand other kinds, religious, moral, serious, futile, very general and very specialized, large and small. Americans group together to hold fêtes, found seminaries, build inns, construct churches, distribute books, dispatch missionaries to the antipodes. They establish hospitals, prisons, schools by the same method. Finally, if they wish to highlight a truth or develop an opinion by the encouragement of a great example, they form an association." This is still true. We even have an association for keeping statistics on our associations, which reports that Americans have registered 1.5 million nonprofit organizations, one for every two hundred citizens. (Of course, many more associations, clubs, and groups of friends and neighbors have not incorporated or registered.)

Associations connect parents to improve their children's schools or join lovers of the ocean to clean up trash from remote beaches. Through their memberships, people influence one another, often in shared spaces too small for corporate journalism or government to notice, drawing information from life experience and their neighbors while discounting politicians or the news media. They change the country by changing the baseline values of the entire society, new realities that reprogram the norms of behavior for the powerful and the obscure alike. Recently changing norms are easy to list: littering became unacceptable, homosexuality acceptable; large families less common, recycling more common; privacy less important, safety more important. In a century the social status of African Americans rose from barely human to worthy of the presidency. Governments and corporations led none of these changes, but even the largest and most powerful of their officials must avoid violating such basic social beliefs. The essence of American democracy, such as it is, consists of our ability to define these baseline values, which the marketers of political and cultural products then endeavor to feed back to us for their own benefit.

For this system to work—for human relationships to remain intact and capable of establishing norms—national power must be kept at bay, outside the space self-governing communities need to solve their own problems. That space is easily invaded. Profit-seeking corporations or benevolent governments can displace these loosely affiliated associations and choke off their culture-molding ingenuity. American radio broadcasting is a handy example: the federal government grants the public resource of frequencies to large companies that maximize profits by standardizing music and news nationally, with little access for local voices. The antifederalists always feared that kind of interference in their communities. But Alexander Hamilton discounted those fears because he couldn't imagine why national leaders would want to meddle in local affairs once they possessed the power to make war and build a capitalist empire.

Hamilton's genius is incredible. He envisioned the large-scale market economy that would become the nation's defining feature—even though no such thing yet existed—and used down-and-dirty political skill to create the conditions that would bring it into being. But the antifederalists proved to be right about the hazards of this new system and Hamilton wrong. The rich and powerful who controlled the government were not satisfied only with great issues but intruded into everyday life as well. A creature's appetite grows as its size increases, and with each measure of self-determination that Americans lost, their central institutions became hungry to take more.

Examples of this phenomenon abound in daily life. American towns used to support family-owned hamburger stands and hardware stores, the kind of businesses whose owners would donate to school activities and volunteer for PTAs and school boards. Enormous corporations took over and standardized these businesses, removing decisions to distant headquarters. Publicity creates the impression that corporations are major charitable contributors, but in fact they give only one dollar of every twenty donated to charity in the United States. Individuals give more than 80 percent of the total. Government also displaced local autonomy, consolidating schools into larger, bureaucratic districts and passing state and federal laws to standardize their educational goals. Federal courts must protect individual rights—including desegregating schools—but community decisions about teaching children are the essence of culture making, not a bureaucratic function. Meanwhile, inequality of wealth weakened the ability of communities to resist these trends by dividing people and driving their ambition. Americans who already had everything they needed dedicated ever greater effort to earning more. The loss of community diminished their ability to generate fulfilling social and environmental relationships.

The cost comes not only in lonelier lives but also in less competent government. Those closest to problems almost always come up with the best solutions, and when the federal government addresses local issues it usually behaves like a bumbling idiot. The same is true of enormous corporations that span nations. Together they overrode local experts to send supertankers through the rocks and icebergs of Valdez Arm without double hulls or effective measures to prevent or clean up an oil spill. Likewise, while local groups, towns, and states pushed for a decade for solutions to climate change, national corporate and government authorities actively thwarted their efforts. The direst predictions of the antifederalists came to pass. Any constitutional covenant protecting different forms of order was broken long ago.

In fact, the covenant didn't last long. In 1885, a century after the Constitutional Convention, Woodrow Wilson, then a young academic, wrote that power had all flopped to one side: the federal government had assumed all authority from the states. This failure was built into the Constitution's system of contending interests. Since the federal government alone judged the extent of its powers, it made itself omnipotent, regardless of the plain words of the document itself. The Supreme Court functioned as a political body, centralizing authority with the growth of the market economy and national expansion. Hamilton had begun this process when he used the implied authority of the commerce clause rather than explicit permission in the Constitution to enact his economic program. The Supreme Court progressively expanded the commerce clause through the nineteenth and twentieth centuries until it was construed to give the federal government power over virtually every aspect of American society. For Wilson, the Constitution was a dead letter by its first centennial. The federal government already came to everyone's doorstep. The pragmatic response he recommended in his widely read book was to sweep away the rest of the checks and balances. If an all-powerful national government was to run everything, at least make it work efficiently.

Theodore Roosevelt's muscular presidency and the professional efficiency of Gifford Pinchot's administrators answered the concern Wilson had raised, as did Wilson's own progressive administration. Franklin Roosevelt subscribed to Wilson's ideas, too. They intended to protect the public—and stabilize society—by corralling exploitive corporations and the rich. A powerful permanent bureaucracy would reign in the role of Hobbes's supreme sovereign. But before the Progressive Era ended in the United States, the fallacy of this idea was already evident—power corrupts progressives, too. Without it they wouldn't have proposed regulating every aspect of citizens' lives, even their genetic heritage. Robert Michels, a disillusioned German socialist, writing in 1910, likened reform in a complex modern democracy to the rhythmic roll of ocean waves breaking on the sand and receding to the sea. Elections change the nameplates on the doors, but the leaders of change movements inevitably become members of the ruling class themselves, as willing to hold power over their subjects as their predecessors. Since the ruling class never surrenders power, reformers end up strengthening the system of power they start off hoping to change. Michels wrote, "They perhaps contribute to this class a certain number of 'new ideas,' but they also endow it with more creative energy and enhanced practical intelligence, thus providing for the ruling class an ever-renewed youth."

Reforms did happen. Social Security, civil rights, the major environmental laws. But history exaggerates the role of politicians in those acts. It can't see the cultural changes that rose before the legislation, the norms that spread invisibly through community networks. Social movements always preceded national reforms. The periphery lifted the center. Politicians like to say activists prepared the groundwork, as if society merely moved around dirt upon which government could build structures. But really the important work was done before the first vote. The making of a law is the final step, as the governmental system adjusts to an underlying social norm it can no longer resist. In ten years we'll probably hear that Al Gore's *An Inconvenient Truth* created the sea change on global warming; in fact, the movie was the foam atop a cresting wave making itself visible.

The relationship of a community to an ecosystem is a matter of the heart, something we share in the most intimate way as we project natural rights onto trees, animals, and mysteriously meaningful places. Norms we barely understand link us to our sustaining commons. Here we must be free. Even well-meaning outside authority diminishes our sharing and takes away, with our autonomy, the dignity of our gifts. Vincent Ostrom called this tendency the "central-government trap." He believed the words of the Constitution, if accorded meaning, could have kept the sharp machinery of powerful institutions at bay, protecting this space for us. "The fundamental condition for avoiding the central-government trap is to have mutual respect for each other's freedom, to rest all political experiments on the capacity of mankind for self-government, and never to view oneself as the master of others," Vincent wrote. He believed people working as self-directed equals could learn to cooperate, avoid the pitfalls of counterproductive relationships, and turn conflict into mutual support. "This is what it means to live in a self-governing society. Those who view themselves as political masters are trapped in a vicious form of servitude that denies them access to what it means to be free."

I began this chapter asking if a society could take advantage of the efficiency of competition when appropriate while also protecting space for people to share a sustainable commons. It doesn't look good, and most people who have thought about it seem pessimistic. Our shared space gets smaller all the time. But I believe it's possible. At least, I'm certain the community spirit remains a part of us as organisms. We're built with a sense, in some form, that nature is worth saving regardless of its material benefit to humanity. That doesn't mean, however, that our deepest hopes will prevail. My fear is that the self-expanding institutions founded in the eighteenth century have become too strong for us to overcome.

The ecological conquest by the powerful continues. I saw it myself: oil spreading over the smooth water of Prince William Sound in 1989, poisoning sea otters and burning their eyes. I pulled unrecognizable lumps from black pools that had collected on porous cobble beaches—dead cormorants and loons. No human being wanted this. No one had the ability to stop it or mitigate the damage. No one could stop even the corrosive dysfunction that tore apart communities for two decades after the disaster. But the *Exxon Valdez* oil spill was not an accident. It was the visible manifestation of the system by which we live.

Part V

Exxon Valdez:

Blackness Visible

27

Finding Oil

On Easter morning, 1989, when I was twenty-five, I drove my beloved red Toyota four-wheel-drive pickup through the mountains north of Prince William Sound, where the snow piles so deeply by spring that some never melts. Bruce Springsteen was blasting on the stereo and my friend and *Anchorage Daily News* coworker David Hulen was riding along by my side. In back, under an aluminum shell, another writer and a photographer rode in the pickup bed. The cloudless sky glowed a blue supercharged by the light reflected from the white land. We descended the cliffs of Thompson Pass, back and forth, like an airplane circling downward, and landed, after seven hours, in Valdez (pronounced Val-*deez*), the little industrial town at the end of the Alaska pipeline, where tankers loaded. Seven of our reporters and photographers covering the nation's big oil spill shared a motel room, balancing hamburgers in Styrofoam take-out containers on our knees. On Friday I had been covering a suburban chamber of commerce and now I was here, on the biggest story on earth.

David walked the streets, talking to cabdrivers and waitresses, tracking the tanker captain's last evening ashore, cocktail by cocktail. Reporter Larry Campbell followed the deliberations of the dithering corporate and government officials arguing over what cleanup techniques to use—mostly hypothetical arguments, since they didn't have resources at hand to pick up, burn, or disperse the oil except in token amounts. Other members of our team documented the neglect of the pipeline operators, the years of cuts in their spill response staff, the equipment that had been out of order or buried in snow when the *Exxon Valdez* went aground early on Good Friday, March 24, 1989.

At the Valdez convention center I attended mass press conferences along with scores of other reporters and ordinary people who saw their world ending. Miraculously, by Easter Sunday the oil had lain essentially stationary through three days of rare dead-flat, calm weather, spreading into an area of merely one hundred square miles without making landfall. Frank Iarossi, the president of Exxon Shipping Company, a remarkably candid man handling his first oil spill, admitted that his people, still arriving from all over the country, hadn't accomplished much oil recovery, but said the next day, Monday, would be different. "We weren't even up and running until twelve o'clock yesterday," he said. "We have been so frustrated today. . . . By tomorrow, we will be going all-out."

There wasn't any point sitting in press conferences with everyone else. I made my goal to snag a seat on an airplane or helicopter out to the sound. Monday morning I found Dan Lawn and asked for help—he was the local pipeline regulator for the State of Alaska who was punished by his superiors for pointing out the inadequacy of oil spill preparations for years before it happened. Dan had been up late Sunday night piling sand bags on the skids of helicopters at the Valdez Airport to save them from a ferocious north wind rushing down from the mountains. Not atypical weather in March, but nothing was flying. In the afternoon, however, someone at the pipeline terminal offered a ride—as strange as that seemed—and photographer Erik Hill and I strapped into a helicopter that leapt into the air and then bounced on stormy winds as if hanging from a spring. My first helicopter ride. There was the *Exxon Valdez* on Bligh Reef, in foul waters known since Captain Cook gave up exploring Prince William Sound there. The wind had turned and twisted the hull and torn floating containment booms from around it. The slick now trailed away toward the south farther than the eye could see.

We landed on a north-facing cobble beach on Naked Island. Wind had whipped the oil into a black-brown emulsion with water, a nightmare substance called mousse, as sticky as oil-based paint. It came halfway up our rubber boots. The pilot pulled a shape from the glop. A cormorant. The oil hid its true, graceful shape: the long neck and the angular wings that cormorants air out, half-furled, when they stand on rocks. Iridescent blue reflected from its oil coating. Surely there were more here, but they were hard to find in the muck. A biologist later reported, on Green Island, putting something down on what he thought was a rock that then moved, a head rising up from under a wing—a bird buried alive in oil like the living horses underfoot on the trail to the Klondike.

I stuck my hand in the oil and Erik took a photograph of it dripping from my fingers. The oil is still smeared on my notebook and it never came off my boots. Or off my heart. The nauseating chemical odor seeped in and I can still smell it. We stayed only a few moments because the helicopter was short of fuel. Lifting off, the islands of the sound lay to the south, receding into the distance without apparent end, fringes of rock puncturing the surface of the water as if through ragged fabric. The oil had already gone that way; the front flowed on the prevailing current toward Montague Strait. Nothing could stop it. No human force could contain the oil or defend those coves and headlands, whose infinitely branching, hidden crevices made them immeasurably vast.

In 1977, when the pipeline was completed, my family visited Valdez for the festivities surrounding the arrival of oil. My father, an attorney specializing in municipal bonds, had helped the town participate in the financing of the oil tanker terminal, the biggest deal of his career (after the spill my mom found his commemorative photograph of the terminal in a garbage can). Pipeline construction brought the gold rush back to Valdez: huge brothels, organized crime, wild drinking and drugs. For the oil-in celebration the town held a wet T-shirt contest on the back of a flat-bed truck. That night on the beach with my seventeen-year-old cousin I watched a guy in the back of a pickup truck spray gasoline from a barrel onto a towering bonfire, the truck getting stuck in the gravel as the flames shot higher. My cousin and I stayed late at the fire, passing a jug of cheap red wine. At fourteen, I poured it down my throat fast, emulating the more experienced drunks around me, until the world spiraled and the ground seemed repeatedly to rise up and collide with me. Later, in the motel room, wine and stomach acid erupted from my mouth, a toxic black liquid that seemed to spew forth inexhaustibly, as I struggled, disoriented, poisoned, unable to think, crawling, astonished and wondering if it would ever end.

Twelve years later, those in charge had handled the oil like a fourteen-year-old with a jug of red wine. The drunk captain, a relapsed alcoholic, had recently lost his license to drive a car, and the corporate heads knew about his problem and ignored it. A lot of officers drank in the cave-dark Pipeline Club in Valdez where he had loaded up on vodka—a waitress there fretted to me that the oil spill would hurt long-term business as the mariners began pacing their shots. The Coast Guard had reduced its Valdez manning so deeply that the vessel traffic controller hadn't been properly trained or supervised; he gave the tanker permission to leave the shipping lanes, but he didn't monitor the ship's passage on the radar because he didn't know how to adjust its range. On the ship, Exxon had cut staff to the point of exhaustion. When the captain abandoned an unqualified third mate on the bridge to do the navigational job of two men, the mate struggled on alone because the other officers had already been working around the clock. The night the tanker left port, fishermen and other local environmentalists had been meeting to discuss setting up their own oil spill response system because the industry's preparations were so inadequate and because their concerns had been brushed aside for more than a decade. Yet when the oil spilled out, those in power were amazed and reacted as if no one could ever have predicted such an accident.

They cleaned up like a hungover fourteen-year-old, as well; I remember dabbing the motel room's ruined carpet with a hand towel. And equally absurdly, they sought to hide the magnitude of their error and incompetence. They: the oil industry, the federal government, even the state government. As I flew in the helicopter back to Valdez from Naked Island that day, the shape of the story started to clarify in my mind—the inconceivable immensity of this horror and the tiny people who let it loose and talked of recapturing it. They were so small they disappeared in the landscape. The oil slowly flowed onward, out of the sound, along the Kenai Peninsula, to Cook Inlet and Kodiak Island, killing hundreds of thousands of birds and thousands of otters, sinking irremovably into beaches, and spreading across an area so large it took much of a day and a change of planes just to fly from one end to the other.

Those in charge knew it was hopeless when they looked. The next day, Tuesday, I slipped onto a helicopter with Alaska's governor, Steve Cowper, and with the administrator of the EPA, William Riley, and the secretary of transportation, Samuel Skinner, and their retinues. Without knowing who I was, Riley and others used my notebook to pass messages back and forth, as the Coast Guard helicopter was too loud to carry on conversation. We saw oil floating across vast areas of the sound with no human presence in evidence. Riley handed a note to the Coast Guard commander across from him: "If cleanup ongoing, where is it?" The officials seemed shell-shocked and fled from the media after the flight. Cowper despaired. He told me, in his car, "That's the end of marine life in that area for a long time . . . I don't know what you could do about it at this stage."

As I spent more time on the sound, in oil, the press conferences and carnival of activity in Valdez seemed increasingly irrelevant and disconnected from reality. Exxon officials always announced numbers—miles of boom, numbers of skimmers, millions of dollars spent—facts that, if they meant anything at all, couldn't be checked. And they announced a lot of other operational information that could be checked and often wasn't true. State officials, fishing groups, and the like pointed out Exxon's faults, lobbing impotent verbal shells from bunker to bunker. People debated questions such as whether or not the sound had been "pristine" before the spill. The Coast Guard sent a series of admirals to take charge, issuing commanding statements to once and for all get the situation under control, to marshal forces and solve the problem at hand. For the most part, they came across as self-important fools, without impact, always retreating to the key imperative of protecting their own image of effectiveness, like the other agencies.

Everyone adopted the metaphor of war. We were an army in rout and we needed leadership and aggressiveness to meet the enemy and start taking back ground. But who were we fighting against? You can't make war against a stain. On the other hand, the situation was like war in some ways. Workers who picked up dead animals grew hollow-eyed with grief, but couldn't stop. Nothing else in our lives felt so important, the vividness of crisis bringing experience to an addictively sharpened focus. The outside world seemed pale and people there could never understand what we had seen. Decades later, certain images and objects could bring veterans to unexpected tears—a fragment of floating boom or oil-absorbant pompom might be enough. And the press conferences were like the Vietnam War that I had read about, with generals optimistically counting up numbers and making diagrams in Saigon while the hell of futile combat slithered through the jungle without discernible order. My job was to resist their image making: to stay out there and record the filthy truth. I did it fiercely, without caring who I hurt.

The newspaper bought a boat for us to get out and cover the story and rented an apartment in Valdez, where I spent the summer. Another gold rush came to town. Exxon hired street drunks as beach cleaners for $1,750 a week—anyone who could stand—at least until their piss tests came back. People seeking work flooded in; the woods filled with camps where the newly employed drank up their earnings. Businesses lost their staffs to Exxon. Trying to order in a busy restaurant one day I finally realized the last member of the waitstaff had just quit to work for Exxon and no one there would be able to order. But staff stayed in the bars, where servers made huge tips and cash-flush workers sometimes lit their money on fire. At the waterfront Club Bar the staff turned down big spenders who wanted to buy rounds for the house—it was too much work and no one cared anymore. Crime increased sixfold. The main street became a midway of stalls selling velvet paintings, suspicious home electronics, and T-shirts—"Let's go to Naked Island and get crude." The Alaska economy pulled out of a four-year depression with the infusion of money and the unemployment rate fell to the lowest level since the construction of the pipeline. Town boosters who had felt betrayed by Exxon became its defenders.

About two weeks into the oil spill I called my editor and said I wanted to write a story without any people in it—just about the rocks. Despite his skepticism, the next morning I got a seat on a helicopter from a temporary employee for the State of Alaska who was gathering up journalists and putting them on choppers to bill to Exxon as state oil spill business, at $1,600 an hour. The pilot had no instructions and most of the reporters on board knew nothing about the area, so I directed the flight to Green Island, where I'd heard there was good tide pooling.

The helicopter descended toward a narrow tidal isthmus of gravel that attached a pair of vertical rock islets west of the main island's low-profile forest. The 1964 earthquake had thrust these diagonal shafts of rock upward like crystals, shards of land amid slivers of sea. In the twenty-five years since the earthquake, kelp, urchins, and popweed had recolonized the rubble collars at the foot of each tower. At the top of each, crowding tiny patches of green beyond the reach of salt water, uncombed shocks of little spruce and hemlock trees were hung with moss. Sun off smooth water warmed our faces as the helicopter drilled downward, as I imagine a spaceship would do, committing our footsteps to this one world from among the many above which we had hovered.

As the helicopter's engine spun down and quiet fell I worked around the slippery rocks on the outside of the island by myself, hearing a raven cry and the distant voices of the photographers—they had found an oily bird, hollowed out by scavengers, in the beach grass. I heard barnacles clicking, drying under a sheet of sun-warmed tar. The place smelled of sulfur rather than salt and, in places, stank of rotting animals. I remembered an Independence Day weekend party on Yukon Island, in Kachemak Bay, when we climbed over rocks like these and picked mussels to cook in a pot over a beach fire—a warm, sunny evening, sipping wine—the mussels steam open in just a few minutes, and then each shell is a tiny plate, rich with butter and garlic and that tender, delicate flesh. The mussel beds on Green Island were as plentiful, but they didn't need steaming to open them—my fingers could easily do it, and inside the meat was black.

The tide pools varied, as always, worlds within worlds with their own stories of birth, sustenance, conflict, and death. I remembered being eight years old on a similar shoreline with my mother, staring down through the glassy roof of each pool, pulling aside forests of eel grass to find tiny creatures squirming, and scooping water in a clam shell to catch inch-long fish, watching hermit crabs scamper away from sunflower stars with twenty-four legs and fifteen thousand feet. Some of the Green Island tide pools had ceased to exist: they were full of syrupy sludge, smothered and obliterated. Other pools were poisoned, but still visible. A flowerlike sea anemone had turned brown; another looked nearly normal but didn't respond to a touch. A sunflower star's thousands of feet were smeared with thick oil. Limpets and chitons, which hold hard to rock, fell away at a touch, and even barnacles could be swept away by a fingertip. Down under the eel grass, in the crevices, creatures without common names—white, shrimplike animals—curled up dead, in piles.

Another pool was still relatively healthy, but the tide was coming in. A slick that met the shore rose with the water. A seal swam, like a ghost, under the oil, then popped its head up and looked at me. As the flood met the tide pool, the oil swirled in.

The helicopter roared. The rising water had neared its skids, and I had to run back to get on board.

28.

Costs and Values

Within the first ten days of the oil spill, lines of young workers in hard hats and rain gear marched each morning down the ramps at the Valdez small-boat harbor, some with shovels on their shoulders like rifles, all with eyes straight ahead and as wordless as members of an honor guard. They boarded tour boats and sailed off, presumably to clean up beaches. But when the boats left the harbor mouth they just drifted aimlessly in Port Valdez: managers had no plan for how to use the workers and sent them out only because it was all they could do. When the workers began landing on oiled beaches they were given oil-absorbent rags to wipe off individual rocks. I spent a day with them on Naked Island. Those who needed a sense of purpose would pile up the pebbles they rubbed in little mounds—otherwise, at the end of the day, there would be no evidence they had done anything at all.

Like characters in an absurd Samuel Beckett play, young people making the biggest wages of their lives struggled to find meaning in the meaningless task they had been given. I irritated them by asking questions that pointed out the futility. A woman said every teaspoon of oil removed mattered. Someone else had a theory about trying things that didn't work before finding a solution. A man said taking an Exxon paycheck for doing nothing was his only way of punishing the company. A couple of guys simply enjoyed the sunshine and the swimsuit videos they watched on the tour boat, and dreamed of spending their paychecks. They threw their rocks at me. The foreman, George Cowie, a purposeful young oil field construction worker, said, "Exxon is a multibillion-dollar outfit. They haven't gotten to where they are by being ignorant or stupid. And they've got guys making one hundred thousand dollars a year working on this. So I basically trust that they know what they're doing."

Why does God do things we don't understand? Because he controls the universe itself, in which we're only players. Mighty Exxon, godlike in size, wealth, and power, impervious to punishment or coercion, also did many things that we could not understand, including spilling the oil itself. The company's actions led inevitably to the wreck: cutting payroll, defeating calls for precautions, saving pennies—certainly not rational choices. But its policy made sense in the theater of the absurd, as an act of profound freedom, the omnipotent one arbitrarily devastating the wilderness and the lives of thousands of people simply because doing so was easy. Beside such a large legal person as Exxon, the biological people and animals smothering in black excrement barely registered, like insects drowning in a drop of sweat. For Exxon's finances, the spill was no more than a blip; other than the captain, those responsible got raises and promotions. As mortal prisoners of this universe, we would all be wise to do what we're told and stop asking questions. Cash your paycheck and use the money to buy machines that burn Exxon's products. Believe that freedom consists of a powerful car; that's as much freedom as you can hope for.

But what if the inmates, the victims, discovered they could change the shape of the universe that imprisoned them? Exxon's wealth depended on its ability to take collectively owned resources without paying for them— the atmosphere, the earth, and the ocean its ships crossed without adequate safeguards against a disaster. The spilled oil made visible, for once, the cost of a system in which competitors can grow to unlimited size by consuming the tangible and spiritual sustenance that belongs to all. For a moment, the world looked on in horror. Perhaps the possibility sprouted of a new collective belief, a new social norm giving back the sea to the fish, the otters, and whales, the fishermen and kayakers—or at least that Exxon shareholders and not subsistence hunters should absorb the cost of the company's actions. The cleanup responded to that possibility. In theory, it would reduce the costs Exxon had externalized upon the rest of society. The cleanup was more than a public relations gesture. Exxon tried to shore up a worldview.

In physical terms, it failed: Exxon's cleanup did little to reduce the spill's cost. In the real world, Exxon was not a god, and its money, technology, and logistical skill were impotent to diminish the harm caused when its shrug oiled Prince William Sound. The possibility to mitigate the harm ended with the wind that blew on Easter night. Oil spill experts who converged on Valdez gave clear, accurate predictions of the fate of the oiled shores. Oil had been spilled on cold rocks before. Each tide would remove and redeposit it until every shore was coated from the low tide line to high water. On shores exposed to large waves the oil would last only a season until it weathered and faded into the ocean. But on miles of protected beaches, emulsified oil would sink in deep and stick in dark, interstitial spaces for at least a decade, probably more. From there it would slowly burp its toxins into the intertidal ecosystem. Removing the oil would require destroying life with scalding water or chemical solvents and would create collateral damage with its toxicity and physical force. Adding fertilizer to hasten biodegradation would help where oil-eating bacteria enjoyed conditions conducive to growth. Otherwise, nothing would work.

The experts were Cassandras, doomed to be right but unheeded. The people in charge treated the *Exxon Valdez* oil spill as the first ever, experimenting with increasingly harsh cleanup techniques as their frustration grew. Exxon offered to do everything possible, and the Coast Guard pushed it onward, with public and political opinion at its back. The Chugach people wanted the oil gone, regardless of the cost, so they could trust their subsistence foods again, and they influenced Alaska officials, who also were motivated by their own indignation and perhaps by a sense of guilt. Given a couple of months to get started, Exxon's know-how and some $2 billion created wonders. In a remote bay on Knight Island I found a floating heliport anchored next to a 438-bed hotel for beach cleanup workers newly built on a huge steel barge, where a brightly lit cafeteria served tender steaks for dinner at night and pies were displayed in a glass case. A fully equipped floating hospital offered care for the workers' scrapes and bruises at a cost a doctor estimated at sixty times a shoreside clinic. At night hundreds of lights reflected on the smooth water of Green Island's rock-guarded anchorage, a mobile city of boats and barges full of Exxon's new workers. Most of these people backed up the forces on the front lines, who operated barges equipped with massive boilers and pumps capable of spraying beaches with nearly boiling water through a water cannon. The high-pressure hot water released oil that crews then caught in floating booms.

The day after a successful hot-water treatment, the stink began. The scalding water cooked every living thing. Sea stars, shellfish, and seaweed rotted in the sun. A clean start for life—or so the officials told themselves when they checked these beaches off on their clipboards. Visible oil was gone, but oily sediments flowed on the hot water down below the tide line, killing ecosystems and contaminating seafloor that had been relatively clean. And even on the sterilized rocks, life didn't come back the way it had been. Opportunistic species of algae and barnacles came first, the way grass grows after a forest fire, and crowded out recovery of the seaweed species that support the community of seashore life. A natural succession of seaweed did develop, but intertidal ecosystems collapsed again in 1994 when the popweed, *Fucus gardneri,* all died of old age at once. Worse, the high-pressure water re-sorted beach sediments. The sea arranges these beaches with smaller grains below and armoring rocks on top that protect the sand from waves. Clams and other burrowing creatures thrive in those conditions. Two decades after the cleanup, researcher Dennis Lees found that colonies of clams had not recovered on beaches washed by high-pressure hot water. Re-sorting had barely begun, as it had in places uplifted by the 1964 earthquake, and in 2007 he predicted it would take decades more.

Did the cleanup as a whole do more harm than good? The question is

unanswerable without defining good. It benefited Exxon and its competitive corporate-government paradigm. Cleaning eliminated most visible
evidence of oil. The effort took so long world attention turned to other issues and anger faded. A diminished Prince William Sound became the
new baseline for the next generation of people. Today the spill has passed
into history and Exxon still rules its world. Good also, perhaps, from the
point of view of human users of beaches, since the cleanup may have has-
tened the time when they felt safe eating clams, fish, and seals again. Cleaning probably shortened the time active contamination affected some
species. But if good is defined as the total health of the ecosystem, it's prob-
able that much less cleaning would have been better. Oil would have dispersed anyway, more slowly but without the cleanup's many environmental
costs. And more of the visible black asphalt would have remained, biologically inert but a powerful warning about the cost companies like Exxon
impose on our shared birthright. Cleaning removed the evidence, but not
the damage.

The attempt to clean the animals produced an even clearer case of
harm and a deeper chapter in the story of hiding spill costs to avoid reck-
oning with outraged human values. Everything comes back to these
questions of values and cost.

Individual animals suffered horribly. Kelly Weaverling, the Cordova
bookstore owner who had drawn animals on years of kayak trips, and
knew individuals as friends, organized a flotilla of boats out to Knight Island in the first days of the oil spill. He felt he had to because no one else
was doing anything. We were together when morning first showed how oil
had buried the branching, smooth water of Herring Bay—snowflakes lay
briefly unmelting on the black surface as they fell—and the animals were
all missing. Tears streamed down Kelly's dry, lined face. We chased birds
around, their oil-weighted wings flopping on the water, but even the small
ones easily avoided our powerful motor boats and dip nets. A couple of
fishermen grabbed a shivering otter that had hauled out on a rock, and
one of the men was bitten in the process. A seventy-pound otter is a tube
of muscle, its teeth and claws sharp enough to open clams. The otter's eyes

told stark panic. Its fur was glopped with a thick gel of oil mousse, which it frantically licked. Otters survive cold water by keeping their coats perfectly clean.

News photographers in Valdez converged urgently on the first animals we sent back by helicopter. A fight broke out among them. Combatants knocked down a worker holding an oiled bird, nearly crushing it. Without adequate facilities, workers stored birds and otters in stacks of cardboard boxes and dog kennels lent by pet owners. Exxon quickly built treatment centers, remodeling a school gymnasium in Valdez with a false floor that concealed plumbing like a big-city aquarium's. Volunteers adopted individual otters as their personal charges. Some sat day and night in their enclosures, stroking and cuddling animals that were near death. When the otters did die—the agonizing fate for a third of the rescued animals—managers whisked the carcasses away so volunteers and the media wouldn't see. As spring turned to summer, more centers were built in other communities and more otters were rescued, less oiled, more likely to survive, and indeed with better chances if they had been left where they were found.

In a refrigerated truck trailer in a back parking lot I met Cal Lensink, who was dealing in obscurity with the far larger number of otters and birds that came in dead, which, in turn, were a tiny fraction of the hundreds of thousands of birds and animals that died and were never recovered. Cal, recently retired and volunteering, had spent a long career in the U.S. Fish and Wildlife Service; he began counting otters in the sound in 1959. He looked like a 1950s outdoorsman, with a gray crew cut above a V-shaped face and a pipe bit in the teeth of a squinting smile. With another scientist he patiently pulled apart bagged lumps of animals rendered shapeless by oil, identifying them by wiping off a webbed foot or a beak, snipping heads from long-necked birds with a pair of scissors for later analysis. He pulled teeth from dead otters with pliers to learn their age and cut purple fetuses the size of small cats from the bellies of the females. As he worked he reflected on the illogic of the animal rescue. Lensink said, "With many hundreds of otters dying, we try to save just a few, and most of them, we're not successful, and they die anyway. In not being successful, you're prolonging the agony for birds and animals that will die anyway."

A married pair of state biologists persuaded officials to prohibit rescue of harbor seals. By killing some seals and analyzing their brains, they established that seals were dying from neurological damage caused by oil fumes, not poisoning or oil on their coats, and so could not benefit from cleaning. They also pointed out, as other scientists did, that releasing treated otters back to the sound posed a serious risk of carrying domestic diseases into the wild that could rip through populations without immunities. Spill otters had been held in those used dog kennels and in close contact with volunteers. In one center all otters picked up herpes sores. But U.S. Fish and Wildlife Service officials didn't see a choice. With two hundred otters on their hands, too many for the world's aquariums to accommodate, the animals either had to be released or put down. Exxon had already paid about $90,000 for each otter captured and cared for, and the emotional investment was much greater.

Here is an apparent contradiction. I have said we value wild nature deeply because of the sense of personhood—the spirit—which our human nature detects within certain animals and places. We need the heart to do good, to care for land and ocean and to join in community among ourselves, to rise above the lonely fate that awaits us as purely self-interested beings. But a certain kind of love for sea otters led people to betray them with killing kindness. "Given the disease and all that, I'm not sure they should have brought *any* of them in," said biologist Charles Monnett. "It's very hard not to, but the ones that were going to die were pretty much going to die, it turns out, and by bringing them in all they did was create other complications in the process, and risk a very large healthy population.

"Everybody was trying to put some positive spin on this, and not just the people who were directly involved, either. Certainly Exxon and the government wanted it to end well. But there were a large number of people that just, I think, couldn't emotionally deal with the idea that we hadn't solved this problem."

Monnett and his wife, Lisa Rotterman, had been in the sound tracking otters with implanted radio transmitters before the spill, under contract with the U.S. Fish and Wildlife Service, flying Monnett's own small plane. Afterward they expanded their project with boats and another plane to measure the effect of the oil and—after losing the fight to stop the release—studied the fate of the animals treated in the rehabilitation centers. Half the released otters died in their first winter. The couple also continued tracking its never-oiled otters in the northeast corner of the sound, away from the spill, near where the treated otters had been released. Fully 40 percent of those unaffected animals died after the captives entered their area, compared to a 6 percent death rate distant from the release sites.

As a percentage of the study population, that die-off of unoiled otters destroyed as many as the oil spill did where it hit. The follow-up questions were obvious: How many otters died and in how large an area? And was it caused by the release from the treatment centers or some other cause? Monnett never got the chance to find out, because his research contracts were not renewed. The data from the study received scant analysis and were written up without mention of their implications. Monnett's career came to a standstill and never fully recovered. He was unable to get a job or research contract in his field for nearly five years—he says a supervisor told him he was blacklisted—and he had to leave Alaska to find an entry-level position and begin working his way back up. A former coworker and a different supervisor confirm that Monnett became unwelcome at the Fish and Wildlife Service, but they said the personality clashes were partly caused by his intensity, or arrogance, driven by his intense love for wildlife and the habitat where he worked. No one doubts how much he cared: he twice crashed his plane while tracking otters during that troubled period, the second time swimming forty minutes in the frigid Gulf of Alaska from the sinking wreckage. After so much time, it's impossible to know exactly what happened to his relationship with the agency, except that Charles Monnett was convinced that rescuing otters tormented the animals and threatened their population, and he was probably right, and many scientists agreed, and yet the agency still plans to do it again the next time there is an oil spill. "We were blunt," Monnett said. "It's very unpopular to be blunt about something like that."

Few moments of outrage in the modern era have been as potent as the public sentiment after the *Exxon Valdez* oil spill, and probably none concerning the environment has ever been so strong. Nationally, all but 6 percent of Americans were aware of the spill; by comparison, 26 percent didn't know the name of the vice president (Dan Quayle). Briefly, the truth lay exposed, undeniable, along a thousand-mile coast, as if a sewer had backed up with the sick, dark stuff we prefer to keep hidden. Killing the ocean slowly is much less visible, much more comfortable. Americans were ready to act, potentially to change the basic rules of this relationship. Such changes can happen; that year, 1989, the Berlin Wall fell. But, at this time for facing difficult truth, kindhearted television viewers were instead dealt the happy ending of seeing treated animals released, as if cured, rather than images of terminally disabled otters being euthanized. By denying them that truth, the Fish and Wildlife Service compounded the harm of the oil spill.

The power of the moment received precise scientific measurement in a study by leading academic researchers commissioned by the Alaska attorney general. The study rested on an interesting idea about how to value environmental injuries in lawsuits. By traditional legal measures, an otter might be worth only the resale value of its pelt, about $200. Prince William Sound obviously was worth much more, but since its intrinsic value could not be bought or sold, the market could not put a price on the damage. The concept of contingent valuation came into federal law as a way around this problem. Economists believed they could find the value of environmental goods by asking ordinary people how much they would pay to avoid losing them, with surveys that concocted plausible scenarios to make the respondents believe their answers would amount to a real vote for a tax or other authentic cost they would have to pay. The contingent-valuation study team sent interviewers to 1,599 doors, in every state, with information explaining the damage of the spill and a supposed foolproof plan to prevent another spill, without which, respondents were told, the disaster would inevitably be repeated within ten years. The interviewers drastically understated the damage of the oil spill, both to make the study defensible and because some of the harm wasn't yet known. In their statistical analysis, as well, the economists minimized the results at every opportunity. Thus, the study's outcome represented the absolute least Americans were personally willing pay to prevent another Prince William Sound oil spill, a figure a half to a fifth of what it might reasonably have been—yet that price, $2.8 billion, was so high the lead state and federal attorneys discarded it as incredible and unlikely to succeed in court.

They turned out to be right. Although the government settled its cases for much less (as we'll see later), private plaintiffs fought on. In 1994, a federal jury valued the case similarly to the survey respondents, although entirely independently, penalizing Exxon $5 billion. But Exxon fought that verdict through a series of appeals that concluded nineteen years after the spill with the U.S. Supreme Court arbitrarily reducing the amount to $507 million—nothing to Exxon, the most profitable company in history, which had paid its CEO, Lee Raymond, $683 million over the fourteen years the case was on appeal. As James Madison intended when he created the constitutional system, the justices acted as supreme sovereigns to protect the private property of the wealthy against the democratic will of the majority, in this case making new law on their own authority to do so.

Exxon's victory was complete, but I am left to wonder how it could have been different without a fundamental change in society. The contingent-valuation concept doesn't really make sense. By finding that each American household would pay $31 to prevent another spill, and multiplying that number by all the households in the United States, the economists had come up with a value of $2.8 billion for the damage to Prince William Sound. But why should only Americans have been asked—why not all the people in the world? Surely we all own the ecological integrity of the earth itself. Or, if the economists intended to measure only value for the legal owners of the land, what of the fact that the beaches and territorial waters belonged to the State of Alaska? Surveying only Alaskans would have yielded a much lower value, because they are far fewer in number. But the economic value of an object isn't supposed to depend on who owns it.

As I consider what I would have paid to have prevented the oil spill, I realize the problem is much deeper, down in the roots of what value means. If the survey takers had asked me, I would have said I would pay everything I have, but that wouldn't be honest—I wouldn't give away my children's education or my ability to manage my other obligations. Just as there is no amount of money I would accept to allow the harm, I'll never have enough money to pay what I would offer to prevent it. The law puts dollar values on a severed arm or a dead daughter, losses that no one would accept in a willing exchange. These values have meaning only to corporations: they can use them to determine the cost of killing or dismembering people with their products, measuring the downside, for example, of selling defective merchandise.

Punishing a corporation through the legal system is like striking a chair against which you have stumbled. Exxon absorbed the hatred of a generation of coastal Alaskans without flinching, impassive, because it is not a person—it is a thing, ultimately just a set of symbols, an interchangeable subsystem in an economic machine that functions without regard to individuals' wishes. A higher cost in dollars, even one high enough to bankrupt the company, would hardly change the system of relationships that governs our lives, and through which we are inexorably killing the oceans. Financial penalties would create stronger incentives for care, but only so long as government agencies meting out punishment remained uncorrupted by the corporations they regulate—which they never do for long.

Patrick Norman, the Chugach chief of Port Graham, said protecting the environment requires no more than that each corporate board member possess a conscience and a sense of responsibility, putting their duty to shareholders below their duty to humanity and nature. The oil companies that continue to dump petroleum waste legally in Cook Inlet north of his village use the law as it was intended to be used, as did Exxon. Justice, instead, would require them to decide not to pollute the ocean with their toxins, because it is wrong. "Even in Western culture there is a set of values people live by," Patrick said. "Maybe the Western culture needs to remember them, or re-create them."

Monetary penalties couldn't give Exxon's board a conscience. That kind of learning is more personal. Communities sharing the commons rely on shame and exclusion to punish those who abuse shared resources, such as those Native Mexicans who ban outsiders when scallops are scarce, and deter overfishing within their group using embarrassing sexual jokes. We would need similar tools to change the behavior of Exxon's leaders, including the smug chairman in 1989, Lawrence Rawl, or the well-compensated Raymond, whom the company promoted to Rawl's job after he led the U.S. subsidiary through the spill and cleanup, or the other executives who made the sloppy decisions that brought so much harm and grief. We would need to punish them as one would punish a child who needs to learn the difference between right and wrong. Then, instead of paying other people's money to clean beaches and animals, they themselves would futilely wipe the rocks with rags. They personally would put down the oiled animals, ending the pointless suffering of each otter with quick, merciful death, as any decent person would do.

But they were not members of the Prince William Sound community and they were immune to its shaming. A real sanction for the oil spill would require a return of power to local people as they haven't had for centuries. Give the community a tool with human meaning. Let it exclude Exxon from the sound, forever.

29.

A Community Collapses

Each April, when the herring returned, Cordova would awaken from winter dormancy like a tide pool at the first lick of the rising sea. Fishermen geared up at the boat harbor, processors brought on their first seasonal workers, and spending money began to flow at the hardware store, the bar, and the bookstore. Bank accounts that had dwindled since fall, nursed through frugal months, began to grow again. Winter coats and skis went up to the attic. Musty boathouses and net lofts opened to spring air, which carried the odor of sprouts piercing the muddy snowmelt. Quilting and music groups lost interest as the dirt baseball diamond above Orca Inlet dried out and members of the women's softball league limbered their arms. At the high school, teens prepared for graduation and another season of work on their families' boats.

On the fishing grounds, a competitive dance of seiners and their spotter planes angled for a rich set of their nets, pilots dodging each other while gauging the thickness and color of the schools in the water below—an indication of herring roe content and value—while each crew dreamt of wrapping up a legendary $100,000 ball of fish in a single hoist. Captains elbowed their vessels into position like basketball players fighting for a rebound—or like the huge male sea lions that bark and roar on the rocks in symbolic tussles, all blubber and musk, before swooping under the water and catching fish like swallows snatching flies from the air.

Back in town, the joy of spring would rise with news of the size of the catch. Everyone would benefit from the health of the herring run and the skill of the skippers.

Cordova's economy was an extension of the sound's ecology, and it supported a tight community of equal, self-reliant people. There was no big boss, no outside corporate office making decisions. Businesses mostly were owned by families who knew their customers intimately. It was possible to build a house entirely by bartering, trading skill for skill. But there was no locksmith. Some never locked the front door—didn't even have a key. Cars sat along the street with keys in the ignition, and sometimes a friend would just hop into one and borrow it. Children could roam freely without fear of strangers, cautious only of bears and taking care to wear life jackets in their boats. Cordova, more or less, was a self-governing community sustained by commonly owned renewable resources. Until the oil spill.

With news of the spill and the lack of response, dozens of fishing boats set off from Cordova to help, informally organized by the fishermen's union to guard the Armin F. Koernig Hatchery they had cooperatively built in Sawmill Bay, which lay in the direct path of the oil. Governor Cowper dispatched a state ferry as a work platform, and the fishermen, volunteers, and state workers successfully strung a series of floating booms across the bay and kept the oil out. In town, the community gathered in big, emotional meetings at the high school. People worried about the upcoming herring season and what it would mean if they could no longer fish. They recited poetry and one woman read out the Declaration of Independence, in tears. Fishing families worried they would lose their season, or more, and their permits and boats would lose value—everything they owned. The town drew together in solidarity to get its message out to the world, dispatching representatives to Valdez to talk to the media. Exxon officials arrived, at first not even aware that Cordova existed. The corporation's Alaska representative famously declared to a packed gymnasium of angry people, "We will do whatever it takes to keep you whole."

The money began flowing, but not from fish, and that made all the difference. Exxon paid fishermen for the lost herring season. When Cordova merchants complained they would suffer from the lost season, too, Exxon began paying them to make up the difference from what they had taken in the previous year. Exxon also sent checks to the vessel owners who had volunteered in the cleanup—huge checks that fishermen could hardly believe. An average boat chartered for $3,000 a day, almost all profit, since expenses were paid separately by Exxon. As the boats stayed out for weeks and months, often with hardly anything to do, life-changing sums of money accumulated; owners of big boats, or of more than one, became spillionaires. Fishermen who rushed to the Cordova union hall to get their boats hired complained of favoritism and collusion if they didn't get a contract, or if they got only one boat hired on while others had three. Those they accused were their neighbors and fellow fishermen. Others refused to take Exxon's money on principle and resented those who did, resentment growing deeper as they saw friends earning enough money to send kids to college or buy new houses or boats while they got nothing. They observed that some of those who had denounced Exxon the loudest were now anchored in remote coves doing next to nothing and collecting huge charter fees under contracts that bound them not to speak publicly. Old friendships ended.

Mike Webber, a successful, aggressive young Native fisherman—and the grandson of Mae Lange (whom we met in Chapter 19, describing her childhood in Katalla)—charged out of Cordova Harbor an hour after being called in the first days after the spill, fighting a seventy-knot wind in the middle of the night to get to Sawmill Bay. The work was chaotic. Everyone was in charge, but no one would take responsibility. Then the money showed up—he was paid retroactively as a volunteer—and the suspicion and jealousy started. "It didn't settle that well with a lot of people, and that started tearing our little community apart," Mike said. "The whole picture, being out there, wasn't healthy. It wasn't worth the money. I chose fishing over it. It was healthy. Healthy for our minds."

But with his fishing income and his spill charter—fifty days' worth—Mike received more money, all at once, than he had ever dreamed of, and he and his wife went on a spending spree. Soon, their relationship fell apart. "We weren't used to that money," he said. "Money doesn't do good for people by any means. It makes them greedy."

The town went crazy for the summer, and it didn't seem to get better. Businesses couldn't stay open because their employees quit to work for Exxon. The radio station went off the air because there was no one left to run it. After months of crisis, the local government fell upon itself: a pro-Exxon council member sued anti-Exxon council members for supposedly holding secret oil spill meetings, comparing them to Hitler, while they said she was paranoid, demented, and a stooge for Exxon, and one told her at a council meeting he wanted to vomit on her. Legal fees of $1 million to defend council members depleted the town's savings, although the plaintiff sought nothing more than an admission of wrongdoing.

Patience Faulkner, a stepdaughter of Mae Lange, had moved to Anchorage for schooling toward becoming a lawyer—she was already a paralegal—when her fisherman brothers in Cordova called and asked her for help dealing with the attorneys who had come to town to sign up plaintiffs against Exxon. Cordova had no attorney of its own. As the community spun into chaos, she advised fishermen, helped them file claims, analyzed the deals offered by various law firms, and organized them for cleanup work. Anxious wives of fishermen dropped by needing to talk, and she knitted with them. Eventually she became a local representative for the law firms, compensated by a monthly fee, free office supplies, and dinners out with the lawyers when they came to town. She also worked at the liquor store and took odd jobs.

"People would bring me food to eat, and fishermen brought me fish," Patience said. "I got spoiled. I got richer in all these intangible things, all these relationships." But she abandoned her studies, as taking classes would have required her to leave town. "People said, 'If anything happens, the first thing we're going to do is call you. You can't leave.' And I didn't take that as a joke. I took that seriously." The paid work lasted nine years, and she went on as a volunteer for another ten years after that. By the time the case was over she was closer to retirement age than college and becoming a lawyer was no longer practical.

In August of the spill year a pair of researchers showed up in Cordova and began asking people to fill out surveys about their emotions. Friends called Patience to ask if it was okay to answer. She and others took pity on the researchers, fed and sheltered them. The professors, J. Steven Picou, a sociologist from the University of South Alabama, and Duane Gill, then at Mississippi State University, had never been to Alaska before. They had spent their grant money on plane fare and didn't have proper rain gear for the autumn downpours, and were subsisting on crates of Beanie Weenies (canned beans with hot dogs). Long-term friendships grew between them and the Cordovans they studied to learn how the disaster had affected their mental health and community connection. Unlike research paid for by the government or Exxon, which seemed mostly to reach conclusions favored by its sponsors, Picou's group worked independently, with funding (adequate after that first year) from the National Science Foundation and other independent entities. And Picou used random sampling methods and a control—the southeast Alaska town of Petersburg, which is similar to Cordova but unoiled—to develop a psychological profile of the community in the midst of a disaster.

Picou found broken people and a broken community. Surveys disclosed signs of stress such as intrusive thoughts and emotions, bad dreams and sleepless nights, unfocused anger and soured family relationships—markers that scored as high in Cordova as in personal catastrophes like rape or the death of a child (although Steve emphasized he doesn't equate those events to the oil spill). As the years passed, the feelings got no better, or even worsened. Pink salmon prices crashed to the point the fish weren't worth catching. Herring came back with lesions, and then stopped coming back in numbers large enough to fish. Then the pinks disappeared for a time as well. A fisherman like Mike Webber who had owned boats, gear, and permits for fisheries that kept him busy around the calendar now lost one season after another, until he was fishing briefly and making as little as 10 percent of his former income. The value of his permits dropped even more than his income; a salmon seine or a herring seine permit had sold for nearly $300,000 each, but the salmon permit fell as low as $12,000, and there was no market for a herring permit at all, because the fishery remained closed year after year. No more April surge of money into the town from big balls of herring. The head of the fisherman's union lost his boat to the bank, but when it came up for auction there were no bids. The community became corrosive, in Picou's description: sorrows compounded themselves in conflict and a sense of lost connection, lost faith in the sea, the town, and the institutions that had failed to protect them. Patience spent long evenings listening to a fisherman nearing retirement who had lost everything and didn't know what he was worth anymore. Everyone could count the names of the suicides on their fingers, and she didn't want another.

If you believe that a town can be part of an ecosystem, it makes sense that the human community would sicken along with the natural system upon which it depends. But even as the environment later stabilized, Picou found that the litigation itself kept the pain alive. The verdict took five years, promising the $5 billion punishment that the Anchorage jury imposed on Exxon to be divided among thirty thousand fishermen, processors, Natives, landowners, and others—enough to pay for retirement or a new boat for many. But Exxon kept fighting for fourteen years more. With unlimited resources, its strategy was unlimited litigation—deposing everyone involved, seizing scientists' notebooks as they came in from the field, fighting every point on appeal to the very end, even suing Picou in an attempt to force release of his confidential mental health interviews. The legal war went on and on, up and down between the district and appeals courts. In Picou's surveys, it emerged as a greater stressor than the event itself, with its confusing issues, the offer of money held always just out of reach, and constant reminders of the painful losses of the past. The hollow feeling that justice isn't real. Eighteen years into the process, a fisherman stopped me in a Cordova restaurant and told me Exxon had already won—although the Supreme Court's decision remained a year away—because no one would ever dare challenge the company again. Picou by that time had found some recovery in Cordova, primarily by replacement of the victims: some had moved to escape their feelings, about 20 percent had died waiting for the case to conclude, and many others had been born and grown to adulthood since the spill. Members of that new generation suffered the litigation stress only if a family member was involved. "The litigation is kind of like a disease," Picou said. "It afflicts all of those who come in contact with it."

Mike Webber broke his back in a fishing accident and fought back to be able to walk stiffly, in constant pain. He succumbed to a drinking problem and checked himself into rehab. One winter, while caring for a family member who was dying of cancer, he began reading about his Tlingit heritage and took up carving cedar logs he picked up from the beaches. To his surprise, complicated traditional carvings came from his knife smooth and sleek on his first attempts—people said the spirit must have been in him already, from an ancestor—and he began sitting long winter hours in his workshop, among his nets, carving happily. He developed the dream of visiting the world's museums to copy the great old art of the sound and bringing it back to Cordova in his own pieces. He showed me his work and his workshop with pride. But, eighteen years after the spill, he remained bitter for all he had lost.

"It'll never go back to normal—the fishermen, the people, will never be normal again," he said. "The oil spill has not been buried in the back of our mind or soul. It's like a lesion on the herring. It's always there. It will always affect us."

Mike carved a totem pole in the Haida tradition of the shame pole, an ancient form, long out of use, with which a person of high status would be called to account for an obligation or misdeed. It now stands in the Ilanka Heritage Center, a living museum representing each of the cultures that interacts in Cordova. Mike covered the pole with symbols. Dead birds and animals—individual creatures Native people see in themselves. A fishing boat sinking with a family on board. A tombstone for friends who took their own lives. People of different races holding hands, each with a hole in the heart. A scale of justice tipped in the direction of money rather than the earth. He made the dollar signs with his own blood. The head at the top represents Exxon, a cartoonish shape, upside down, with dollar signs for eyes and the long Pinocchio nose of a liar. From the figure's mouth flows a huge black blob, running down the pole among the other symbols, and, spilling out with the oil, the words the Exxon executive said at that meeting in the high school gymnasium so many years ago: "We will make you whole."

30.

Rick Steiner and the Fight for the Trees

Rick Steiner arrived in Valdez from Cordova in the first days of the oil spill, walked into the second floor of the hotel at the harbor where Exxon was setting up its command center, and struck up a conversation with Frank Iarossi, the president of Exxon Shipping. With a musician friend, Steiner joined the command group, whose other members were Iarossi and the top officials from the State of Alaska and the Coast Guard. He pointed out the sensitive areas that should be defended from the oil flowing down the sound, especially the salmon streams and the fish hatcheries, which Exxon didn't know existed. Iarossi listened to his advice. Rick became an intermediary between fishermen, the oil company, and the government agencies, praised all around for a cool head in a crisis and a calm, personable style that put at ease men like Iarossi, who faced so many angry people. Rick didn't mind meeting in private and making clear, practical demands, while holding just out of view the threat of letting loose the dogs—the fishermen, with their red hot fury, who trusted him to represent their interests.

They were unlikely allies, the oil executive who arrived in a business suit and Rick, with flowing hair, sandals, a smiling voice—he looked nearly capable of flight, with his tall, thin body, elastic stride, and shirt sleeves that hung down from his wrists like wings. On more typical days Rick worked from his University of Alaska marine extension office on a dock in Cordova. At lunchtime he would climb into his Zodiac skiff, round Spike Island out of sight of town, strip and swim in Orca Inlet, then warm up with a cup of hot coffee and return to work to advise fishermen. But Rick had grown up talking to powerful people. His mother, Fay, served as a White House staffer from the Eisenhower administration through most of Reagan, and as a child Rick would answer the phone when President Johnson called her at home to chat. Among his first jobs was opening mail in the Nixon White House. He recalled shooting a question at Secretary of the Interior Rogers Morton as a eighteen-year-old, asking why Alaska oil couldn't be piped through Canada, as environmentalists preferred, rather than being shipped out of Prince William Sound (Morton's answer was cost). Rick grew up with an ego for handling power—or, to put it another way, with the naïve assumption that the powerful are obliged to heed the reasonable requests of the people they serve.

Steiner had studied to be an ecologist, but he became impatient with scientists who were satisfied to obtain information but do nothing with it. "I realized once you see the reality and the severity of the environmental problems, that really there are ways of changing human behaviors," he said. "You can see clearly that there are easier, better ways of doing things." The *Exxon Valdez* oil spill appeared to be an epochal event in that respect. "It nailed a lot of people," Rick said, including Alaskan politicians who previously had fought environmentalists reflexively on every issue, but who now, at least for a brief, historic moment, partly blamed themselves for the mess. Rick worked thirty days with the command group before leaving Valdez. "I realized there was no such thing as oil spill cleanup. So what else can we get out of this thing?"

Rick and his remarkable network of politically talented friends in Cordova already had an agenda before the spill. They had been meeting in his office for several years talking about bringing a science center to town to house research on the sound and return the knowledge it produced back to the locals. They also had studied a solution to the complacency of the oil industry and government agencies in preventing spills: a citizen's watchdog group like one at the Sullom Voe Oil Terminal in Scotland's Shetland Islands, which was funded by a levy on oil shippers, giving that community a strong voice for better practices. Two years before the spill Rick unsuccessfully pushed the idea for Prince William Sound with the state government and the Alyeska Pipeline Company, the owner of the Valdez terminal. And there was the Native corporations' tax-loss logging. Rick and his friends were desperate to buy the forests back for conservation. Considering the amount of money required, that had seemed a pipe dream. But now almost anything was possible, as the national media and the government focused on Cordova with that political attention that flicks on like stage lighting—that intense, capricious attention—here once again illuminating the melodrama of a little town in distress, just as it had in 1911, when the stars of the show were Gifford Pinchot and the Guggenheims rather than Rick Steiner and Exxon.

Rick's friend R. J. Kopchak, a fisherman and Cordova's vice mayor, helped rush through articles of incorporation for the science center and landed start-up grants and a building amid the tumult of the spill summer. Chuck Monnett and Lisa Rotterman, the otter biologists, joined the board and relocated their work to the center. R.J. also selected Mead Treadwell, the Republican strategist, through a corporate head-hunting firm, to be the city's oil spill coordinator—a position to be reimbursed by Exxon—choosing him in part because of his close ties in Congress and to the first Bush White House. That was the year Mead brought Carol to Alaska, on that first trip when they fell in love (described in Chapter 6). One of their stops was Cordova, where they briefly helped chase otters for Monnett's project. Later, Mead propped up the science center's wobbly financial legs by persuading Senator Ted Stevens to endow it with permanent funding through federal legislation.

Legislation for the oil shipping oversight group followed after Rick and his confederates brought a new friend from the Shetlands to Alaska and worked the media with the story of the superior environmental controls there. The message got through. Oil executives agreed to the concept at a meeting Rick set up with fishermen. Mead received a call from the White House demanding he set up a meeting between Steiner, Kopchak, and the vice president when Quayle made a symbolic visit to view the spill. (Mead, still doing advance work, helped stage-manage the visit, which I saw from the other end—on a troop ship in the sound, where obsessive efforts were made to prevent Quayle from falling down in front of the cameras, including construction of a boardwalk on the cobbled beach of Smith Island, where he could appear to walk among cleanup workers without running the risk of stepping on oily rocks.) Rick, with a shaggy beard and given to wearing baggy trousers held up with a leather thong, introduced himself to Quayle as Rumplestiltskin; but the meeting went off well and the idea for the citizens' committee won administration support. Legislation sailed through Congress to create citizens' committees for Prince William Sound and Cook Inlet, paid for by the industry but run by the communities. Senator Frank Murkowski pushed the bill along. A Republican normally known for a gold-rush-style environmental outlook, he called Rick in excitement and with a sense of shared accomplishment when the legislation passed. These citizen committees, with their paid staffs, have proved effective, especially the group in the sound, monitoring pollution prevention and fighting for expensive improvements from the oil industry. They stand as a rare victory in the fight of communities to win control of their waters, even though the committees' only weapon is information.

The third item on the agenda, saving the trees, depended on forcing Exxon to pay for the Native corporations' land or logging rights as part of its obligation to restore the ecosystem its oil had damaged. Pollution law recognized the concept of habitat protection as a legitimate way of using money the government won in such cases. Often, the only way to "restore" the environment is to prevent further harm. And no one could think of a better way of spending Exxon's reparations—no amount of money could put back what had been lost. Clear-cut logging on the sound's steep, wet slopes and shallow soil not only altered the coastal ecosystem dependent on old-growth trees, but stood a chance of erasing it permanently because of the damage to the land itself. But the litigation might last for years and the trees were already falling.

Rick went to Iarossi for help. They met in a coffee shop near the White House the January after the spill. Rick suggested that Exxon volunteer $150 million to tie up the timber rights until lawsuits could be settled with enough money to buy the land outright. Iarossi lent no encouragement to that idea, but toward the end of the meeting he said Exxon was negotiating with the U.S. Justice Department for a deal, a criminal plea bargain, that could provide a lot of money to buy trees. He made Rick guess the amount—$500 million—and left it ambiguous as to why he was telling such an important secret: to enlist Rick's help in structuring the deal, which seemed improbable but was his stated purpose, or to leak the information because the deal was a bad one for the environment. Rick investigated and ultimately decided to tip off the *Wall Street Journal* and Alaska's attorney general, Doug Baily. The settlement fell apart, Exxon was indicted, and Iarossi left the company within days for a job outside the oil industry. He never disclosed his true purpose in leaking to Steiner. Rick likes to think his friend was atoning for the spill by killing a sweetheart deal which, Baily said at the time, would have benefited no one but Exxon.

The incident won Rick prestige as an insider, but made the trees no safer. Clear-cuts were growing around Captain Cook's anchorage on Port Gravina and on Montague Island, and now in the rainforest paradise around Cordova itself, and, beyond the sound, in the big trees of the Kodiak Archipelago and on the Kenai Peninsula. Steiner kept working as if still in a crisis, with that intensity that comes in bursts but is difficult to sustain—yet it went on for years. He wrote his own plan for a litigation settlement to buy conservation rights from willing Native corporations, allowing them to continue owning the underlying property. He rounded up a thick sheaf of letters of support from organizations—communities, environmental groups, tourism outfits, and some Native landowners— and courted famous people as well. Over the years he won backing for the timber buyback from former president Jimmy Carter and actors Robert

Redford, Paul Newman, Ted Danson, and Alec Baldwin. Baldwin did a press conference with Rick and Patience Faulkner, who stood in as a Native representative. At breakfast before the appearance, while talking about the spill's impact on Cordova, Patience broke down sobbing, until Rick guided her out of the restaurant. Rick said, "This all led to Baldwin really laying into Exxon at the press conference—it was truly a thing of beauty."

At the end of 1990, a new governor took office, Walter Hickel, who, besides serving as Nixon's first secretary of the interior, had been Mead's mentor since his arrival in Alaska. Hickel said he wanted to settle the litigation and use the money to buy timber to protect the sound, where he had cruised for decades on a yacht named for his wife, Ermalee. Mead had first explored the sound on the *Ermalee,* too. After a year of drama, the state and federal governments finalized a settlement with Exxon for $1 billion. But then more years of delay began as the governments used the money to pay themselves and Exxon back for expenses, to set up offices, employ scientists, and establish a six-member so-called Trustee Council of federal and state bureaucrats to spend the money, which began lengthy rounds of planning, hearings, and studies without any action. It wasn't until 1994, almost five years after the spill, that the trustees approved a restoration plan, with policies for habitat protection. Loggers had already ripped through the area around Cordova. Chuck Monnett and Lisa Rotterman had lost their research contracts and were gone. Pink salmon prices had crashed and the 1992 run had failed.

A sense of dread hung over Cordova as that summer of 1992 ended. No one had made money and the winter loomed ahead to drain family assets even further. The town's moment on stage had ended; now its survival was in doubt. Rick's manner surprised me—no longer buoyant enough to fly, but angry, dark, and earthbound. "At every turn in the road it seems like they try to screw with us more," he said at the time. "They really don't want to do anything with this money." He had various theories about why this was happening. "I've become probably more cynical, more bitter, and more angry over the years. And it's not just that the oil spill happened, but that nothing constructive is going to come out of it. I've just been amazed at how bad our government can be. And, you know, I don't want to spend my life writing letters to federal judges or standing up yelling in Trustee Council meetings, telling them that they're all breaking the law—which they are."

But even in the sound not everyone wanted the trustees to buy the Natives' forests. Edgar Blatchford, then president of Chugach Alaska Corp., which had built a money-losing sawmill to process the trees, said Steiner was selfish. He said environmentalists had always been the enemies of the Chugach people, and the main harm of the spill for the corporation was that it had made environmentalists stronger. In Chenega, where no logging had ever been planned, Gail Evanoff believed selling the trees would be a setback to the rebirth of the village, to which her family had devoted their lives: "You're selling your identity. And we fought so hard for the land. So hard." Sylvia Lange, Mae Lange's daughter, had been a member, with Rick, of the Pursuit of Happiness Puppet Theater, but disagreed with him now on the timber buyout idea. She feared her children would lose their birthright: "They will not have the feeling of ownership," she said. "Land is forever. We were only just getting a feeling of pride and ownership, and what our grandparents went through. . . . We just received it back, and then they dangled money in front of people when they desperately needed it."

Cordovans faced losing their boats and homes. Rick, who had invested in a seiner and permit with partners, now feared bankruptcy. His relationship with the massage therapist he had lived with for years fell apart and she left town, getting away from its toxic emotions. The fishing got worse. The 1993 season brought herring too scarce and deformed to harvest and no pink salmon at all. On the grounds, talk boiled over on marine radios, and fishermen decided to block Valdez Narrows and shut down the pipeline. Tankers had to turn around. The Coast Guard threatened to arrest everyone. Rick took the role he'd had during the spill, calling oil company presidents and mediating between them and the fishermen (although Exxon wouldn't participate). The fleet's demands were for more science to explain what had happened to the fish—after spending $100 million on studies, the Trustee Council had no answers—and financial help from the oil companies. Rick advised the fishermen to back down when the companies seemed to meet part of their demands, but after the blockade was gone, those promises came to nothing.

Rick became isolated. Fishermen began opposing him and his timber buyout because they wanted the money for science or to shore up their industry. Organized environmentalists, who had gained influence at the Trustee Council table with the election of Democrats as president and governor, began to resent Steiner as too shrill and even counterproductive. He believed they cared more about the council's process than the habitat they had committed to save, trees falling as the bureaucracy ground on. They treated him as a crank. Loggers left threatening messages on Rick's answering machine. One called him out for a fight. Another confronted him face to face and demanded he leave town because he was trying to take food from his family, while the logger's wife and child looked on. A member of the university's Board of Regents with logging interests tried to get Rick fired. His mother died. He became ill. "I was ostracized," he said. "All throughout the town I became a demon. It had been such a wonderful place to live before all this, and it became a nightmare." He withdrew and avoided leaving his house, where he now lived alone.

When the oil spill occurred some of us felt profound, dizzying grief. For a moment, in shock, many seemed to share that high and terrible emotional space. But later, lines diverged, as some found a way to assimilate the losses into their daily lives, while others labored on, pain and indignation clinging together, as if to lay down the burden would betray the sound and the self of the past that had endured so much agony. At first, "getting over it" was for people who had been bought off by Exxon, and seeing it happen aroused new anger. But, later on, good, healthy people recovered, refusing to let one event define the rest of their lives. They included Rick's friends who had come to Cordova and stuck immovably there because of their love for the sound, and who had fought at his side, but who now found and followed strands of life that could lead them through the pain and on to new days. Even years later, his close friends didn't really know how hard a time he had.

Those who couldn't let go of the weight of the spill, as their numbers dwindled, learned to recognize one another. A writer, Marybeth Holleman, saw it in Rick. Her marriage was failing partly because her husband couldn't share that same tumultuous, unquenchable sadness. In an Anchorage convention hall she saw Steiner call out angrily among those who, with professional reserve, were making a career of science paid for by restoration money, while trees continued to fall. Eventually, she left her husband and began spending time with Rick. On a trip to Cordova, Rick picked up Marybeth and her son at the airport, thirteen miles out of town, in his jeep, which he drove so little that buying gas was a form of annual maintenance. On the way down the highway the car began to shudder and wobble. Someone had removed the lug nuts from one of the wheels.

An unlikely person in Cordova recognized Rick's depression and isolation: Margy Johnson, who owned the biggest hotel in town and led Cordova's prodevelopment faction, with strong connections to conservative state politicians. Margy invested nearly hyperactive enthusiasm in the town and her business and resented that the spill had robbed her hotel bar of its laughter. In May 1993, a beloved former mayor committed suicide, partly over the collapse of his businesses in the dead economy, and partly his regret that he couldn't do more to help the town; his successor in office, Kelly Weaverling, the bird rescuer, likened the entire town to wounded war veterans. But Margy rebelled, demanding people let go of their anger. She told a reporter, "The ultimate cost of anger is death." She won the next election over Kelly by a single vote (he refused a recount) and tried to organize a jazz parade to symbolically bury Cordova's blues. But she dropped that idea after an angry backlash. "I thought people were going to burn my building down," she said. "Vicious, vicious comments came out. People were just brutal. Absolutely brutal."

Margy knew Rick received comments like those all the time. She called the university chancellor to help him keep his job and took him out to eat regularly to prove to others he was still okay to associate with.

She and others organized a testimonial dinner for him—not held at her hotel, because some people refused to go there—keeping the plan secret from Rick to make sure he'd show up. Margy said, "It seemed abhorrent to our values that we were mean to one another. How can you be mean to Rick? He has such a pure heart."

Margy left Cordova in 2002, planning never to go back. Rick left in the mid-1990s, married Marybeth, and settled down in Anchorage, without his big circle of friends—he came to rely instead on his relationship with his family. With Marybeth, he shared the sense that he lived once in paradise, and now it was gone and impossible to recover, the falling trajectory of tragedy without the redemption of life's circle of rebirth. "There was this wonderful life there, and then there wasn't," he said. "All the things you take for granted when you're a happy person, all coming undone. . . . Families broke up, brothers were against brothers, sons against father, over who took Exxon's money. And then, when it all broke up, it was every man for himself. As Steve Picou says, it was a corrosive community. I like to fix things, and I couldn't fix it."

I asked Rick to go with me to Cordova in the fall of 2007, but he said he couldn't. He hadn't been back in more than ten years and even avoided taking planes that stopped at Cordova's airport. He rarely went out in the sound anymore, either, except to lead journalists to lingering oil, which remained in 2008, on the day the Supreme Court made its decision reducing Exxon's punishment. But when the decision came down, Rick sent Margy an e-mail, saying it might be for the best—although he had stood to gain financially by the punitive damage award—because now people would realize that companies like Exxon cannot be punished, and so must be stopped before they do their damage. I wondered if he was finally circling back to when we first began discussing how to restore Prince William Sound, almost twenty years earlier, back when he still thought the oil spill could change society for the better. Back then he had a broad idea of what restoration means. He wanted to buy trees, but he also wanted to spend some of the money on science education, so children wouldn't repeat their parents' mistakes.

"[The oil spill] was a symptom of a pathologically unsustainable society, and psychologically, for you and me, we need to do something about this," he had said. "We're asking Exxon to do the things we haven't been able to do ourselves, and that's create a sustainable lifestyle. We're asking Exxon and Alyeska [Pipeline] and the oil companies to restore us and make us whole. And they can't do that. We have to do that ourselves."

31

Understanding Life as a System

When the sound's herring and pink salmon fisheries failed in 1993 and fishermen blockaded Valdez Narrows, Exxon spokesmen were able to claim without contradiction that science couldn't prove oil was to blame. Yet the sound had been the focus of one of the most intensive government biology projects in history. The federal General Accounting Office investigated and found the Trustee Council was working without an overall plan, its members had allocated science funds to their own agencies for routine work they would have done anyway, and they hadn't required timely or accurate reports on past studies before commissioning new studies on the same subjects. Most projects had focused narrowly on damage to individual species. The passage of four years confounded those assessments by bringing changes in weather, ocean temperature, currents, and nutrients, changes in the many factors affecting the mix and abundance of predators and prey, as well as unrelated human disturbances, disease, and the chaotic fluctuations caused by the internal dynamics of ecosystems—in other words, the complexity of ecology—which escaped the frame of the two-dimensional, single-species picture. Linking cause and effect would require a holistic understanding of the ocean system. With the problem embarrassingly evident, the trustees resolved to change, and ordered a huge project, the Sound Ecosystem Assessment, to gather the pieces and start providing answers.

Fourteen years later, I joined a pair of scientists from the National Oceanic and Atmospheric Administration (NOAA) on a Prince William Sound science cruise funded by the Trustee Council's restoration fund. They were still trying to understand the basic ecosystem relationships around herring. One calculated how many herring were eaten by humpback whales. The other measured the fat content in the fish to track their energy use over the winter. They were energetic young scientists, new in their careers, driven by the excitement, which fishermen and marine biologists share, to reach into the mysterious community of life hidden in the other world below. We pursued humpbacks through the mountain corridors near old Chenega, dipping nets for their feces and the scales of partly consumed fish, and we jigged lines into herring schools to catch for fat content analysis. A deckhand fishing along with the researchers muttered, "This is science?" The trip was fun. But I felt old. The questions hadn't changed. The goal of understanding the entire ecosystem was only a little closer than when I first heard biologists preach for it soon after the oil spill—at which time the principal investigator on this herring fat study was in middle school writing about the *Exxon Valdez* in a composition book. Trustees now had spent roughly half their $1 billion settlement fund on the conduct and administration of scientific studies and still couldn't say for sure what was wrong with the herring.

This matters aside from the oil spill—which is, after all, beyond anyone's ability to change—because comprehending marine ecosystems and being guided by their limits is fundamental to saving the ocean. We use the ocean to produce food and disperse pollution, we appropriate ocean habitat, and we rely upon the ocean as our greatest buffer against climate change, as it absorbs heat and carbon dioxide. We cannot live on Earth without affecting the ocean and being affected by it. If we choose to share existence with other free-living organisms, we will need to define and fit within a sustainable ecological role.

Managing individual species doesn't work. Fishing depleted many of the world's fish stocks even though catches were limited based on scientific studies of the targeted species. Traditionally, managers allowed fleets to catch excess fish, defined as the percentage of the total that could be removed without affecting future fishing productivity. But precise measurement of fish numbers—and thus the supposed excess—is next to impossible. And even if abundance were knowable, the percentage that can be called excess changes according to oceanographic and ecological conditions that combine unpredictably in an infinite number of ways. The rules often change without warning.

The demise of Prince William Sound's herring illustrates this challenge—at least, if you accept the newest and most plausible hypothesis about what happened. Although obviously exposed to oil and in some cases deformed by its toxicity, herring continued returning in quantities considered adequate for fishing during the spill year of 1989 and for three seasons more. Then the stock crashed in 1993 and it has not recovered. Most scientists assumed Exxon was not to blame because of the four-year lag from the spill to the fishery's collapse—they looked for the cause instead in 1993 and the years that followed. But now it seems more likely that oil reduced the hardiness of the herring while warmer waters caused by climate change made it harder for their fat reserves to last through the semihibernation of winter. Skinny, oil-weakened herring would be more vulnerable to disease, starvation, reproductive failure, and other threats, changing the rules about how many fish it was prudent to harvest each year. Even if total numbers of fish seemed adequate, the excess number had invisibly declined. In four years the weakened stock was fished out.

But by the time ecosystem studies started, it was too late to know for certain if that explanation was correct. Scientists had called for broad ecological study of the sound from the start, but during the litigation, before the resource agency trustees took over, the decisions about oil spill science were made by lawyers. Their roles in the adversarial system led to a poorer outcome for all. To support the lawsuits, attorneys wanted studies to measure the immediate damage of the oil spill and vetoed long-term studies or ecosystem approaches. The scattershot science program further suffered from being confidential—because of the litigation, specific findings and even general information about the spill's harm were for years kept secret from the public or outside scientists who could have theorized, planned further work, or critiqued the process. Then it was too late. Without coordination in collecting the data it was impossible to pool information to find patterns or test new hypotheses. With these flaws in the initial work, it's hard to see how the oil spill science program could ever have developed a comprehensive picture of the ecosystem.

While the cost of this secrecy and poor planning were enormous, the benefits for the lawsuits are hard to identify. Lawyers made little use of the data the scientists gathered for them. For their purposes, measuring the spill's harm was always less important than how to put a price on that harm. To get big numbers, the attorneys needed contingent-valuation studies, the economic surveys asking Americans what the lost resources were worth. However, in preparing those surveys, the attorneys, in their caution, used almost none of the results of their scientific studies—for example, they told respondents the number of dead otters that had actually been found, not the true casualty count revealed by the science, which was eight times higher. Then they discarded the resulting contingent-valuation numbers as well, choosing to settle the case for an essentially arbitrary figure of $1 billion. We don't know how much of that money would have been sacrificed if scientists had been allowed to design a public, ecosystem-spanning coordinated program from the start. Probably none, and certainly not half—which is the cost of the studies that were so crippled by the secrecy.

Rick Steiner believed little should have been spent on science at all. We already knew it was harmful to dump thousands of tons of toxic crude oil into the marine environment. But advocates said the government had a duty to track the recovery of species, if only to respond to Exxon scientists who quickly reached the conclusion that everything was fine. The government's researchers also had to defend themselves against Exxon scientists' unfounded accusations of bias and impropriety, and to dispel improbable theories propounded to cast blame away from the company. (Scientists on both sides had good credentials and worked for respectable institutions—the difference in their conclusions correlated mainly with who was paying the bills.) After the litigation ended, the trustees' science was supposed to guide restoration plans, but there don't seem to be many examples of that happening. Of the many shelves of reports produced, a small minority are of enduring interest. At a recent symposium a new project was proposed to mine the useful information from that vast trove.

Science itself could be a powerful form of restoration if it produced results that allowed humanity to interact more harmoniously with the marine ecosystem, and if people then heeded those results. Some science like that did happen. The most important new knowledge to come out of the spill emerged from the understanding of how low levels of chronic pollution could reorder and diminish an entire ecosystem through a cascade of effects on many species. The critical initial discoveries came from researchers led by Stanley Rice, of NOAA's Auke Bay Lab in Juneau, including Ron Heintz and Jeff Short. Rice goes by the name Jeep, and he is jeeplike: solidly built, rugged, practical. Four years after the spill, he and his colleagues set out to test whether oil was holding down pink salmon numbers in streams where so little contamination remained it wasn't measurable in the water—but where mortality remained so high that people had begun doubting the data documenting their demise.

The accepted understanding had been that the most toxic parts of spilled crude oil, such as benzene, last only days, because their light, one- or two-ring molecular structure causes them to evaporate quickly. The heavier, polycyclic aromatic hydrocarbons, called PAHs, with three, four, or five rings, stayed around longer, but were thought not to be very poisonous. Heintz and Short designed an extraordinary experiment to test this assumption. They sprayed a tiny bit of weathered oil into gravel tumbling in a cement mixer. When the rocks came out, they had no perceptible oil on them. The team then collected and fertilized pink salmon eggs over the treated gravel. The tiny fish burrowed down into the rocks before growing large enough for release into the sea. Two years later, when the salmon came back, they were small, abnormal, and less likely to

reproduce. A significant number didn't come back at all. At concentra-
tions of five parts per billion—the equivalent of less than a tablespoon
of oil in an Olympic-sized swimming pool—the fish suffered a range of
sublethal effects that weakened them enough to affect their population.
"We didn't believe it," Jeep said. "That's why we repeated it three times."

The toxicity of long-lasting PAHs helped solve the mystery of why
various species failed to recover as expected, vindicating those who insisted that the sound didn't seem as abundant as before the spill. Poisons
dragged down the Prince William Sound ecosystem for many years. Herring eggs sustained the damage after sixteen days of exposure. Mussels
and clams concentrated the contamination from the gravel around and
beneath them. Otters and birds got it from the mussels and clams. Those
animals, even if born after the spill, tended to be smaller, to die younger,
and to underpopulate their habitat. Rice's team went looking for the
source of the old oil and found that natural weathering and dispersion
had slowed after the first few years, especially under mussel beds and in
other protected spots. A survey led by Jeff Short in 2001 estimated sixty-
one tons of oil still in the beaches. At that rate, oil would remain for
thirty years.

Targeted cleanup to expose pockets of old oil to the air probably sped
up recovery. But a more lasting outcome of Rice's work is its implication
for other estuaries, such as Puget Sound or Chesapeake Bay, where urban
runoff constantly introduces PAH contamination as serious as the *Exxon
Valdez*'s buried oil. Most urban estuaries are struggling biologically. The
oil that runs off a parking lot in a rainstorm may be enough to prevent an
ecosystem's recovery.

I last talked with Rice in a cafeteria in the Juneau federal building during a lunch break in yet another meeting on ecosystem science. A federal
funding agency had drawn researchers together to discuss creating a computer model of the entire Gulf of Alaska ecosystem. In theory, the model
would simulate the gulf as one, so that fishery managers and other resource
agencies could test how its entire fabric would be affected when they pulled
on any single strand—the goal of ecosystem management. I'd first heard of

ecosystem science at a similar meeting seventeen years earlier, also with Jeep, the first gathering of oil spill researchers, which was convened in Cordova by its new science center, then still being born. That meeting had been full of passionate calls to link studies of the water column to plankton, to sandlance and herring, and up the food chain, spanning times long enough to find patterns. Scientists reported that their lawyer bosses were canceling projects that could do that, and the discussion itself was stunted by gag orders. Now, two decades later, in Juneau, with all that money spent and the new effort underfunded, getting excited again about the same topic wasn't easy. Jeep seemed disengaged and, although he was under orders to attend, he would likely slip back to his office when the lights went out for the next PowerPoint.

Problems remained that had existed all along: agencies funded research of commercial fish species and marine mammals, while far less was known about climate, oceanography, plankton, and forage fish. The barriers remained of bureaucracy and interest-group politics, scattered research goals, and the sense, as during the oil spill, of a disconnect in scale between anything that happens in these meeting rooms and the immense mysteries one senses watching schools of fish roiling the surface of the water. Most of the scientists' presentations were pleas for more money for their own work.

One researcher had been trying to build a computer model of nutrients and phytoplankton production along a line in the Gulf of Alaska, and he gave an unintentionally funny talk. A simple plankton model made a good place to start, with a constrained area and fundamental relationships—the bottom level of food production—and real data from along the line that would allow checking the model's results. The computer ran numbers representing physical aspects of the ocean—the temperature, salinity, nutrient runoff, and so on—and spat out results for plankton, which came out all wrong. Where was the problem? Could it be the penetration of light? The stability or density of water layers? Each time the researcher had jiggered a number, the output went out of whack in a different direction. Other scientists at the meeting started calling out suggestions, like amateur mechanics trying to fix a car.

I doubt a control-room view of a complex marine ecosystem will ever be possible. Software is for sale to build ecosystem models, but the most advanced research still gropes to understand the theory of food webs with simplified, idealized systems. If the drivers of change are themselves unpredictable, then predictive modeling is bound to fail. For example, a top resource user could change food preferences for its own reasons—and not only human beings. Sea otter numbers crashed in the Aleutian Islands beginning in the mid-1980s, with only 10 percent remaining in some former strongholds, and many scientists believe the missing otters were eaten by killer whales that switched their favorite food.

Craig Matkin, the killer whale researcher who paid for his own studies in Prince William Sound by fishing, had shown how a few important whales could affect an ecosystem. He became a scientific star during the oil spill because his familiarity with individual animals allowed him to make definite statements about how the oil had affected their pods. After the spill he could have had all the funding he wanted, from the government or from Exxon. He remembered being followed around the sound by a large Exxon-chartered vessel with an expert chef on board and repeatedly invited over from his seiner, the *Lucky Star,* and plied with doughnuts. When he received an unsolicited check from Exxon for $280,000, he sent it back with the message "to shove it as far up their butts as possible." He kept autonomy from government research agencies, too. Despite holding contracts that funded four boats at once and gathering a team of researchers in various subspecialties under his North Gulf Oceanic Society, Craig continued seining for salmon for five more years.

Matkin and his team found that elder females in two pods disappeared during the spill. Without those matriarchal leaders, their killer whale tribes fell into long-term decline. The social breakdown of the AT1 pod damaged its reproduction and recovery permanently, far beyond the actual number of animals potentially killed by oil. Chemical contaminants and the scarcity of harbor seals probably played a part, too. The pod no longer inhabits its former hunting grounds. Only eight of twenty-two are left. Killer whales eat sea otters, sea otters eat sea urchins, and sea urchins trim the kelp forest, which is a critical habitat for a range of creatures. Theoretically, the deaths of a few individual killer whale leaders could alter an entire coastal ecosystem by allowing more otter survival, just as killer whales' increased appetite for sea otter could decimate them and swing change in the other direction.

If such small changes bring huge effects, the system is chaotic and fundamentally unpredictable. Ecosystem management therefore is doomed, at least if its purpose is to allow us to continue using the ocean as a machine to produce as much material wealth as possible. To succeed, ecosystem management has to mean something different than giving new tools to the same political institutions that serve money and power. We need new social and economic relationships for humankind to fit adaptively into the network of species as a part of the ecosystem, guided by the voices of communities and individuals with values that cannot be monetized, and ruled above all by caution and humility before nature's complexity. Everyone is responsible; everyone should be empowered.

An intellectual movement expounding such a way of living already is afoot. Success may seem improbable, but even cultural transformation is more likely than achieving ecosystem management under the old paradigm, with scientists comprehending everything and bureaucrats turning the dials of fishery management, pollution control, habitat use, and every other human input to precise settings. In the words of a pair of resource scientists, Hanna Cortner and Margaret Moote,

> Under ecosystem management, the roles of scientists and managers are redefined from expert to educator, public relations specialist, technical advisor, or some combination of these. Correspondingly, the role of the citizen also includes resource management, for under ecosystem management, all

citizens take responsibility for achieving ecological sustainability.

We don't need to model the marine ecosystem to treat the ocean better.
More knowledge would help, but the basics are simple—as Rick Steiner
explained it, put less bad stuff into the ocean, take less good stuff out. We
need to be gentler, to listen, and to adjust our actions as we see their results.
To do these things, we need to restore our sense of touch, moving close
enough to nature to appreciate its scale and our role there. Stop poisoning
the fish and animals, for example, with tanker spills or oil drips from under
cars. Stop stripping coastal rainforests of old-growth trees. Government
bureaucrats running computer models can't do this for us. The change
must come first in our hearts, then in our relationships, and finally in our
politics.

32

Hearts Connect to the Coastal Forest

In the fall of 1989, the oil spill year, after Mead Treadwell brought Carol to Cordova to track otters, they next flew by float plane to Loonsong Lodge, in the mountains above the south side of Kachemak Bay, in that steep, private forest valley with its deep lake and its glacier, surrounded by great spruce trees robed in dark branches. They were alone together for days, gazing from bed through the big window Mead had helped install with his friend, the owner, Michael McBride. The tall spruces seem so knowing: from a distance like ancient monuments whose carved faces lie just below the bark, waiting to be revealed, but close up, among their trunks, emitting a scent of sap, of growth, as vital and irrepressible as a rising sexual climax. And below that scent, the odor of rot, as the ecstasy of love inevitably pairs with premonitions of mortality. In such a place, amid the continuity of age-old wood and the bursting tips of fresh needles emerging, the idea of mating for life acquires profound emotional logic. And there Mead and Carol's hearts confirmed their unbreakable attachment.

But after they left, the trees remained under threat of clear-cutting. Although the forests of Kachemak Bay State Park were protected according to Clem Tillion's plan in 1971, when the Halibut Cove patriarch's power was still rising in Juneau, the Alaska Native Claims Settlement Act later opened the area for selection by a corporation representing the bay's indigenous people. Seldovia Native Association chose 23,800 acres on the south side of Kachemak Bay from within the state park. A village lay in this forest once, somewhere near China Poot Bay, a wide basin of shifting sandbars and tide rips, indistinct in silver light, populated by salmon when the water is high, and, when drained by the ebb, by birds with narrow beaks picking their way across the mud. Russians kidnapped the entire village and forced its people into service elsewhere. A collector for the Berlin Museum passing through in 1883 stripped the remains of the abandoned site without recording its location. By the time of a more enlightened archaeological survey, in 1930, the village had slipped into legend and no one could remember where it had been or what had become of its people. With the weird justice of the Native claims' role reversal, Seldovia Natives, probably unrelated to the lost village, now planned to level the forest and extract its economic value, to the dismay of relative newcomers for whom the trees and seashore had become engrained, with fervent wonder, in their lives' sense of meaning.

One such was Michael McBride, who arrived in 1966, made friends with Clem and Diana Tillion, and like Clem and Will Langille before him explored the bay in an open boat, cove by cove, finally settling on a wooded isthmus that forms one side of China Poot Bay, with huge, mossy trees and a square depression amid the ferns where an ancient Native home once stood. Michael's father had been a World War II hero and he himself arrived as a member of the air force, not a rich man, but he had the skill and sensitivity to build a lodge with his own hands that integrated into its setting as gracefully as the huge, slowly rotating mobiles of driftwood, shells, and feathers that he hung from the trees above. Michael's wife, Diane, gave her warmth, taste, and practicality to the business, and he contributed his charisma—his joyful, enthusiastic personality, and the sense of deep significance he brought to every conversation with his eloquence and steady brown eyes. Instead of bringing guests to hunt and fish, like almost every other wilderness lodge in Alaska, Michael tramped them through tide pools to inspect odd little animals and eat seaweed, and over forest trails to pick berries and learn the names of plants. A *New York Times* travel piece in 1978 and international press that followed made the Kachemak Bay Wilderness Lodge famous. The McBrides added the mountain lodge, Loonsong, and a bear-viewing camp. Michael led the CEOs of the world's largest corporations through the tide pools, and also powerful politicians and European royalty. He returned visits on some of his new friends, taking his family to Europe to stay in their historic palaces. He also used his contacts to become a nationally known environmentalist and lecturer, a member of the Explorers Club and of the Smithsonian Board of Trustees.

But the McBrides seemed unable to stop the impending destruction of the forest. The original creation of the park and the buyback of Kachemak Bay oil leases in the 1970s had come about through the political magic of Tillion's legislative power and Jay Hammond's governorship, aided by the money from the early oil era and the public reaction to the excesses of uncontrolled development; but no such fortunate moment blessed the forests when the tree cutting on the south side of the bay progressed and doom approached land within the park. The Seldovia Native Association didn't want to log the park: they offered to trade the land to the state government in return for property elsewhere of equal value, but state officials never offered what the Natives or local park activists thought fair. Later McBride and his allies lobbied the legislature to buy the land, but failed. Oil prices had crashed and the state was short of money. McBride organized people around Kachemak Bay and his wealthy associates from outside Alaska, raising enough money to hire a full-time staff person to work on the issue. The oil spill hit, tar balls rolled into the bay on the tide rips, and people started talking about using Exxon money to buy Seldovia's land. But time kept passing and nothing happened—just secret science and legal fees, a million dollars a month spent on lawyers on the state's side.

The Seldovia Natives finally sold their holdings to a timber company that had made huge clear-cuts all over North America's Pacific Northwest. Brightly colored tape showed up on big trees, marking the future cuts. Chainsaws fired up, demonstrating the company was serious, felling trees near the McBride's lodge. The loggers' plan called for stripping the area around China Poot Bay, with a road to be punched in from the shore.

"It was like this war that was being waged against this peaceful nation of people who had never done anybody any harm, which is the nation of trees," Michael said. "In a way, I felt like my father at Bergen-Belsen. There was this horrible thing happening, and no one seemed to get it."

Mead and Carol married the summer after the spill, 1990, and that fall the sponsor of Mead's career in Alaska, Walter Hickel, suddenly decided to run for governor under the flag of a fringe political party, just six weeks before the general election. Hickel was already an old man and often pegged as an outdated, gold-rush-style boomer, but his huge personality tended to overwhelm his politics. He had been born a poor sharecropper's son in Kansas, dyslexic and always reliant on a prodigious aural memory, and worked his way to Alaska before World War II as an itinerant boxer and brick carrier. After failed starts, he hit it big in real estate and construction in Anchorage, repeatedly plunging everything he had on larger and larger bets, and winning each time because the new state was exploding economically, until he became Alaska's biggest rich man and Republican politician. He was elected governor in 1966 and Nixon appointed him secretary of the interior in 1968, perhaps expecting a typical anticonservation Westerner, but Hickel's sense of destiny and love of grand gestures made him more than unpredictable: among other actions, he listed the great whales as endangered species and he wrote Nixon a public letter taking the side of protesters against the Vietnam War. He was fired and sent home, where he and Hammond fought each other for the soul of Alaska through the 1970s, Hickel for bigness and wealth, Hammond for slow growth and conservation. Mead, a recent graduate in 1978, served at Hickel's side in a titanic battle for the governor's office, which was decided in Hammond's favor by a few votes in the Alaska Supreme Court. When Hickel finally did get back to the governorship in 1990—through the force of his inspiring grandiosity, his fortune spent on a blitzkrieg campaign, and his extraordinary luck—the state's culture and politics had transformed, and no one really knew what might result from intermixing Hickel's volatile chemistry.

Certainly, he would have no patience with drawn-out litigation. Hickel adopted the idea Rick Steiner had been peddling of a quick legal settlement that could be used to buy trees. He knew McBride and had originally put Mead together with him. Hickel felt a deep link with the sound, which he had explored on his boat—he told me of a psychic connection he had once experienced with a black bear, which moved to safety from hunters seemingly at a warning sent by the power of his thoughts. But as governor, Hickel appointed a cabinet of men without his imagination for the wilderness, men from his own generation, the generation that had made Alaska a state, that had settled Native claims on the corporate model and pushed through the Alaska pipeline to Valdez, and that didn't believe in buying the Natives' land back again. In his own office, Hickel brought in budding young Republicans like Mead had been twelve years earlier; one of them used the governor's personal letterhead to write a snotty letter to Steiner suggesting the town of Cordova itself be bought out and reclaimed as well as the forests.

Mead refused the role of a young aide in the governor's office and ended up in the second tier of authority, as a deputy commissioner handling oil spill issues. In an administration thick with intrigue, he had the job of approving millions of dollars of oil spill response spending, but he didn't have control of the levers of power—his access to the Trustee Council was through his boss, the commissioner of environmental conservation, who was one of the six council members. The attorney general, Charlie Cole, was the real policymaker, responsible for settling the lawsuits. Cole, a brilliant and charming small-town lawyer, resident of Fairbanks since 1954, had left a one-man office on that town's skid row to negotiate the $1 billion deal with the attorneys from Exxon and the federal government, and then dominated the Trustee Council as its most opinionated member. It was Cole who insisted on a careful plan before buying trees, a slow, cautious approach, and driving a hard bargain with the Natives corporations.

As winter ended in 1991, Mead told Michael that his new wife was pregnant, and Michael suggested using the clear water from the spring at his home on China Poot Bay for a baptism. Michael wanted the ceremony in the cathedral of old-growth trees. When the Treadwells' son, Tim, was born in May, Mead and Carol brought more than two dozen guests from all over the country to the lodge. Michael was a godfather. Carol's spiritual guide also came, a Catholic priest from Chicago, who agreed to perform the baptism among the trees despite the pagan implications of the location. The ceremony took place on deep, damp moss in front of a spruce so large it would have taken at least three people with outstretched arms to encircle its trunk, and surrounded by ferns, tiny raspberry blossoms, and the huge, hand-shaped leaves of the devil's club. Carol held the baby and beamed an enormous, toothy smile. The next day, Michael and Mead went for a walk alone in the same woods, carrying mugs of hot coffee, and Michael revealed that the tree where they had stood had been marked for cutting with a ribbon that read TIMBER HARVEST BOUNDARY.

Michael said, "It was sort of a trick on my part. It was very deliberate. I really wanted it to happen here, under the trees."

Mead said, "That ribbon is what made me realize how special old growth is."

This is how it has to work: when we're moved to action by a love beyond ourselves. The economic value of these places or their material services to humanity will not protect them. Forms of value recovered from the deep past instead come to us in personal moments of recognition, when something more important than fortune or power makes itself irrepressibly evident. This is the alchemy that can dissolve the chains of selfishness and create new social norms for sustaining wild nature.

Mead became a campaigner within the Hickel administration to hurry up and spend *Exxon Valdez* money to buy the trees in Kachemak Bay State Park before it was too late. Early in 1992 a bill worked through the legislature that included about half the $23 million needed to buy the land on Kachemak Bay, and also money Steiner wanted for land near Cordova and Tatitlek, and smaller appropriations to tie up parcels all over Prince William Sound, and besides that, a grab bag of little projects, some marginally related to the oil spill, which legislators inserted through the process of exchange that goes on while rounding up a majority to spend money. Mead's work behind the scenes made him notorious among administration colleagues. Although Hickel wanted to buy the land, and the administration nominally backed the bill, those around the governor didn't. Charlie Cole asked members at a cabinet meeting who at the table favored buying habitat with oil spill money to stop clear-cuts and only the governor himself agreed. When the bill did pass, Hickel's advisers shut Mead out of discussions about a possible veto. A letter from Cole to Hickel savaged the legislation, mixing legal and policy arguments, stating, "While it may seem superficially plausible to purchase habitat," any such purchases should be minimized and subjected to the kind of exhaustive review the trustees were still carefully planning.

After a dinner at Mead and Carol's house in Juneau that summer, a young Hickel aide let slip, while waiting for a cab, that the bill would be vetoed the next day. Mead stayed up all night and met his boss early the next morning at the Shrine of St. Térèse, a lovely Catholic chapel built of beach rock that sits on a tiny island north of Juneau, among huge rainforest trees and tide pools. Together they went to the governor's office, but were blocked from entering by Hickel's chief of staff, who held up a hand saying, "Mead, don't even try, the decision has been made." A furious argument erupted on the spot, voices raised. Hickel emerged. Mead recalled, "We were standing in front of the receptionist's desk having this fight, having this catharsis, and the benefit was to have the governor say, loudly and clearly, 'We're going to do this [purchase].'" The veto stood, but the administration's old men got their marching orders from Hickel to find a way to buy the trees, although how they would do it remained in doubt.

The news of the veto hit coastal communities like a bomb: in Cordova, Steiner and his friends were devastated and the mood turned more bitter; in Kachemak Bay, disappointment and fear, as the cutting had already begun. But with Hickel's reinforced backing, Mead participated with Cole in planning another oil spill legal settlement that produced an agreement that would direct a third of the money needed for the Kachemak Bay timber buyout. The next winter, with that secret agreement in mind, Cole suddenly changed his position with the Trustee Council, proposing it immediately allocate another third of the cost of the purchase, regardless of the lack of studies—shocking the other trustees, who nonetheless quickly adopted the reversal of policy. The legislature provided the final third of the money and the sale was completed in the summer of 1993.

The story ought to end happily there, but this is real life—not so simple. Mead and Carol had another child, Jack, and planned to christen him on the shore of Kachemak Bay, as they had with Tim, but he died in his crib at nine days old. They went anyway, with all the friends who had already planned to attend, and held a memorial weekend instead among the trees that Mead had helped to save. And then most of the trees died. An infestation of spruce bark beetles that had started at the head of Kachemak Bay swept down, over several years, through the length of the state park. The beetles had always been there, but a two-decade series of exceptionally warm springs and summers, unprecedented in weather records, had allowed their reproduction to explode. The massive blight became one of the first documented biological impacts of climate change. (Not that the buyout was a waste: it did spare the park the building of roads, erosion, and the other wounds of logging. Also, the beetle did not affect the damper forests of Prince William Sound.)

Intrigue continued in the Hickel administration, but elections put Democrats in the presidency and governor's office, and their pro-environment perspectives reoriented the Trustee Council. People who wanted to buy habitat took control and, although still saddled with an unwieldy system—for example, every decision on the six-member council had to be unanimous—large habitat purchases slowly but steadily went forward. The trustees discarded Cole's policy of requiring Natives to give up all rights to their land, instead purchasing only conservation rights to protect the forest. With less tumult, more got done. Bureaucrats who worked huge land purchases through thickets of rules and problems deserve credit for many more acres than were saved during Hickel's term.

The land sales again split Native communities. Native corporation shareholders approved the sales by huge margins, but many villagers didn't have stock, or votes, because of where or when they were born, and often opposed the loss of traditional lands. One village on Kodiak Island kicked out its corporate board and split up the proceeds from the sale of its lands, producing $200,000 per-person payments that some blew on short-term purchases. Most corporations put the money in trust or used it for business development, sometimes with spectacular success. Tatitlek finished clearing the trees on Port Gravina around Captain Cook's anchorage, and for miles farther, before selling the ruined land to the trustees. Rick Steiner's priorities came last. A deal finally took ownership or conservation rights for 75,423 acres around Cordova, including the habitat trustee scientists had ranked the most valuable in the sound, concluding the transaction almost a decade after the spill—long after some of the area's most beautiful vistas had gone under the saw, and long after Steiner had moved away, planning never to return. Mead Treadwell likewise won a bittersweet victory. Although he helped save the forest where they fell in love, Mead couldn't save Carol, who died in 2002. Five more years passed before he again convened some of the friends who shared her final days, and their children, at a wilderness lodge in Prince William Sound.

Commercial logging has ended in the sound. In total, the spill money bought protection for about 650,000 acres covering 1,400 miles of coastline, most of the Native land in the region, an area comparable in size to some of the famous national parks and an accomplishment perhaps as extraordinary as creation of any of them, since it required expenditure of real money, not merely setting aside lands the government already owned. The money tangibly declared the value of nature. But not unequivocally. The trustee's expenditure of $370 million for habitat (augmented by $56 million from sources such as Exxon's criminal restitution) was just over a third of the $1 billion recovered from Exxon, which was itself a fraction of the value Americans had put on the spill-damaged resources in economic surveys. The rest of the settlement money had been spent on science and administration or reserved, with relatively little to show for it.

Forgetting the money and politics, however, and forgetting the absent herring, seals, and shellfish, and the rest of the screwed-up ecosystem, I look upon these wooded islands—looking upon them as Will Langille did, a century ago, and perhaps as the ancients once did, when they regarded the wily raven in a high spruce tree—and the knowledge that the forest will not be destroyed gives me a profound sense of safety and of hope. The money ultimately doesn't matter nearly as much as the heart. The heart heard. The heart survived. The human heart, sensing the meaning in this place and sensing its spirits, whatever their names—it survived the maze of history, the wounds of oppression, the numbness of wealth. Pulsing within many different kinds of people, suffering endless obstacles and regrets, it continued to love the sea, the forest, the creatures that cannot, really, feel as we do. People who never knew each other formed a community around this place, and gave. We were still alive, and we still had hope to stop short of hollowing out the oceans.

Part VI

CONCLUSION: THE PROBLEM
AND HOPE

33

Choosing Good (Beach Cleaners Part 1)

Penetrating spring sunlight picked out the ripples and miragelike undulations on the broadest part of the sound, east of Peak Island, where the water is so wide the mountains on the far side shrink into the horizon even on a cloudless day. Morning had already warmed the shore's large, round cobbles when a dozen middle-aged men and women climbed from skiffs, wearing life jackets and backpacks and carrying yellow plastic garbage bags. Most likely they were the first people to land here this year, or perhaps in several years. This rocky amphitheater resembles innumerable others that scallop the sound's shores, cobble beaches bounded by bedrock cliffs and backed by thick spruce forest, and there's rarely a reason for anyone to land. Voices thinned against the big quiet as the people spread, picking their way over a wreck of silver drift logs above the tide line, each finding his or her own spot. Antonia Fowler lowered herself into a pocket of tall beach grass in a dip among the logs, the lower half of her body seeming to merge with the bleached stalks. She looked like a berry picker setting to work in a thick patch, reaching rhythmically into the dark gaps among the logs. She came up with a fishing buoy, a chunk of Styrofoam, an empty water bottle—just trash—and filled a garbage bag.

Within an hour a huge pile of bulging bags accumulated, large items stacked nearby—plastic buckets, hunks of tangled rope, a tire, a television set. Which was extremely odd, because, from the water, this beach had looked pristine. The debris had been nearly invisible even standing in front of it. Over a lifetime plastic garbage had worked its way into the mind the way it settled into the gravel and mussel shells: a yellow line entangling a tree branch or a Clorox bottle encrusted with barnacle husks became as much a part of the landscape as the dried seaweed. Only when the plastic was gone did it become visible, in its absence, like an irritating noise that lasts so long it fades from consciousness until, suddenly, it stops. Muscles relax; the memory of silence renews. The beauty of the beach springs out again, brand-new.

Antonia rose from her patch with effort and with a huge smile. Cleaning a wilderness beach feels like an act of creation—that's how different it looks when the plastic is gone. For those who love this feeling, the work is addictive. Antonia had sought out the organizers five years earlier when she first heard about their annual spring outing and had never missed it since, each year bringing with her friends who she knew would work hard. At the dock this spring they met others seen once annually, with hugs. A boat dropped their group on Naked Island on Friday; they cleaned a beach there, camped out, sat around a fire, and Saturday morning were brought to Peak Island to clean. Antonia is an outdoorsy woman, with big brown eyes, freckles, and a body that looks strong, but she already seemed tired. She has multiple sclerosis. Sitting on a log, she wrote in a field notebook to record the location of the garbage pile so a U.S. Forest Service boat could pick it up later. Word came over a handheld radio that our crew would climb over to the next beach down the shore rather than waiting for the skiffs, which were redeploying other volunteers on other beaches. Someone led the way, beyond the grass into the mossy trees, where a steep, crooked game trail led toward the top of dark, bedrock cliffs standing like the island's buttresses.

Antonia climbed, but then stopped and settled back into the moss, letting others go ahead. She had chosen a lovely spot, where sparkles of sunlight shot up from the water below the cliffs, through the tree trunks and shadowy green. But in fact she hadn't chosen—her legs had stopped working. A friend asked, "Did you take your meds?" Antonia said, "Yes, I did. Did you?" She tried to wave people on, but several stayed with her. She is a physical therapist and some of her friends were nurses and other medical professionals. Eventually she emerged at the next beach, with help getting over steep rocks and slippery vegetation, but those going before had already gathered up the junk there—a sleeping pad, more fishing floats, a rope as thick as an arm, buried and intertwined with logs, and, as always, plastic water bottles. Antonia passed that pile to catch up with where the work was still going on.

A swarm of kayaks with younger beach cleaners worked around the southern end of Naked Island. A doctor and his friends cleaned in the passage between the islands. A couple of boats picked along the east side of Naked Island, led by a marine biologist and his family. The fifty-four-foot *Johnita II,* the flagship, floated in the channel, Zodiacs and skiffs touching at its aft deck like bees at a hive. Chris Pallister stayed on board, directing the fleet and shore parties by marine radio, frequently losing contact, struggling to communicate about the complicated, unnamed geography, tracking who was stranded where and for how long, trying not to waste volunteers' time with waiting, although that was usually impossible. He was thin and physically stiff due to a back injury, which he said wouldn't allow him to sit for long, but he also seemed completely uninterested in sitting, as he paced his deck and concentrated on the radio and chart. Pallister quit practicing coastal management law to run his own beach cleanup nonprofit for a salary of $2,000 a month. He had organized one hundred volunteers for the weekend—around eighty actually showed, although no one seemed to have an exact count—and he had lined up the boats to carry them forty miles from the nearest town, making sure each worker had safety gear and everyone pitched in for fuel and food, and he had mapped the messiest beaches so they could work productively. He had coordinated with land agencies, funders, environmental scientists, and once anchored between Peak and Naked islands, he dispatched boats, counted bags, and fixed broken equipment. He seemed to thrive on constant aggravation.

Pallister and a few friends started picking up marine debris on spring weekends in 2000. After four years they had done 70 out of 3,500 miles of shoreline in the sound and calculated that finishing the job would take two hundred years. That was when Pallister and his friends, Ted Raynor and Doug Leiser, decided to devote the rest of their lives to the project. They raised money to hire Pallister and Leiser's college-age sons for the summer and began working week after week without breaks, in all weather, beach to beach, island to island, sleeping in the close quarters of a small plywood boat called the *Opus,* so named because it looked like the ungainly penguin from the comic strip *Bloom County.* In one summer they covered 350 miles. But the ocean is full of plastic, from towns, fishermen, and boaters, from ships on the high seas, from big-city storm drains, and from mysterious Asian sources: many items have Cyrillic, Korean, or Japanese lettering. A beach that's cleaned once picks up a water bottle or fishing float with the next tide.

"I'm not going to be discouraged by that—we just have to make this international," Pallister said. "By and large the people on this thing are optimistic, and they really love this place, and they want to do something."

Everyone who cleans the beaches and who catches the addictive, creative feeling emanating from a newly clean beach also experiences contradictions entrained with these good emotions, problems that probably feel different in the chemistry of each personality. To stop drinking bottled water and eating Styrofoam cups of instant noodles? Revulsion is hard to avoid after picking up so many of the empty containers, seeing the absurd waste they represent and how they last nearly forever in the marine environment. But using such resources is inescapable: it takes fuel and plastic-wrapped food to get to a place like Peak Island to clean it. What's the balance of these environmental debits and credits? In pure money terms, the garbage collected costs sponsors and volunteers around $2 a pound before it gets to a landfill. Antonia has worried about the environmental price of cleaning, especially when the young men spend hours digging up Styrofoam fragments deposited by bears, which for unknown reasons tend to attack and atomize the material. Is it worth it? The biggest contradiction arises with believing in what you're doing. Every float and oil jug recovered is one less to poison or entrap an animal: the pleasure comes from that sense of purpose. Yet for every piece of plastic that's gathered, an infinite number remain in the ocean, gently landing in soft summer swells or flung by the wind into the trees on stormy winter nights—and that fact reduces your accomplishment to a futile gesture. Is this all just to feel good? How do you reclaim the sense of purpose once you have the sense of scale?

For Pallister, Leiser, and Raynor, the sense of purpose is everything—cleaning the next beach and dreaming of cleaning them all. Doug Leiser is a gentle, cautious man with eyes that scan a room from behind thick glasses before he makes a remark. He acted as Chris's lieutenant, driving boats or managing people as needed. He wondered at the odd way in which the ocean sorts the garbage, leaving commercial fishing debris on one beach and recreational litter on another, or dividing the plastic and Styrofoam to opposite ends of a single beach. He didn't like the word "environmentalist." That wasn't how he was raised. His late father, Mann Leiser, was famous in Anchorage for fighting the EPA over pesticides. Doug spent his adult life working in Mann's greenhouse, from back when Anchorage was a cold, dusty boomtown—a stream ran through the building and tropical birds sang, an oasis in the Alaska winter—but after Mann died and the big-box stores came to town Doug found he could make more money by closing the business and selling off his dad's pack-rat collection of toys, magazines, and junk. When not out on the water he worked on eBay four hours a day. He preferred to be out here, with his boys, accomplishing something.

Ted Raynor believes Pallister's project saved him. I've known Ted many years. During the *Exxon Valdez* oil spill he burned with a special anger—it seemed to pressurize him into hyperactivity—and he organized a remote oil cleanup for volunteers, those like himself who wanted to remove the pollution but wanted none of the $2 billion Exxon was throwing around that summer. But soon after that project started at Mars Cove, on the south side of the Kenai Peninsula, Ted became disgusted with meddling from bureaucrats and the media and he quit. Seeing each other for the first time in the eighteen years since, we sipped boxed wine by a campfire on a placid lagoon miles from where the other boats were anchored. Sea lions barked occasionally from the rocks at the lagoon entrance, unseen in an indigo dusk. The Pallister and Leiser boys had told me admiringly about Ted's tenacity and daring in extricating nets from steep, slippery boulders, but he likes to stay on his own, sleeping in his sturdy little aluminum boat with his pit bull, Bryn. He volunteered that he wouldn't kill himself because he needs to take care of the dog.

In a way, Ted was like countless other former fishermen whose lives were ruined by the oil spill: still waiting for his share of the lawsuit, never able to find a job that satisfied him, despite trying work as a sportfishing charter captain and getting training as a computer help-desk technician. But unlike some others who did commit suicide, Ted found a new purpose in cleaning up marine debris. He plans to clean every shore in Alaska, all forty-seven thousand miles, and then to sail to the middle of the North Pacific, to the legendary garbage patch at the center of the gyre of world-spanning currents, a patch said to be the size of a continent, and somehow clean that. The first summer of the paid cleanup Ted brought in garbage bags piled so high they hid his boat. Tourists gathered around and brought him Cokes or cups of coffee. The admiration affected him.

"They say, 'How much is out there?' And I say, 'Like stars in the sky.'" The very impossibility of the job was his motivation. Ted said, "After all the decades out here, I'm still not jaded by this beauty. I need to be out here. This is the last chance for me."

Pallister, Leiser, and Raynor began planning a way to leave the relatively sheltered waters of Prince William Sound and begin cleaning beaches on the wild outer shores facing the open Gulf of Alaska, where ferocious weather slams fog-shrouded rock and almost no one ever goes. They would need a new scale of equipment and money. Crews and boats would have to voyage a long day just to get there—and then only during breaks of calmer seas. But without that step their grand purpose of cleaning it all up would remain merely metaphorical and the work done would always be partly symbolic.

I had joined the volunteer cleanup to find out what sort of people engage in such a seemingly altruistic venture. It's the mirror opposite of competition driving down a common resource. If overfishing and pollution can be explained by self-interest, then nature lovers' gifts of time and money to improve a common resource must have the reverse motivation. But Raynor doesn't see himself that way. He pointed out that he doesn't give money to causes or volunteer to do anything but pick up marine debris. A schoolteacher coming home from the volunteer cleanup laughed and said I was all wrong: she enjoyed the weekend too much to take any credit for altruism.

While everyone takes selfishness at face value, generosity is discounted, suspected as fulfillment of some other, more complicated desire. We can easily define altruism out of existence: any action that a person chooses to do is necessarily what he or she prefers, and therefore self-interested. By that thinking, good acts are possible only by those who don't want them—and only the bad can do good. Nonsense. Suppose, instead, we interpret these debris-gathering volunteers in the same ecological terms Geerat Vermeij used to examine the actions of stockbrokers and snails, by their external relation to the world. He found humanity rushing toward the limits of our ecological niche by competing for power, food, space, all the things that enhance our individual survival, dominance, and reproductive success, to pass on our genes and culture. But the marine debris workers weren't doing that—the benefit of their work went to everyone equally, present and future, litterers and cleaners alike. Spending a weekend or a lifetime gathering plastic bottles and fishing floats from the storm berms of wilderness beaches excuses you, at least a little, for your part in destroying the ocean—even if, while working, you listen to the ravens and smile at the sun.

There are different kinds of people. The real workers know each other and recognize lazy ones easily. Chris Pallister puts Antonia Fowler's group on the dirtiest beaches. He doesn't bring back people who take advantage of the ride out into the sound—worth perhaps $500 a person at going rates—but who don't do their part. Antonia knows whom to bring among her friends even before seeing them at work. When sweating to unearth an *Exxon Valdez*-vintage oil containment boom, you can turn your head to see what sort of people you're with—those pulling and those watching.

Antonia wanted to be like her mother, who took in homeless, troubled kids and called everyone "honey." She felt empathy and equality with animals, explaining, "It's just being calm, and allowing yourself to think about what they might be going through, and not trying to overpower them." She had planned to become a veterinarian, but she tripped up on chemistry and found her vocation instead working with blind, deaf, and otherwise disabled children at a therapeutic horseback riding clinic. She came to Alaska in 1982 for an internship in physical therapy. She was an avid rock and mountain climber, a dedicated berry picker and beachcomber, a ballet dancer, and she explored the western sound by kayak. She'd prefer to be outdoors all the time.

When Antonia's first son was born, MS symptoms that had been undiagnosed since college became clear. Her foot wouldn't come up to level, her vision narrowed to a pinpoint, and she needed a cane. She got better, but with each cycle of the disease there was something that didn't come back: in 1989, loss of feeling on the left side of her hand and arm; in 1993, spasticity in her bladder and bowel. She doesn't dance anymore and she carries extra clothing with her everywhere. She can't work, but still counsels newly diagnosed MS patients and helps a few paralyzed spinal-cord patients with swimming pool therapy. The weightlessness of floating gives her paralyzed patients more ability. They trust Antonia to handle their bodies in the water, even turning them facedown. She rejoices with them over major accomplishments, like moving a thumb. She said, "It's so exciting to see that. To help facilitate that change in someone is so rewarding. And I felt if I have the ability to help someone do that, then that's what I should do."

When her legs gave out on Peak Island Antonia didn't say how badly she was faring; she kept working when she was able, but looked tired and talked of this maybe being her last trip. When I pulled out my notebook and asked for her last name she got angry and told me a story about a newspaper reporter who had embarrassed her in the past. Later she apologized several times. When we met again back in town she greeted me with a hug. She told me many intimate things over coffee and said she trusted me to write anything. For the first several years she had kept her disease a secret, but now she didn't remember why. When I asked, across the table in the coffee shop, if she felt she was one of the special kind of people who prefer to do things for others, for animals, to cooperate—people contrary to the competitive, consumptive system we live in—she hesitated for a long moment before deciding to risk sounding immodest and told the truth.

She said, "I wish more people were that way. And I think there are a lot of people who are that way. And I think the more other people see it, the more they will become that way, too."

She had succinctly stated the basis for hope.

34

How Change Happens

Is there hope of changing people? Can individuals transform society? It does happen. Beforehand, change often seems impossible; the greatest changes seem to be the most difficult for experts to predict. Scholars and commentators failed to foresee the fall of communism, the rise of international religious violence, or the arrival of a level of racial harmony that would allow election of an African American president. All were transformative events, and all had in common that they were preceded by unseen social currents driven by individuals and communities influencing one another. People changed and then the world changed. That's the hope for those of us who imagine a new relationship to the ocean and the rest of nature.

In 2002, I watched the Iñupiat Eskimos of Barrow, on Alaska's Arctic coast, struggle with the warming climate, fearing constantly that the ice upon which they hunted would break from shore, as it did one day, carrying away more than ninety people who could be rescued only by helicopter. That year, leaders of the Iñupiat came to accept that global warming was real and affecting them. Americans were already twenty-five years into the climate change debate and fifteen years beyond its explosion as a national issue, but as a culture they seemed little closer to readiness to act. Lecturing after my book about the Iñupiat came out in 2004, I still prepared for skepticism and the assumption that concern about the climate was a liberal ruse to undermine environmental foes.

·In 2005, a researcher came to Alaska to study how the people directly affected by climate change perceived the risk they faced. He was Tony Leiserowitz, who later went on to direct climate change work at the Yale School of Forestry and Environmental Studies, which Gifford Pinchot founded. I gave Tony a variety of reasons why mostly white, urban Alaska didn't believe in the change yet, despite the dramatic evidence in the environment around them. I even published an article that summer in which I interviewed politicians, pollsters, and activists about why Alaskans didn't care. Then Tony did a scientifically designed public opinion survey and found virtually all Alaskans had noticed the warmer temperatures and thought climate change was at least partly to blame. They cared deeply and were ready to act. Almost all trusted their friends and families to tell them the truth about the issue, and most trusted scientists, but an overwhelming majority thought their elected leaders were lying about it.

I couldn't believe I had been so wrong, but Tony reassured me: Elites often hold false beliefs about public opinion. Everyone does, because our sources of information are flawed. We can learn directly what friends think, but hearing the metaphorical voice of an abstract body called "the public" requires intermediaries and imagination. The problem for elected officials is especially difficult, because almost everyone they meet wants something from them. They interact daily with lobbyists, staffers, reporters, affluent supporters, and fellow politicians. Like all people, they develop their beliefs about the world based on the emotional landscape of their everyday relationships. In Tony's research, members of the political elite sometimes discounted polls if the results contradicted their own intuitions about public opinion based on narrow personal experience.

Historically, the federal government lags behind local and state governments in assimilating social change. Global warming is an outstanding example. In Barrow, you could see the weather changing individual minds and the community's collective mind, which then helped power a national movement. An academic team was there to study Barrow's environmental perceptions and adaptation. One of its members, Ron Brunner, a policy scientist from the University of Colorado, had predicted more than a decade earlier that political acceptance of climate change and the will to act would come up from the bottom, from single human beings, their families, friends, associations, towns, and states—while national and international bodies would continue to spend billions on studies and hold meetings without effect. The problem wasn't just President Bush. In 1997 the U.S. Senate voted unanimously to oppose any climate change agreement that could affect the nation's competitive position with poor countries.

Tony Leiserowitz saw the change coming in his national surveys as people persuaded one another. Hundreds of mayors signed an agreement to address the problem, whether or not the nation or other countries followed. States passed strong laws on the East and West Coasts. Tony helped organize a climate conference of governors at Yale—on the centennial of a meeting of governors held by Roosevelt and Pinchot—where California's governor, Arnold Schwarzenegger, declared in his Austrian accent, "We don't wait for Washington because Washington is asleep at the wheel. America has to lead, and we're doing so even without Washington." Individuals led. They bought hybrid cars, contrary to their economic self-interest. Detroit automakers that had bet against their customers' better nature lost billions on the wasteful vehicles they continued to build.

As the media reported these developments the movement became self-reinforcing. That pattern fits psychological research on how social norms affect environmental behavior. People conserve most when they believe their neighbors are doing so, not under orders, or to save money, or to try to save the world on their own. Simple experiments confirm the phenomenon: for example, a study testing door-hanger messages encouraging energy conservation. Monetary incentives or persuasive arguments were less effective than the simple statement that everyone else was doing it. Beginning in 2005, a new societal norm established itself for conserving energy to protect the climate. Even ExxonMobil and President Bush eventually had to take account of the new reality. Yet Congress remained incapable of passing strong climate change legislation for at least three years more.

Climate change is the farthest-reaching threat to the ocean. The ocean and atmosphere interact constantly over their immense shared surface, passing back and forth heat, carbon dioxide, and water, as vapor or precipitation. The systems are too complex for precise long-range predictions, but it's clear that changes in the temperature and circulation of the atmosphere are influencing ocean currents, surface temperature, ice extent, and the habitat range, behavior, and productivity of organisms. The sea also chemically absorbs most of the excess carbon dioxide we emit, making its waters more acidic and less able to sustain shell-bearing creatures and corals. A large scientific literature documents changes progressing at every latitude.

The arrival of carbon reduction as a broadly accepted social good is a profound first step toward saving the oceans. Moreover, it is proof of the connection from nature through community to social change. For most Americans, the climate threat remains distant in time and space, but their cooperative impulse and goodwill were adequate to form a collective will to act.

Tony Leiserowitz found the connection between community and environmental caring in his surveys. He studied what sort of people perceived risk from climate change and wanted to act nationally and internationally. Statistically combing the data in search of commonalities, he couldn't group these people by race or gender, age or political party. The important common characteristic was a person's value for social equality. People who wanted to save the climate also agreed with the statements "What this world needs is a more equal distribution of wealth" and "Firms and institutions should be organized so everybody can influence important decisions." Only a minority fell into the opposite group, the climate change deniers, and they tended to believe most strongly in hierarchy and individualism. And, unlike the majority, those people did congeal into a clear demographic group. Amounting to 30 percent of white men, they were Republican Party members who listened to conservative talk radio.

In university lab experiments, Lin Ostrom and her colleagues found business and economics students cooperated the least—they were the most likely to fall into the destructive conflict of the tragedy of the commons—while students who had graduated from small high schools had the strongest tendency to work out mutually productive relationships. She asked about high school size looking for evidence of how Americans develop skills of shared responsibility, the social capital that supports democracy—the glue. Lin believed that students in small schools were more likely to have experienced being leaders and would be more ready to help their fellows cooperate in economic games. I think she was right, but I also believe the nature of the educational process could matter at least as much. Humanistic schools that allow students to develop their own interests and collaborate with classmates, without adult intervention, can teach the mutual respect and self-respect necessary for good community members.

To save the oceans, teach children. Education creates the conditions upon which our species interacts with our ecosystem—children who learn to be competitors or cooperators. Schools make our species; they are the agents of cultural evolution. And schools effect change faster than traditional environmental activism. It took twenty-two years for Congress to pass new vehicle fuel economy standards, a law that wouldn't take effect for another twelve years after that—a thirty-six-year effort in all. Today's kindergarteners will be driving in ten years, or choosing not to. Many studies have shown that environmental education works, changing children's behavior and the way they view the world. But fundamental change will require more than taking nine-year-olds on field trips. They must learn as well how to work in a community of equals, how to value the aspects of life that provide meaning and fulfillment, and how to think for themselves. Far more than any technology scientists may invent, their hearts will determine what life survives besides our own species. They'll decide what matters—what's worth saving—and the right and wrong of how people relate to the earth, the oceans, and each other.

35.

Chenega's Next Generation

Wannah Zacher dug clams in the sun on the beach below her house in Chenega with a couple of family dogs at her side. The sounds of her children's play seeped out from the trees above the shore. I introduced myself and my unusual purpose in the village, as a writer, and she accepted my presence as gracefully as if she made new friends that way every day. While looking for a shovel so I could dig with her she missed most of the low-tide period suitable for finding clams. Coming in from clam digging, Wannah sat for a moment at her kitchen table while her husband, Dennis, paced and devoured leftovers. The love and duty of catching food for his family seemed to energize Dennis's wiry body into constant motion. With bright spring sunlight piercing through the trees from the water, they talked about the season. Was it time for goose eggs yet? That depended on the late snow and Wannah's birthday—that's when you go.

Living from the land—subsistence, the Alaska Native tradition—is a hell of a lot of work. The differences from sport hunting and fishing are fundamental. Success really matters, as families committed to a subsistence lifestyle invest the time in food gathering that would be spent earning cash in mainstream society. Also, getting the amount of food needed to sustain a family requires long, intense hours of processing, teamwork, and efficient harvesting. The sportsman's conventions of hook-and-line angling or fair-chase hunting make sense only in the context of leisure, not subsistence. Dennis worked odd jobs for cash, but most of the time he was thinking about the beach for clams, chitons, and octopus, the bay for salmon to cure in his smokehouse, the big garden and greenhouse out back for growing vegetables, the mountain behind the village to stalk deer, the wet ground all around to pick blueberries by the gallon. A dedicated subsistence family such as Dennis and Wannah's harvests as many as seventy species of wild plants, animals, fish, and invertebrates in a year, perhaps half a ton of food per person. Dennis learned the skills and the patterns of nature as he grew up in Tatitlek doing these things with his own family. Now five children are growing up at his side, helping Wannah and him in the woods and on the water.

What Larry and Gail Evanoff planted when they built the first cabin in the new village long ago now grows strongly in their grandchildren, Ray, Joey, Melonie, Toni Rae, and Tyler, ages six to fourteen when I was in Chenega in 2007. The children know about the rest of the world, but they're completely of this place. They have cable TV and the Internet and they've been on plenty of trips, but in the sound they learn Alutiiq, carving, hunting, and all their traditions. The whole family attends Heritage Week in Tatitlek and spirit camp at Nuchek. The children go to school in tiny, individualized classes and practice traditional dance. They roam the village, its beaches and woods, the way Wannah did as a girl. Wannah said, "Their sense of living is language and culture and dance. That whole program they're in. And that's set them in this whole awesome attitude of being proud of their culture and life, and being who they are."

Wannah's first job, as a teenager, was on the oil spill in 1989, and she worked the beaches during the summers through her youth. She got together with Dennis on oil spill work and they began their family with that income still coming in. The spill had hit near the twenty-fifth anniversary of the earthquake and five years after villagers reestablished Chenega. The fishermen-initiated battle to save the Armin F. Koernig Hatchery from approaching oil happened here, a chaotic invasion of workers and money, obliterating village stability and privacy. Social workers and researchers intruded with intimate questions. Ignorant international journalists produced news pieces full of stereotypes. Exxon's splurge of spending that first summer brought seventy thousand pounds of free groceries to the twenty households in the village, enough to fill every inch of storage space and send some off to relatives. The outside interference reminded anthropologist Nancy Yaw Davis of what she'd seen in villages after the earthquake, but the flood of money also caused new, unfamiliar stresses, as subsistence stopped and many people began working long hours for wages they had never dreamed of. New resentment, conflict, and loneliness. People stopped helping each other. Davis said some Native villagers blew their new income as fast as they could in order to restore the balance of financial equality that had always glued their communities together.

No one wanted to eat wild foods that first summer, with so much oil on everything: sea stars fell apart to the touch, brain-damaged seals swam right up to hunters. But confidence didn't return the next year, either. The advice of the experts seemed thin and changeable. As the years passed, the oil became part of the landscape—people knew which beaches to avoid—but spill science undercut that recovery of confidence with the new evidence that PAHs could affect health even when undetectable to human senses. Fifteen years after the oil spill, Chenegans could easily find oil by turning over rocks on many nearby islands. Some were afraid to dig clams. Researchers led by James Fall of the Alaska Department of Fish and Game found a deep new awareness of contamination of all kinds in the villages touched by the spill, a sense that one must always be concerned about whether wild food was safe—a new suspicion where they always used to say, "When the tide is out the table is set." Most believed the traditional way of life had not recovered from the oil. Fall's surveys showed harvests had not returned to prespill levels.

But Wannah said many Chenegans weren't digging clams or gathering much wild food because they never got used to it. Some grew up on store-bought food in Anchorage, followed by only five years living in the village before the oil spill disrupted the recovering subsistence tradition. Wannah's mother, Gail Evanoff, posted a handmade diagram on the wall of her office in the village clinic showing Chenega's journey through time as a ride down a river: from the big sea of culture and identity in the past, through the narrow, rocky passage of the earthquake and tsunami that destroyed the old village, widening again with the tributaries of the land claims and the reestablished village, and then back into dangerous rapids with the oil spill, lost subsistence, increased drugs and alcohol, but with a hopeful future ahead, a broad new sea, labeled "self-determination, sobriety, wellness." She is a behavioral health counselor in the immaculate, well-lit clinic building, where Wannah works part-time as a receptionist. Checking in patients isn't a difficult job in a community with only sixty or seventy residents.

Old Chenega was a subsistence and commercial fishing community, but there are no fishing permits left in the village today, and anyone can order groceries from the air services that fly from Anchorage to a big, modern airstrip built during the oil spill recovery process. The sale of the village's lands to the Trustee Council brought the Chenega Corporation $34 million. It used that wealth and Senator Stevens's bidding preference legislation to become spectacularly successful as a partner in contracts for military, security, and intelligence services for the Department of Defense and other federal agencies. In 2008, with 3,200 employees all over the world and $700 million in revenues, the corporation had a standing offer of work for any of its 167 shareholders, whom it paid annual dividends of up to $30,000 each (or so I was told in the village; the corporation won't say how much it pays out). Lucrative jobs and other spending came as well from an organization set up to prevent and respond to future oil spills. Newly affluent families moved away—to be closer to work in Valdez or elsewhere, or simply because they could—but kept their low-cost housing in Chenega for occasional visits. With those housing units unused, there wasn't room for anyone new to move in, and the population dropped to around forty, with the school endangered, since only schools with ten or more students get state funding. Once a village's school closes, the community begins to die. But Dennis and Wannah Zacher contributed five children, Sandy Angiak had her two, and the couple running the Armin F. Koernig Hatchery brought a family, too—so the school scraped by.

The fact remains, however, that while the village has the best of every-thing that money can buy, and plenty of good jobs doing not much, it doesn't have an economic reason to exist. The money and things all come from somewhere else. In the old village everyone was poor, but elders re-called their sense of purpose working to gather their food and build their shelter as a community. Those needs are gone, taking along with them the meaning they provided—a man's feeling of use when his family's survival depends on his hunting skill. Jim Fall's surveys show that Chenegans still harvest as much wild food as most coastal Native villages, but the villagers themselves kept telling me that few live a traditional subsistence way of life, the kind that is linked to the sea and guided by the elders. An anony-mous Native told Fall's researchers the oil spill had reduced the respect given elders because, "The culture is changing to a monetary system and more competition. Young people want money now." A Chenegan in the survey said the traditional way of life was lost forever: "It's never going to recover. It can't happen. Everybody's got a pocketful of money now. You can't go back."

I heard Chenegans criticize one another's commitment to traditional values along with many other bad things they said about one other. Bitter grudges and resentment seemed to lie just below the surface, directed against neighbors seen every day, fellow villagers with more in common than most of us have with our extended families. I'm not repeating what I heard, because I don't want to make matters worse. Besides, the causes are too old and complex to sort out—Chenega has been through so much, with its historic traumas, its division into the corporation, traditional council, and nonprofit organizations, the associated elections and so on, and the disagreements over oil spill money and land sales. Carol Ann Kompkoff, whose father saw two daughters slip away in the tsunami but kept the strength to be the village's Moses—even Carol Ann feels lonely in the vil-lage at times, because she's been sober since the oil spill and finds it socially uncomfortable to visit hard-drinking friends and neighbors. In a human community, as in an ecosystem, there is no rewind function. What's lost is lost, and hope lies in creating healthy relationships that are new. The grief must die for the next generation to rise.

Sandy Angiak's sons James and Ian impressed me by trading Alutiiq words and showing off the carvings and other cultural objects they had made on many visits to Tatitlek and Nuchek. Sandy taught traditional dance at the school. At the session I watched she sat back and let the young people lead themselves, with James, a high school senior, guiding those around him with a stern, serious voice. James would graduate that spring, the first graduation in many years at the Chenega school, where he, his brother Ian, and Wannah's son Ray were the only teens. The entire community was looking forward to the ceremony, all rooting for James to complete his credits in time. He worried about a speech he would be expected to make. Sandy, with streaks of gray in her long black hair, did not try to hide her pleasure at having raised these two strong, proud Chugach men on her own, standing with them as I took a picture. But when I asked about her own history she suddenly began to weep and they stood uncomfortably away.

Wannah and Dennis invited James to go along on a picnic with the family. That sunny spring morning the hummingbirds had arrived in Chenega and villagers were out on the road telling each other about seeing the first of the tiny, hovering birds. Wannah and Dennis could stay home no longer and set off in their little fiberglass skiff to get a seal for lunch and to gather goose eggs. After adding me to their family of seven, James, and two dogs, the boat sank low in the water and traveled slowly. We followed the mountain-serrated shore of still-snowy LaTouche Island, motoring over smooth, undulating water toward a cove where Dennis knew there would be harbor seals. Spring had made the channel new, mountains and water metallic in their brightness. No chance at all of seeing other people that day, although the island once had a mine and a sizable town, now long gone. Melonie sat in the bow with the dogs. The big boys, James and Ray, cradled their rifles. Little Toni Rae and Tyler sat between their parents at the outboard motor in the stern, pretending to help drive. Wannah, with an irrepressible smile, looked over their heads at Dennis, whose eyes concentrated instead on the horizon and his responsibility. Joey, the middle boy, manipulated a handheld Game Boy computer, watching Japanese Pokemon characters rather than southwestern Prince William Sound. I remembered feeling as he did as a child, unimpressed by the overly familiar grandeur of Alaskan scenery, and I respected his parents for letting him think about what he liked.

We landed well short of the cove where the seals would be found and Dennis sent the two boys with guns ahead on foot. He taught without saying anything, letting them find blinds behind two boulders—he had put them where they could have their own success. Dennis stayed in the boat, ready to retrieve the kill. He hadn't even brought his own rifle. James and Ray hesitated. They sited powerful scopes on the heads of the seals, which barely poked through the surface of water halfway across the broad cove. Two shots thundered, but the bullets skimmed over the surface of the water beyond the seals, which were gone as quickly as the splashes. Dennis's voice from the boat urged the boys on to another spot for another shot, but they missed again. Likely their gun sights were out of alignment. Dennis uttered no word of criticism, but James made a rueful remark, imagining the seals under the water now discussing the hazard of men pointing sticks at them.

James lit a damp campfire with a well-chosen clump of grass and moss and a single match. We cooked wieners on sticks rather than pieces of seal. Dennis didn't bother with the stick: he stuck a hot dog in the hot coals and then ate it in his fingers, without a bun, saving time for his work. He walked miles down the shore, fast, checking for goose nests in the dry grass beneath the eaves of alder branches—nests that could not be found because they hadn't been built yet. While the girls played with shells, rocks, and sticks and the big boys lay by the fire, Dennis never stopped moving, and in the evening, after arriving back in the village, he went out again, fishing, and brought food home after all. He always seemed driven about catching food, but became maniacal when he saw a fish jump.

"I wanted to be a seiner when I was young," Dennis said. "That was my dream, to be a seiner when I grow up. Then that oil spill happened, as soon as I grew up. No more seining."

Almost twenty years after the spill, conversation kept coming back to that topic. Dennis believed the otters and seals were still down, and the herring eggs had never reappeared as thick and plentiful as they once were, when he was a boy and you couldn't even see the seaweed under the eggs.

But to the next generation, James and Ray and the others, the oil spill was just something their parents talked about from before they were born. They didn't remember a more plentiful sound. Their own dreams, still rising, included learning to hunt and fish like Dennis, and other, less-formed ideas. Ray began fulfilling his destiny when he took his first deer and shared it with each of the village elders, just as they would have done in old Chenega. What that meant to his grandparents, Larry and Gail, I could hear in their voices, but the meaning for Ray himself surely was only beginning to form. The teens answered all my questions respectfully and as well as they could, but with faces put on specifically for diplomatic relations with an adult. I could no more enter into their world than I could see through the eyes of a seal as it slips birdlike through the water. Free of us, the boys' young muscles surged forward, powerful and unburdened by the grief of previous generations, toward a place that we could hardly imagine, except for our hope that, in the future, the sea will remain alive, able to envelop the human heart as it has since the first mind woke to its spirit long ago.

36

The Hedonic Treadmill

Elmo Saves Christmas would more accurately be titled *Elmo Ruins Christmas,* because the little red monster puppet's magical wish to have Christmas come every day produces the crisis on Sesame Street that is the film's theme. As the father of four children, I've seen and thought about *Elmo Saves Christmas* many times. Elmo's error of wishing for Christmas every day is one any five-year-old can imagine making. For adults, the result is more familiar than the wish: the grinding repetition of the same Christmas songs, the weariness of excessive acquisition, the nauseating surfeit of material possessions. Elmo's Christmas every day looks a lot like shopping centers all over the United States from October through December, followed by the period when gifts are returned and credit card bills arrive listing purchases that have already lost meaning except for the debt, that sticky presence pervading daily life all year. For the other puppets and kindly adults in Elmo's universe, Christmas every day imposes an existential torment, as if his wish alone had robbed them of the will to choose what will come with the dawn, until Elmo saves Christmas by using another wish to reverse his curse and allow the holiday to come just once a year. The rather menacing Santa Claus played by Charles Durning is only too glad to oblige. The rest of us aren't so lucky. Our escape depends on resisting the spell without the help of such a deus ex machina.

This challenge—of escaping the trap of unfulfilling consumption—

stands in every path potentially leading to a new human relationship to the oceans and wild nature. If everyone needs things without limit, we will surely fill our ecological niche, leaving no undomesticated space for other species. Can we escape without Elmo's magical intervention? No, according to some serious thinkers in economics and biology, who believe we're built to seek wealth and power, even if the ultimate result is individual despair and collective destruction. I've argued that this view is erroneous: we may be programmed for selfishness by an intellectual paradigm dating to the eighteenth century, but ample evidence exists of other ways of relating to each other and the environment. Escape is our choice.

Even if escape is possible, however, many economists doubt it would be desirable. The doctrine of growth holds that competition for escalating wealth has raised humanity as a whole to greater health, longer life, and higher intellectual attainment, and will continue to do so. These advances have real value. But if they're the result of economic growth, it's reasonable to balance them against the environmental costs, which are immense—60 percent of the services the earth's ecosystem provides are degraded or unsustainably used, some with catastrophic consequences—and it's also reasonable to consider the equity of these costs, which fall overwhelmingly on the world's poor, who receive the least benefit in health, life expectancy, or education. Even admitting, however, that economic growth often is good, still I don't believe it is inherently good. If increases in economic output are wasted, growth is more likely to be bad. As Elmo learned, economic goods that overshoot sufficiency—things made and bought that are more than enough—become waste and diminish the lives of the acquirers.

Psychologists and behavioral economists have an eloquent name for the process of always wanting more but receiving no satisfaction from receiving it: the hedonic treadmill. We seek money and things in pursuit of happiness, but quickly adjust to improvements in wealth and material possessions, returning to a baseline sense of well-being that leaves us again seeking more money and things in further hope of feeling better. Research over decades using increasingly sophisticated tools has confirmed the treadmill phenomenon many times. Those with high incomes are not happier than those who make half as much, and the wealthy are more likely to feel angry and tense. Social connection and purposeful leisure make us happy; intimate moments and long-term emotional links such as marriage create well-being; but increasing income and material possessions do no good once the necessities are provided. In controlled experiments with real money, subjects predicted that spending on themselves would make them feel good, but giving away the windfall produced much stronger positive feelings. Volunteering for charity scores near the top on scales of life satisfaction, behind only dancing. On the opposite end of the list, our least happy everyday hours are spent on work and commuting. Financially successful people spend ever more of their time that way, and less on family and friends. Pursuit of excess wealth displaces the activities that really satisfy us. Money doesn't buy happiness. Everyone knows this; moreover, it's an established scientific fact.

These findings create a profound challenge for a competitive society. The costs of running the treadmill include the work of earning goods as well as the cost of producing them for the biosphere and the rest of humanity—the cost, for example, of creating corporations such as Exxon that spill oil in places such as Prince William Sound. Since 1950, the world's economic output has doubled and redoubled almost three times. We've spent most of these gains on increased consumption. In fifty years, the size of an average single-family house in the United States doubled; since families shrank, each person occupied three times as much indoor space, far outstripping improvements in energy efficiency. Still, every room filled with stuff and we had to rent millions of storage units, enough to enclose every person in the country. Since 1982, development in the United States has consumed land at twice the rate of population growth, even faster on the coasts, and miles driven in vehicles to cross the sprawl have increased four times as fast as population. But during these fifty years of growth, measurements of general life satisfaction have not increased in the United States at all, nor in other developed countries. For all the spending and building, we're no happier than we were when we started. So what was it all for?

The break point between enough and too much is not difficult to find. Surveys of people's sense of well-being have been repeated in dozens of countries. The world's poorest nations, full of people lacking food and clean drinking water, uneducated, short-lived, and with inadequate health care, and often subjected to political repression, are the world's unhappiest places. But once a nation's wealth has alleviated those conditions, additional income no longer yields greater well-being. In a 1995 study, citizens of those nations whose per capita purchasing power was half to three-quarters of Americans' reported the highest life satisfaction; the United States came in eighteenth, behind many less affluent countries. The detailed reasons for these differences may be complicated, but the basic message is simple: sufficiency is real—there is such a thing as enough.

Why doesn't the economy stop growing at "enough"? In part, because it values goods according to the psychological state of our wanting them, not their true utility. If a man pays $5,000 for a watch, that decision alone sets the worth of the watch, not its benefit to his well-being. Another reasonable way of measuring the value of the watch would be to consider its use and the satisfaction it yielded. If the man buying the watch knew himself well, then presumably the two ways of valuing—the want and the need—would produce the same result. In a society of people like that, measures of economic activity and happiness would closely match, because consumers wouldn't waste money on unfulfilling things. The economy would grow more slowly, but would still produce goods of true utility, such as music for dancing or candles for romantic evenings, greater health and education, and leisure time to work for others, perhaps picking up plastic bottles from beaches—all of which, research says, yield happier lives than the hoarding of material possessions. That would be a state of sufficiency.

Besides inducing waste, the treadmill also devalues what we already own. The creation of new wealth and new things to buy makes old things cheaper. For example, we used to have only one world; now we have virtual worlds as well, within networked computer games. Players in these unreal spaces compete for objects such as magic wands that yield prestige and power in that frame of reference. For those who want to advance faster in the virtual world, online markets developed where players can buy these imaginary things for real money. Asian sweatshops opened where paid game players produced the items for sale. Interacting people in the game space produced new economic goods, things with tangible value in money but no tangible existence. Economically, these electronic symbols are equivalent in value to goods that feed hungry people.

Within the market system, this equivalence makes perfect sense. Virtual worlds integrate easily with the real economy because both function as games. Both are governed by rules that encourage competition but that set boundaries to keep order. In both, only the field of play counts—things outside the game are not valued or subject to the rules. Our legal and political system works on the same basis, as a contract among players. As a concept of justice, it has enabled the conquest of indigenous people's lands and continues to confer legal protection on careless oil shippers at the expense of the marine environment. The included players and their exchanges establish what has value, and everything else is worthless.

But a problem arises when players ask the purpose of the game. Suppose we assume, as some do, that the purpose of economics is to improve human well-being, and that the utility of goods depends on the benefit or satisfaction people get from them, rather than on their wish to acquire them. In that case, the period of economic growth since World War II has been one of deepening failure, as goods multiplied exponentially without increases in the happiness or well-being of their owners. By that measure of value, the worth of each thing we owned became progressively less. No wonder imaginary objects are as valuable as products taken from the earth or sea—nothing is worth much anymore. Seen in this light, the onward march to domesticate every living thing seems all the sadder, as what we receive in return diminishes to nothing.

An economist suggested to me that things like connection to community or wild nature could be valued along with everything else. If we want those things, we can buy them. Many leading environmentalists have adopted this thinking. Wild lands are goods we lease from the economic system by deferring the use of their resources. The job of those who love these places is to increase their "want" value by extolling them with words and photographs. That approach has preserved about 1 percent of the world's landmass as designated wilderness. Accepting, then, that the other 99 percent will be domesticated, and seeing that the great majority of land already is, this point of view calls for enlightened domestication by identifying ecosystem services that benefit humanity and quantifying their economic worth. For example, coastal wetlands would be protected because they provide storm protection; watersheds would be kept clean to provide drinking water; marine ecosystems would be kept intact because their diversity yields more biological productivity. Unfortunately, biodiversity has yet to prove its economic superiority, and there's no guarantee someone won't invent technologies that perform any of these services better than natural systems do. Besides, to sincerely accept this point of view, you would have to approve of the outcome of the *Exxon Valdez* oil spill as a whole, as there hardly ever has been an incident with environmental costs more thoroughly adjudicated by the combined efforts of the scientific, political, legal, and economic systems. After twenty years, these institutions settled on an exact dollar amount for the damage—its sale price to Exxon—but I haven't met *anyone* who would do it all again. Those who truly believe that nature's value can be established by the economic system should have been happy when the transaction was completed, and likewise be willing to oil more birds or let climate change destroy more coral reefs if the price is right.

The ecosystem services argument betrays all those people—and many they are—who by their actions and their devotion of time choose to place their relationships with people and the natural world over their economic roles. By opting out of our maximum earning potential, we sacrifice influence in the primary form of exchange that counts in the game. The control of power and money instead go to those others, the avid competitors, who work and trade to the exclusion of all other forms of human meaning. They determine society's goal of unlimited economic growth. The connections that a healthy mind makes with other minds, both human and animal, and that it projects as a sort of spiritual life into trees and oceans—those connections have real, intrinsic value, the kind that makes living purposeful, fulfilling, and happy. But generosity can't be bought and therefore has no economic worth. To partake generously in the world means you give away the means to pay. You're out of the game and you don't count.

The puzzle lies in the wanting. Why are material desires so much stronger than is justified by the happiness derived in achieving them? Why do so many exchange happiness for money? Daniel Kahneman, the 2002 winner of the Nobel Prize in economics, argues that a psychological illusion fools people into thinking increased wealth will make them happy when it will not. Mentally focusing on getting rich inflates its appeal to exaggerated proportions. Once you've got more money, you still worry about your children, your commute, and many other things, with less time for anything but work. It's easier to be preoccupied with getting ahead than with your own good fortune once you've got it. Partly, that's the result of immaturity and lack of self-knowledge; wisdom brings more realistic understanding of one's own state of being. More broadly, the hedonic treadmill runs on human adaptability and our use of social norms to establish a sense of the sufficiency or shortcomings of our material circumstances. We tend to adapt to new luxuries, transforming them into necessities. We tend to judge our success compared to those around us. Making more money and rising into a higher economic stratum only produces new standards for comparison and a new range of wants added to the horizon. The outcome is easier to see than the process: who thinks a corporate executive alone in his mansion is happier than penniless young lovers sleeping under the stars?

The treadmill generates norms as it is driven by them. Marketers arouse new wants to create demand for new products. I stayed in a hotel room recently in which the coffeemaker had been rendered largely disposable along with the cup. Advertising helped create a national phobia about germs, a pathological social norm for an antiseptic human environment. Single-use products became necessities for those driven by this fear. These products, such as disposable coffeemakers and plastic water bottles, consume resources to create them, move them, and get rid of them, and when just a small percentage are lost into the marine environment they pile up on beaches and in the guts of animals. Users' wants drive the economic value of the throwaways, but they receive no benefit in happiness or well-being—they only gain brief respite from a manufactured fear and, incidentally, contribute to the extreme hygiene that appears responsible for the epidemic of allergies in westernized countries.

But the influence of social norms is not a cause for despair. It's an opportunity. Some environmentalists avoid talking of sufficiency because they don't want to call for Puritanical self-denial. But stepping off a materialist treadmill isn't denial, it's freedom. If social norms power the treadmill, then changing social norms can derail it. We know we're biologically capable of such cooperative norms, because they have existed at many times and places, including among many people even in this, the most powerful and consumptive society in world history. In fact, we all live on a continuum of how much we respond to the economy's generation of extraneous wants. There's nothing original about opposing excessive materialism. Whenever shame is attached to neglecting family for work, driving a fuel-wasting car, or buying disposable, single-use products, there social norms for sufficiency are at work.

Simply changing the menu of wants is not enough. The fate of the oceans depends on changing the social, economic, and political system that values wants. We are built to be cooperators and altruists, too—givers, not only wanters. We are capable of joining in communities that elevate our loves instead of our drives.

The world I imagine is one in which people do work that is meaningful to them. Schools don't prepare children just for jobs, but for fulfilling intellectual, creative, and social lives. Long-term communities govern their own pieces of land or ocean, with the power to shame their own members who abuse the commons or exclude visitors who do so, such as multinational corporations. National governments and international organizations protect the realm of true democracy, devolving power to self-governing communities and protecting the interests of local institutions, a true federalism that allows groups to associate to solve problems at the small scale. The nation limits itself to truly national concerns, such as ensuring the rights of individuals, restraining the power of large economic entities, and preserving the integrity of ecosystems that lie beyond the scope of any community. This world is not unrecognizably different from the one in which we now live. The fundamental difference can be summed up in the prosaic term "local control."

The human will for cooperation and sufficiency is strong. It rises despite the seeming impossibility of success; it is rising worldwide, in thousands of individual decisions and countless environmental groups in countries rich and poor. All those people must be told that the world belongs to them and that its economic and political institutions should serve and empower them—that the spirit of kinship they sense in their neighbors and in the natural world is real. Only their wishes can ultimately reform governments and economic institutions to allow the many gardens of community to grow.

37

Making a Difference
(Beach Cleaners Part 2)

Jim Miller, the carver and counselor from Port Graham, looked on with amusement as I filled my arms with so many salt-washed plastic bottles and floats that pieces bobbed up through my embrace and back onto the beach. We had come in my aluminum boat to gather washed-up logs and boards for him to carve, but I found trash and couldn't leave it. Jim crouched in the beach grass, slowly and carefully examining hunks of wood, gifts of the creator, searching for the physical and sacred properties that would enable him to shape forms—masks, paddles, boxes—based on tradition and his own inspiration. It was a rich July afternoon, all life out in full force, salmon jumping and a pair of black oystercatchers with bright orange eyes defending their nest. The flatness of Cook Inlet gently accepted the warm weight of the summer sun. A perfect place.

I'd been cleaning beaches in Prince William Sound with Chris Pallister and his crew, an experience that rendered shoreline garbage visible and unacceptable to me. But full arms gave me a dilemma. Why take any if you can't take it all? Or, if taking only some, which pieces? Jim's mind saw the beach differently. By choosing wood and carving it, he would bring meaning out of the vastness of the ocean. His counseling functioned similarly. Jim talked with men while making things with them, believing men needed purpose to find meaning and needed meaning to heal. The things I gathered were purposeless, meaningless—identical, disposable, as numerous as droplets in a cloud—and the act of removing one from the ocean's infinitude was as futile and insignificant as scrubbing a single pebble on a thousand miles of beach. When a plastic bottle slipped through my arms it didn't matter if I picked it up again or not, and that made me feel foolish.

A paradox faces everyone who wants to protect the environment. As individuals, our efforts don't matter, because each good act is overwhelmed by global materialism. On the other hand, *only* individual efforts matter. Only individuals can determine social norms, such as the one that tells us it is wrong to throw garbage in the ocean. Picking up a plastic bottle from a beach, even in private, helps create that norm, and ought to generate a sense of purpose and meaning, something like selecting a piece of cedar to carve into a box. But reason says otherwise when the particular act seems senseless—this bottle on this uninhabited beach. At least, that's how it felt with Jim that day.

Chris Pallister and his friends needed meaning—they weren't dedicating their lives to cleaning beaches for symbolism. Each had his own need. Chris, with the hyperactive drive of a doer, needed an escape from the practice of law into fixing marine engines and commanding his little navy. Doug Leiser, whose family greenhouse business had been driven down by the big-box stores, needed relief from long daily sessions on eBay selling off his late father's trove of knickknacks and toys, which provided his income. Ted Raynor, the former fisherman who still blamed Exxon for ruining his life, needed to return to the one place and function that had ever given him satisfaction, with the dramatic declaration that to clean up all the shores of Alaska had become the one purpose of his life, without which he no longer cared for his continued existence. They had crossed a line from trying to do some good to trying to do something that appeared impossible—the line crazy inventors cross when they stop caring what others think of them. Alaska attracts men like these, but usually they try to transform the wilderness rather than restore it.

I ran into Ted Raynor on the dock in Homer that July as he was headed for a long day's run around the ocean side of the Kenai Peninsula to the no-man's-land of black cliffs and incessant storms and mist, where they had chosen to make their stand. Ted had just learned from a newspaper website that Governor Sarah Palin had vetoed the group's funding, a $150,000 grant from the State of Alaska intended to match a grant from NOAA's marine debris program. In fact, the NOAA grant money wasn't in hand either, owing to a late federal budget and paperwork complications Chris was trying to untangle. Yet Ted and the others never considered calling off their attack on the garbage of Gore Point. They had scoped it out in the spring by helicopter, taking pictures of enormous mounds of buoys and bottles, ropes and fenders, and every other kind of persistent mess, which had piled up one hundred yards back into the trees all along a crescent-shaped beach that reached like a hook into the coastal current. They had fixated on redeeming that one wild spot and proving their work was real.

With a couple of boats anchored behind the point, in relatively sheltered waters, Ted and the young men, Leiser and Pallister's sons, worked like fiends to bag up the mountain of trash. The boys had been promised $200 a day, well under what oil spill cleanup laborers had earned here twenty years earlier. One hundred volunteers had offered to help, but only five actually came when the remoteness and hardship of the trip became clear, and some of those five couldn't hack it. The men worked long days on the isthmus of that curved beach, constantly wet, slipping on stones and logs, whipped by the wind and fog constantly flowing over the shore. Clouds ominously concealed the tops of the fantastically severe ice-capped mountains and the dark tower at the point's end. The men carried full garbage bags at arm's length, like zombies, because when the bags touched snags they easily broke open, pouring out all the trash. In a quiet moment, gathering plastic amid the trees, I asked Chris's son Ryan how long he could do this. He said, "Not forever. At least, hopefully not forever. I'd like to do it more at some point in the future. But you'd like it to end. But I guess the trash won't end." After working a while longer he added, "It does start to seem kind of hopeless."

In the prehistoric past, Gore Point was the site of a village. In 1778,

Captain Cook sailed by and named the point for Lieutenant John Gore, who irritated him on that part of the voyage by insisting on trying each possible route to the north in search of a passage to Europe—by which side trips Cook discovered Prince William Sound and Cook Inlet, both wastes of time, in his view. In 1989, Gore Point caught the flow of Exxon oil, which gathered as deep as a swamp, full of dead animals, but the insistent ferocity of the weather and surf removed that long ago. Now the land is protected, part of Kachemak Bay State Park.

Only one event in human history left a permanent mark on Gore Point: the invention of plastic. Behind the wreck of huge drift logs on the outer shore, behind the black sand and beach grass, the wind and waves deposited a debris field of discordantly bright colors as vast and out of place as a municipal landfill. Ted and Doug fantasized about coming back here when the work was done, when the bare earth under the plastic had regrown its moss and ferns, shaded by the straight, tall trees, which were limbless near the ground so they stood like columns on the isthmus. I could well imagine it. A glorious, vertical space, quiet as a church but for the shushing ocean swells, soft light, and raindrops filtering down from the gray sky above. This, if they could accomplish it, would be creation as well as a restoration, befitting all this effort.

I went in July and picked up trash in the rain, and carried bags at arm's length until my muscles burned. I had no doubt they could do the job if physical effort were the only factor. It was much more complicated. Getting the material off the point would require carrying it to the protected side, where boats could land, but state archaeologists forbade using all-terrain vehicles for that task because they might disturb signs of the prehistoric village. Carrying it all by hand was obviously impossible. Chris thought of moving it by helicopter, but he didn't even have money for the fuel and salaries he had already expended. He was carrying the operation on his credit cards and facing financial ruin if the grants didn't come through. A reporter from the *Anchorage Daily News* had flustered his funders with inquiries about a corrupt state senator who was co-owner of Chris's boat. An environmental group was after him, too, for using their name without permission—his Gulf of Alaska Keeper wasn't affiliated with the national Waterkeeper Alliance. Chris had already formed the Prince William Sound Keeper and then had been forced out by locals from Cordova who saw him as an Anchorage interloper. He was deep in the hole financially with mountains of plastic bags on Gore Point and no way of getting rid of them. A helicopter would cost $1,882 an hour for the full day of work and a vessel large enough to take the stuff back to Homer would need several days at $4,000 a day; and cargo slings called Super Sacks for hoisting the bags would cost another $2,000.

He kept hustling, and as the summer ended he landed a grant of $42,000 from a fishing industry group that had supported the work in the past, the Marine Conservation Alliance Foundation, which would pay for the helicopter and a one-hundred-foot landing craft. Chris arrived in Homer to launch the operation with a couple of cars full of helpers, in the rain. The boat Chris had hired, a rusty relic formerly used in logging, had not yet returned to the harbor from its last job carrying a load of junked cars and such across Cook Inlet. We drove around town as a convoy trying to arrange for barrels of fuel and a float plane flight to carry workers and Super Sacks out to Gore Point, which the deteriorating weather made impossible. At the helicopter hanger the guys learned, theoretically, how to hook cargo to a tether hanging 125 feet below a chopper. None of them had done anything like this before and it all seemed terribly improbable. Chris paced a wet dock with a cell phone on his ear—finding out where the boat was, dealing with the newspaper reporter, talking to spooked donors. "These guys think I'm making all this money, and I'm out here going broke," Chris said. "I might lose everything on this deal. I might have to sell everything I own to pay for all this." I realized the group had nowhere to get in out of the rain. I brought them to my mother-in-law's house to eat spaghetti and sleep on the floor.

The landing craft, the *Constructor*, arrived in Homer the next day. It was as much rust as black metal. The bunks in the tiny superstructure were buried under greasy tools and the toilet was a plastic bucket on the back deck. The owner, Otto Kilcher, also had a ranch and a machine shop; his extended family was a famous group of eccentric Homer pioneers and musicians. A gale warning was up for Gore Point and Otto's skipper was highly skeptical of taking this tired, thirty-year-old vessel into its eleven-foot seas.

"You were never in the military, were you, Otto?" the skipper said.

"No, thank God," Otto said.

"If you had been, you'd know never volunteer. You shouldn't have volunteered to take passengers on this boat."

"I've done a lot of exciting things in life by volunteering. I'm from a family that likes to push the envelope. You know what they say: 'A boat is safe in the harbor, but that's not what boats were made for.'"

"That's also how you end up getting sued in court."

It turned out Chris and Otto didn't have a firm agreement over how much the trip would cost or what it involved (Otto only owned the boat on a handshake). But with time running out, Chris launched the expedition, riding the landing craft overnight to Gore Point, finally able to sleep out of range of cell phone coverage. If the weather didn't clear enough for the helicopter to fly to Gore Point, the *Constructor* and the workers would be sitting out there with nothing to do. If the *Constructor* couldn't make it through the storm and had to hole up somewhere, the helicopter might show up anyway and waste all that money. But when morning dawned, the clouds over the mountains clumped in ragged patches, not a solid block. The forecast was wrong: the boat made it and the helicopter could fly.

For once, the sun was out, the broad Gulf of Alaska barely rumpled by long swells, empty of boats but with a family of humpback whales cruising along the surface. Gore Point stood solid and strong in the clear autumn light, no longer a mysterious dark shape in the mist. Drew Rose, the pilot for Maritime Helicopters, briefed the workers standing on the beach on how to stay safe under 250-pound Super Sacks that would be spinning around their heads. He said, "If you're worried about the hook hitting you at ninety miles per hour, just run the other way."

The dance began with the apocalyptic wind and gut-shaking rumble of the five-passenger Bell 407 hovering above, dipping the tether to the sacks for the guys to hook up, then whizzing the bags aloft as if on a magic elevator, a moment later to lower them to the deck of the *Constructor* to drop off. Chris got whacked on the head by the hook, which sent his hard hat flying but left him essentially uninjured. Over all, the work progressed faster and smoother than anyone had a right to expect. The chopper began bringing three Super Sacks at a time. Ted said, "I'm starting to feel happy again."

Drew, the pilot, looked down through the window between his legs to bring the hook to the hands of the workers, tweaking the joystick so the helicopter slid back and forth on the air like a skateboard, gently whipping the long cable to bring it in on the spot. Then, hooked up, pulling away, the rotors beat heavily in the air, the airframe leaning hard against the weight down below. He remained relaxed, chatting about the many disasters that were possible when lifting heavy loads around obstacles on a wilderness shoreline. Toward the end of the day the biggest piles remained and clouds and wind were bearing down from the sea. These piles were in those columnlike trees. Drew hovered downward among the trunks. The hook lowered smoothly, gently, into narrow spaces, the helicopter barely visible among the branches above. He was in his rhythm, in the zone.

Chris couldn't relax. Standing on the deck of the *Constructor,* waiting for the next load to come down from the sky, he said to one of his sons, Eric, "I'm not sure this was worth it. Worth all the money, and hassle, and wear and tear."

"It's worth it. The first time you do anything is always the hardest," Eric said.

Chris believed this would be his last year.

Suddenly, after a long day, the work was done. The *Constructor*'s huge deck lay buried in a pile of bagged garbage ten feet deep, twenty feet wide, one hundred feet long. High fives all around. Looking at the enormous pile, Doug said, "And all picked up with fingers." Ted walked through the forest, which was back to looking like a forest again, tidying up missed fragments and putting them in his pockets. He said, "This is the most gratifying work I've ever done." Ted and Doug had been on the point for fifty-four days straight. They were elated to be finished. Chris was just tired. He still had a lot of work left to do—the relatively easy physical work of getting the boats back to Homer and the garbage to the landfill, and the harder, political work of getting everything paid for. Someone walked the beach after the work ended to check on what had washed ashore during the day: twenty-two plastic bottles, two Clorox jugs, and a champagne cork.

Ted planned to stay out a while longer with his dog, anchoring his boat in some pretty cove to do some fishing. But before the *Constructor* left he recorded the size of its cargo of plastic trash with his video camera, slowly panning. "Here it is, all of this region's garbage on one boat," he said, narrating. "We've done a good thing."

38.

Celebrations

In conclusion, I'm hopeful, because I see so many people choosing community over competition. We can call this the end of a long, sad story—we're not obliged to continue the mistakes that were handed down to us. We're free to enter fulfilling relationships with people and nature. Many of us are opting out, at least partly, from the programming of more more more. I'm hopeful, as well, because so much science points to our better nature: our propensity to perceive meaning within other people, animals, and special places; our innate preference for fairness and cooperation; and our happiness when we fulfill those parts of ourselves. This promise of joy makes the hope of saving the oceans real. On one hand, we can hollow out marine ecosystems and engineer a biosphere exclusively of human beings eternally battling for escalating wealth; on the other hand, we can choose lives of purpose and material sufficiency, justice and shared membership in the continuum of creation. This is an easy choice.

There are no good or bad people. We all have a lot to learn. We need to gather and help one another. And why not? It feels good to get together, to share food and celebrate what we have in common. That's how we connect as communities, which is the first step to changing society. If you want to save the oceans, organize a potluck dinner.

• • •

A couple of weeks after the Gore Point cleanup, Chris Pallister invited me to an end-of-season party in Anchorage at the home of John Whitney, a cofounder of the beach cleaning group. Everyone was there, including Antonia Fowler, the superstar of beach cleaners and MS patient. I climbed a series of catwalks to the top of Whitney's steeply peaked roof, where a group stood sipping drinks on a sort of crow's nest, which looked out on the forest that surrounded the house. Among them was Gary Luchi, a marine debris gatherer from northern Australia who had come to learn about the Alaskan program. "I've come here because I told them you were years ahead of us. That's why they gave me the grant," Gary said.

"I thought you were years ahead of us," Ted Raynor said.

Gary described working with indigenous Australians, which sounded a lot like working with Alaska Natives, and about the things they found on the beaches, including a dead whale full of plastic. His beaches were like our beaches. Even the names were the same. Captain Cook visited at length inside the Great Barrier Reef, near Gary's home, repairing his ship after running aground, and left behind the same names he had scattered along Alaska's coast, including Hinchinbrook Island and Gore Point. Gary himself seemed like an Alaskan—expansive, ready for anything. Apparently the other side of the world was a mirror of our unique place, and people there were doing the same things we were, walking beaches picking up fishing nets.

Downstairs, Chris was checking the weather forecast and persuading Antonia to join him for a voyage to the sound the next day to show Gary around. It seemed all his talk of quitting had been forgotten.

The next time I saw Chris was the following summer, at the gas dock in Homer, late one night. Harried by complications and business problems, he had persuaded the dock boys to stay open after hours to sell him fuel for a trip back to Gore Point, where his crew had collected one hundred bags of garbage that had washed ashore over the winter.

. . .

I met Craig Matkin and Eva Saulitis again at a swim meet for Craig's son, in Anchorage. The last of the children was leaving home. When I had joined them in Prince William Sound two seasons earlier, looking for killer whales, we had talked a lot about raising kids and what values they would learn. Craig regretted his children's sense of entitlement to wealth and leisure, but added, "I'm the one who served it up, so there it is. These kids feel like they can do anything they want to do." Eva had nagged Craig's girls for years about being too materialistic. Then they went to Central America. "They suddenly get it," Eva said. The girls became active in conserving resources, biking everywhere, which Eva celebrated as a significant success, "with their whole lifetime ahead of them, to get that environmental consciousness."

Craig's killer whale work had opened up dramatically with the use of tags that could broadcast whales' positions via satellite once a day back to his computer in Homer. Besides saving fuel chasing the whales around, the tags produced huge surprises—the whales ranged much farther and faster than anyone ever thought. One whale spent time in the Bering Sea, met up with various other whales, then swam as fast as one hundred miles a day south, halfway to Hawaii. The whales' movements were also more complex, more territorial, and more mysterious than had ever been suggested by decades of watching them from above the sea. They couldn't be generalized; each pod and even each whale had its own personality. As a researcher, Craig was excited all over again. New knowledge had generated new mysteries.

Now that the children had left, they had stopped eating much red meat, for reasons of ethics as well as health. Eva felt differently about eating fish, such as the silver salmon she caught in the sound, unapologetically noting that the distinction was not rational. "When I eat them I feel this incredible connection with them," she said. "To eat something from Prince William Sound, and put it in my body, is the biggest gift."

I told her about my trip to England to meet Nicky Clayton and her brilliant birds. Eva had no doubt that many animals are intelligent and might even be conscious. She felt she had communicated with some of them. But that didn't diminish her friendship with a Chugach hunter. "I admire people who kill animals," Eva said. "It's a superconscious, spiritual choice."

All human food production affects other animals. Even organic farms displace natural ecosystems. Eva couldn't avoid eating, but she made her food choices seriously, with full knowledge of their impacts. I, too, respect those who look free animals in the eye before harvesting them for food. The Chugach believed animals had spirits and made gifts of their bodies to humankind for our sustenance. In that worldview, eating creates a responsibility. And that seems right. Finite energy reaches the earth from the sun; every breath we take is an expenditure of that precious, limited resource. Selecting food wisely and using it efficiently can stretch the supply, but the fate of the ocean may depend more on how we use the energy we derive after eating—whether we waste it, spend it on ourselves, or attempt to repay the debt that each meal implies.

The Native food potluck in Tatitlek faced a serious shortage until organizers approached the village's most active hunter, a young man who I never heard speak but who possessed such self-confidence he seemed more physically solid than other people. He fed the whole community that night: bear, deer, goose neck, seal ribs, sea lion flipper, porpoise skin, and other things we weren't sure about. Jim Miller was at my side, identifying what I was eating. The celebration, part of Heritage Week in the village school gym, honored Gary Kompkoff, the late village chief, who had led Tatitlek with wisdom and strength for two decades, through the years of the oil spill, the litigation, and all the changes, but had died before the lawsuit against Exxon ended. Speakers after dinner cried over their memories. The village was divided and people wondered how they could go on. Sheri Buretta, the chairwoman of Chugach Alaska Corp., said Chief Gary wouldn't have wanted that. "Break down your barriers," she said. "Be part of us as a community, because any of us may not be here tomorrow."

Patrick Norman, the chief of Port Graham, had come to honor Chief Gary. We talked about the decline of the long, lovely bay where his village lies, off Cook Inlet, just outside Kachemak Bay. Another villager had told me the bay was dead. And Patrick and others took responsibility for some of the losses. No seals, because a seal wouldn't last long in sight of the village. Cockles, crab, clams, and chitons missing, too. A graduate student worked with villagers to establish that they had driven down the chitons themselves, which Patrick said occurred because they hadn't followed the knowledge of the elders—traditionally, their people wouldn't have harvested chitons in the summer months. But they had never heard the elders' guidance on that.

Chief Patrick hoped to bring back the bay by bringing back the traditional way of life. The traditional council of which he was chair had been created by the Bureau of Indian Affairs, its membership elected by villagers. But elders wouldn't put themselves up for election—that's not the Native way. So younger, more westernized villagers took the positions and the elders were brushed aside. The village lost the value of their knowledge, wisdom, and leadership. As chief, Patrick wanted to use the village government to reverse those losses. "Our council's main purpose is to re-establish our elders' role in the village to what it was in the past," he said. The first goal would be practicing traditional resource management. Government scientific agencies would be brought in as partners to back up what the elders said rather than taking the lead. The village's lawyers would continue fighting the oil industry, which continued dumping toxins into the inlet from drilling platforms. But within the village Patrick hoped the values of sharing and passing on knowledge would become fertilizer for rich marine life to return to the bay.

The problem was old and couldn't be fixed quickly. Patrick himself, forty-nine, had no one to teach him to hunt as a young man. But the next generation would have teachers, if the village could rebuild families and the tribe as a cohesive unit.

Patrick said, "Long time ago, it was all structured for basic survival. That was the reason for this structure, the basic survival of the tribe. There's going to be changes, and that's inevitable, but the basic traditional values remain the same throughout time. A base set of values that kept the tribe together as a unit.

"That's an area we need to bring back out."

Rick Steiner never went back into the sound to see the mudslides on Tatitlek's clear-cut land. But he accepted my invitation to find the place where Gifford Pinchot and Will Langille made their stands against the fraud and waste of gold-rush-era development, where they stymied William Howard Taft's rollback of conservationism, and where, as much as anywhere, modern environmental politics were born. He wouldn't go back to Cordova, either, but we flew together into its airport and there boarded a bush plane with balloon tires out to a beach a half day's walk short of Katalla. I brought along Pinchot's diaries and Rick brought a file documenting his efforts to buy back the coal deposit Pinchot had fought to conserve, which Chugach Alaska had acquired through the land claims process and had sold to a Korean businessman.

But we ended up storm-bound in a tiny cabin on an immense sandy beach owned by huge brown bears, which was raked by that same merciless east wind that had driven Cook to shelter in Prince William Sound and destroyed Katalla's waterfront in 1907, that had sunk the Klondike gold steamer *Portland* and punished Pinchot himself, sweeping away the illusion that a town could ever survive here. I began the book describing this wild weather. The wind, the blowing sand, and exploding surf overwhelms all else and spins time back in on itself, leaving history in a coil of close-packed layers. Nature spools it all up. Those other conservationists and boomers, Pinchot and his foes, seemed barely out of reach, wrapped up against us, or even present in ourselves, as we sat by our lantern and heard the walls of the cabin creak in the same wind they heard.

My conversations with Rick became his catharsis, as he spilled out the frustration and grief he felt watching thousands of acres of forests fall in the sound while bureaucrats dithered and wasted money on aimless scientific studies. But he had also moved on, working to prevent similar disasters elsewhere in the world. He said, "When I'm doing the international stuff, I really feel connected. I really feel I'm helping. In Alaska, I feel like I'm banging my head against the wall."

Rick's fame and credibility were far greater away from home. Indeed, NOAA's Sea Grant Program and the University of Alaska, Fairbanks, recently canceled some of his funding in retaliation for taking positions against offshore oil development in Alaska, and in October 2009 he resigned in protest. He had testified to the Russian Duma and advised communities and governments in Colombia, Japan, Korea, and Indonesia, traveling to the Baltic nations on behalf of the U.S. State Department. After spending weeks in an obscure district of eastern Siberia, a group he worked with influenced the rerouting of a pipeline from Russia to China to avoid environmentally sensitive areas. The government of Lebanon brought him to see an oil spill caused by the 2006 war with Israel while bombs still fell, and he's also worked in Pakistan, the Niger Delta in Africa, and as a consultant to the United Nations. When we went to Katalla he was fresh back from Papua New Guinea, where the coastal environment was threatened by deep-sea mining. He helped indigenous people there create a citizens' advisory council like the one in Prince William Sound—that's his message all over the world, to replicate the model of local communities being given a voice to influence development within their ecosystems. But there was a misunderstanding on the New Guinea trip: in a traditional jungle village, tribesmen thought Rick's offer to help oppose a mining ship "in any way" included joining them in their dugout canoes to attack it with spears.

Rick's next trip would be to Azerbaijan. There and in Kazakhstan he had previously helped set up citizens' councils overseeing oil development, and those councils had begun to take hold. He said participatory democracy worked better in many places around the world compared to the United States, especially in the former Soviet republics, where Rick found true involvement and enthusiasm in newly free communities band-ing together to protect their own land and water. The year of the *Exxon Valdez*, 1989, was also the year of the Tiananmen Square uprising in China and the fall of the Berlin Wall. Since then, international nongovernmental organizations (NGOs) for the environment had vastly multiplied, become wealthier and more sophisticated. They often paid for Rick's travels. Local environmental organizations had risen up in a vast number of places to take charge of their own environments. Rick compared the shift from national to local power to what happens on the streets of Anchorage during a blackout. Usually, drivers are rude and aggressive at traffic lights, but when the power goes out and the lights stop working, they cooperate and take turns. Governments and corporations formed a destructive complex in many more places than Alaska, but Rick had seen citizens' councils effectively protect local ecosystems from that economic predation.

Rick's strongest message was his example, as he took it upon himself to parley with the powerful. "All these people are human, no matter what position they're in, and they need to be talked to by ordinary people," he said. "There's a growing amount of folks doing that."

The Chugach people call the sauna the *banya* and, like many such things, it means much more to them. Practically, it is the primary way of bathing. But it's also an intimate social space in the evening for losing the day's aches and tensions. Men and women bathe separately, but in groups. It's slow and quiet. A time for talking cures, as nothing can be hidden when naked. Diane Selanoff said women's bare skin shows any signs of abuse. Jim Miller said that for men, the steam is like the womb, and they come out brothers.

Jim taught me much in our conversations, gave me many enlightening ideas, yet also spoke like a prophet, sometimes too blunt to be appreciated, probably not easy to live with. It could be hard to grasp the contradictions. For example, the tension between freedom and purpose. He said he wanted his children to be free to go anywhere, do anything, forget about the racism and layers of intergenerational grief, take with them as much of their Native heritage as they desired. Be themselves. But he also talked of men and women's purposeful lives back when no one had such choices. In the old days, it took two men working in rhythm to cut down a tree with a handsaw that each pulled against the other; now one man with a chainsaw could do the job faster and without a friend. Contractors built houses now, not work parties. Head Start could raise the children instead of grandmothers. People had freezers and fast boats. They could harvest a year's worth of subsistence and put it up; they could harvest until the food was gone from the bay. The changes that brought freedom made freedom less worth having. "There's all kinds of tricky things that go into being a Native in a diminishing world," Jim said. "Because this world that we live in is getting smaller and smaller and smaller."

This is everyone's challenge, not just the Natives'—how to manage our own outgrown size in comparison to the place where we live, our habitat. As Jim put it, "There's a panic—it's a quiet panic. It's subconscious for some, and it's very aware for some. The world isn't any longer able to meet our needs."

When he went into his tiny workshop with those hunks of cedar drift-wood, Jim chose something new to make good. A first set of cuts on the band saw to get the general shape, then large chisel cuts and scooping out extra wood, finer cuts with ever smaller knives, and finally sanding. Some-thing new appeared that was also very old, something he made himself as no one else could, but that also reflected the sea, the forest, and the elders who lived long ago, using their designs. A new carving full of meaning, able to connect Jim to the person to whom he would make it a gift. Free-dom could take many forms. With such acts, the world could grow again.

At the end of a day together in Port Graham, Jim invited me to join him in his neighbor Richard's *banya*. He brought along a towel and washcloth for me. Four of us undressed in the narrow alcove to the shed in Richard's yard, then plunged through a little door into a dark room high enough only to stoop. A huge barrel stove roared in this tiny space. We sat on foam camping mats. Our host sprayed water on the stove and whipped a cloth in the air to mix the hot steam. The heat burst out like a solid, living thing.

This wasn't like my sauna at home. It took my breath away. I covered my ears to keep them from scalding and poured cold water over my head to escape the heat. My mind produced strange, disjointed thoughts—of burning up, of hell—but I didn't want to leave, at least not before the other men, and miss the crux of the experience. Just when I thought I couldn't stand it another moment, we retreated to the alcove and panted in the cool air. Then in again. In and out four times, washing up the last time in the hot room. Almost two hours passed. Afterward, I felt like I'd been seared down to my essence. My skin thinned, my blood whooshing through me like water, the air and light brilliant and fresh.

Walking back to my boat, I encountered the rest of the community at the airstrip, where the totem pole stands that Jim carved for the dead, with faces looking out of the holes of a skybound kayak. The young people were playing Aleut baseball on the flat runway, Chief Patrick Norman officiating as bugs swarmed around his head. The elders and almost everyone else in the village watched, sitting in the grass and cheering. I couldn't quite follow the game. There was one base. The rubber ball bounced this way and that. The batter—man or woman—was free to run or not upon hitting the ball, but once running the other team could get an out by hitting the runner with a thrown ball. A young woman came to bat amid much encouragement. She whacked the pitch, which spun crazily backward. The crowd hooted. She ran while a defensive player tried to extract the ball from between the bumper and tailgate of a pickup. Patrick had called it fair. Everyone laughed in the warm sun, which shone at the low, golden angle of an endless July evening, its light picking out the colors in faces, hats, and spruce boughs, and in the faces peering from Jim's richly colored totem pole for the dead, above the scene, with black-and-white killer whales ushering the kayak of the departed up from the blue sea to the blue, starry sky.

Acknowledgments

Creating this book required contributions from many people for a purpose they could not know, a fine example of the altruism I hope to convince readers is real. Sources invited me into their lives for weeks at a time and confided intimate information about themselves, gifts of trust I have tried to repay with honesty and discretion. For the most part, readers will learn the names of those who helped me when I refer to them in the text; I've also acknowledged those I interviewed or from whom I've learned in the endnotes. Please read each reference as an individual thank-you note. I regret dozens of others cannot be acknowledged lest this become an endless list of names.

Wendy Feuer, to whom the book is dedicated, is my dear friend and colleague through many past endeavors. Her contribution here is large and essential. She brought brilliance and diligence to digging out hundreds of historic documents and sources, unearthing the heart of the book, and contributing ideas, synthesis, and a breadth of understanding that I could not have accomplished without her. The material on the progressives, conservation, and eugenics belongs largely to her. All this work while dealing with cancer. Now she is departing life, a process filled with her usual grace and wisdom and with the constant love of her brave husband, Jeff Rubin, and sons, Nathaniel and Elias. I cannot repay her for all she has added to my life. I will always admire her and cherish our friendship.

My literary agent, Nicholas Ellison, besides his invaluable publishing instincts and skill, also gave me insight, caring, and loyalty through the long and daunting journey of creation. Over the years the book had three editors at Thomas Dunne Books, first John Parsley, then Joel Ariaratnam, and finally Peter Joseph, all of whom made important contributions. Peter's professionalism and sensitivity ultimately proved the key.

Finally, I thank the Anchorage Public Library, where I found both the conceptual skeleton of my book and much of the flesh to fill it out. I relied frequently on Bruce Merrell, now retired but still a friend and help, for his encyclopedic knowledge of Alaskan literature and sources. Dorothy Knaus researched the University of Oregon archives for me, joining in the hunt with gusto and effectiveness. Fred Bauman did similar good work in the Library of Congress Manuscript Reading Room.

Notes

The book relies on sources that include personal experiences, interviews, and records in many forms. Where it would impede the story to cite these sources in the text, I have listed them here. You can locate citations for each fact by using the page numbers and key phrases that match them.

1. The Gulf of Alaska Cauldron

9   **Sand lance:** Martin Robards et al., eds., *Sand Lance: A Review of Biology and Predator Relations and Annotated Biography* (Portland, OR: U.S. Department of Agriculture Forest Service, Pacific Northwest Research Station, 1999), Research paper PNW-RP-521.

9   **Humpback whales:** R. P. Angliss and B. M. Allen, "Marine Mammal Stock Assessment Report; Humpback Whale, Western North Pacific," Report NOAA-TM-AFSC-193, January 2, 2008; http://www.nmfs.noaa.gov/pr/species/ (accessed April 26, 2009).

9   **Herring will gather:** Robert T. Cooney et al., "Ecosystem Controls of Juvenile Pink Salmon (*Onchorynchus gorbuscha*) and Pacific Herring (*Clupea pallasi*) Populations in Prince William Sound, Alaska," *Fisheries Oceanography* 10, Suppl. 1 (2001): 1–13. Cooney also helped me understand ecosystem issues in several interviews.

10   **First the storm:** Robert B. Spies and Alan M. Springer, "Ecosystem Structure," in Robert B. Spies, ed., *Long-Term Ecological Change in the Northern Gulf of Alaska* (Amsterdam: Elsevier, 2007), 18–19. I relied often on this volume, an invaluable compendium of knowledge on the gulf ecosystem.

10   **The runoff from the gulf's:** Thomas C. Royer, Chester E. Grosch, and Lawrence A. Mysak, "Interdecadal Variability of Northeast Pacific Coastal Freshwater and Its Implications on

Biological Productivity," *Progress in Oceanography* 49 (2001): 95–111; also, interview with Royer, October 11, 2007.

11   **biological productivity of the Gulf:** Robert B. Spies, ed., *Long-Term Ecological Change in the Northern Gulf of Alaska* (Amsterdam: Elsevier, 2007).

13   **the population of fifty-year-old halibut:** Royer et al., "Interdecadal Variability." Royer was a great help in explaining fascinating ideas about the gulf ecosystem, most of which I could not include in the book.

13   **Science calls the system complex:** Spies, *Long-Term Ecological Change,* 551.

13   **unrelated herring populations:** Herring fishing in Kamishak Bay closed in 1999 and has not reopened. Some southeast Alaska herring stocks, including those in Lynn Canal, have been depressed for years.

2. Killer Whale Culture and Human Spirituality

17   **Farley Mowat's memoir:** Farley Mowat, *Never Cry Wolf: Amazing True Story of Life Among Arctic Wolves* (1963; New York: Back Bay Books, 2001).

18   **this split among killer whales:** Numerous interviews with Matkin and Saulitis from 2005 to 2009 provided much of my knowledge of killer whales; also Janet Mann, Richard C. Connor, Peter L. Tyack, and Hal Whitehead, eds., *Cetacean Societies: Field Studies of Dolphins and Whales* (Chicago and London: University of Chicago Press, 2000); and Luke Rendell and Hal Whitehead, "Culture in Whales and Dolphins," *Behavioral and Brain Sciences* 24 (2001): 309–382.

19   **their rate of genetic change:** Interview with Hal Whitehead, March 20, 2007; Hal Whitehead, "Cultural Selection and Genetic Diversity in Matrilineal Whales," *Science* 282 (1998): 1708; Hal Whitehead, P. J. Richerson, and Robert Boyd, "Cultural Selection and Genetic Diversity in Humans," *Selection* 3, no. 1 (2002): 115–125.

19   **A National Geographic documentary:** *Killer Whales: Wolves of the Sea,* National Geographic Society, Australian Broadcasting Corp., and TBS Productions, 1993, video.

3. Chugach Culture and the Spirits of Nature

27   **the Feast of the Dead:** Kaj Birket-Smith, *The Chugach Eskimo* (Copenhagen: Nationalmuseets Publikationsfond, 1953), 112–113.

27   **greeting ceremony:** Robin W. Baird, "The Killer Whale: Foraging Specializations and Group Hunting," in Janet Mann et al., eds., *Cetacean Societies,* 134.

28   **they've got the families right:** Interviews with Craig Matkin. H. Yurk et al., "Cultural Transmission Within Maternal Lineages: Vocal Clans in Resident Killer Whales in Southern Alaska," *Animal Behaviour* 63 (2002): 1103–1119; Baird, "The Killer Whale," 132–134, and Peter L. Tyack, "Functional Aspects of Cetacean Communication," 296–300, both in Janet Mann et al., eds., *Cetacean Societies.*

28   **more like the Yup'ik or Iñupiat Eskimos:** Interview with retired anthropologist Nancy Yaw-Davis, October 18, 2007.

29   **resources yielded wealth:** Birket-Smith, *The Chugach Eskimo;* as to warfare, see pp. 95–103; my interpretation of the evidence on the role of chiefs is in a note to Chapter 11 ("lay readers became chiefs").

29   **teachers were discussing consolidating:** Letter of Andrew Malakoff, teacher, Tatitlek, Alaska, submitted as an annual report to the Department of the Interior, Bureau of Educa-

tion, Alaska School Service, June 1, 1909, National Archives, Anchorage, Alaska, RG 75, Box 20, Tatitlek 1908–09. Five villages and population in 1909 from the same file.

30 **rot had moved too fast:** Frederica de Laguna, *Chugach Prehistory: The Archaeology of Prince William Sound, Alaska* (Seattle and London: University of Washington Press, 1956).

30 **informants had met these spirits:** Birket-Smith, *The Chugach Eskimo,* 121–123.

31 **raven calls are culturally determined:** Bernd Heinrich, *The Mind of the Raven* (New York: Ecco, 2002), 195–197.

31 **as many as eight hundred at a time:** Mark Schwan, "Common Raven," in *Wildlife Notebook Series,* a publication of the Alaska Department of Fish and Game, Juneau, 1994.

31 **evening happy-hour gatherings:** Rick Sinnott of the Alaska Department of Fish and Game was kind enough to teach me about ravens and take me to one of their preroost assembly sites in Anchorage, March 29, 2007.

32 **the ravens recognized her:** Interview with Stacia Backensto, April 10, 2007.

32 **In both stories:** For the Iñupiat version, see H. Ostermann, *The Alaskan Eskimos, as Described in the Posthumous Notes of Knud Rasmussen,* ed. E. Holtved, trans. W. E. Calvert (1952; repr., New York: AMS Press, 1976), 24–25; Makari's version is in Birkit-Smith, *The Chugach Eskimo,* 171–172.

4. How Animals Think

35 **apes that could form sentences:** Sara J. Shettleworth, *Cognition, Evolution, and Behavior* (New York and Oxford: Oxford University Press, 1998), 363–364, 554–556.

35 **John Lilly gave LSD:** Erika Check, "Exploring the Dark," *Nature* 443 (2006): 631.

35 **Nicky Clayton liked to get out:** Clayton was very generous with her time over a series of interviews in 2007.

35 **Clark's nutcrackers:** Nathan J. Emery and Nicola S. Clayton, "The Mentality of Crows: Convergent Evolution of Intelligence in Corvids and Apes," *Science* 306 (2004): 1903.

36 **Clayton and a colleague:** The colleague was Clayton's husband, Nathan Emery, who is introduced later in the chapter.

36 **projecting their own behavior:** N. J. Emery and N. S. Clayton, "Effects of Experience and Social Context on Prospective Caching Strategies by Scrub Jays," *Nature* 414 (2001): 443–446.

36 **New Caledonian crows:** Jackie Chappell and Alex Kacelnik, "Tool Selectivity in a Nonprimate, the New Caldonian Crow *(Corvus moneduloides),*" *Animal Cognition* 5 (2002): 71–78.

36 **meerkats teach:** Alex Thornton and Katherine McAuliffe, "Teaching in Wild Meerkats," *Science* 313 (2006): 227–229.

36 **songbirds use:** Timothy Q. Gentner et al., "Recursive Syntactic Pattern Learning by Songbirds," *Nature* 440 (2006): 1204–1207.

37 **cichlid fish reason:** Logan Gosenick, Tricia S. Clement, and Russell D. Fernald, "Fish Can Infer Social Rank by Observation Alone," *Nature* 445 (2007): 429–432.

38 **Humphrey first pointed out:** Nicholas Humphrey, "The Social Function of Intellect," in P. P. G. Bateson and R. A. Hinde, eds., *Growing Points in Ethology* (Cambridge: Cambridge University Press, 1976), 303–317.

38 **Jackdaws pair up:** Nathan Emery et al., "Cognitive Adaptations of Social Bonding in Birds," *Philosophical Transactions of the Royal Society B* 362 (2007): 489–505.

38 **jackdaw will even pick up:** Selvino R. de Kort, Nathan J. Emery, and Nicola S. Clayton, "Food Offering in Jackdaws *(Corvus monedula),*" *Naturwissenschaften* 90 (2003): 238–240.

39   **memories of moments and events:** D. R. Addis, A. T. Wong, and D. L. Schacter, "Remembering the Past to Imagine the Future: The Prospective Brain," *Nature Reviews Neuroscience* 8 (2007): 657–661.

40   **the jays adjusted their schedule:** Nicky Clayton and Tony Dickinson, "Episodic-like Memory During Cache Recovery by Scrub Jays," *Nature* 395 (1998): 272–278.

40   **experiment with children:** J. Russell, D. Alexis, and N. Clayton, "Episodic future thinking in 3- to 5-year-old children," *Cognition* 114 (2009): 56–71.

41   **Philosopher Peter Singer's work:** Peter Singer, *Practical Ethics* (Cambridge: Cambridge University Press, 1979).

5. Thinking and Being a Person

44   **the chimps didn't immediately grasp:** Daniel J. Povinelli, "Can Animals Empathize? Maybe Not," *Scientific American Presents,* 1998, 67–75; Derek C. Penn and Daniel J. Povinelli, "On the Lack of Evidence That Non-human Animals Possess Anything Remotely Resembling a 'Theory of Mind,'" *Philosophical Transactions of the Royal Society B* 362 (2007): 731–744. I interviewed Povinelli in September and November 2007.

45   **taken up by Christian creationists:** For example, the website "Evidence for God from Science," http://www.godandscience.org/evolution/imageofgod.html (accessed May 23, 2009).

45   **"ghosts, gravity, and God":** Daniel J. Povinelli, "Behind the Ape's Appearance: Escaping Anthropocentrism in the Study of Other Minds," *Dædalus* (Winter 2004): 29–41.

45   **contest that division:** Michael Tomasello et al., "Understanding and Sharing Intentions: The Origins of Cultural Cognition," *Behavioral and Brain Sciences* 28 (2005): 675–735; this publication includes both Tomasello's article and a section of "Open Peer Commentary," including an essay by Christophe Boesch, "Joint Cooperative Hunting Among Wild Chimpanzees: Taking Natural Observations Seriously," which states that field studies have demonstrated chimpanzees do share goals and intentions in hunting. I interviewed Tomasello on July 31, 2007.

46   **social thinking naturally led:** Nicholas Humphrey, "The Social Function of Intellect," in Bateson and Hinde, eds., *Growing Points in Ethology.*

46   **the same with the Eskimos:** Kaj Birket-Smith, *The Eskimos* (New York: Crown, 1971), 182.

46   **the child glanced to see:** Susan C. Johnson, "Detecting Agents," *Philosophical Transactions of the Royal Society B,* 358 (2003): 549–559.

6. The Essence of Carol Treadwell

51   **priggish and self-satisfied Christian:** Adam Gopnik, "Prisoner of Narnia: How C. S. Lewis Escaped," *The New Yorker,* November 21, 2005. I agree with Gropnik's critique. The Narnia books themselves suffer from Lewis's narrowness, tinged by racism, relegating girls and women to weaker roles, and allowing the rich necromancy of the author's imagination to be drained by Christian allegory, as the Christ-surrogate lion, Aslan, steps in to solve every problem, leaving the human characters as bystanders. Although I loved the books as a child, I stopped reading them to my own children.

51   **Nicholson's script:** William Nicholson, *Shadowlands* (New York: Plume, 1991).

52   **it can change a personality:** Richard E. Cytowic, "The Long Ordeal of James Brady," *New York Times,* September 27, 1981.

53   **masterfully conduct a choir:** Oliver Sachs, "The Abyss: Music and Amnesia," *The New Yorker,* September 24, 2007.

53 **retained theory of mind:** R. Shayna Rosenbaum, Brian Levine, and Endel Tulving, "Theory of Mind Is Independent of Episodic Memory," *Science* 318 (2007): 1257.

53 **fantasies become less detailed:** Addis et al., "Remembering the Past to Imagine the Future."

7. Connected or Alone

56 **A scientist at Oxford University:** Danielle Egan, "Death Special: The Plan for Eternal Life," *New Scientist*, October 13, 2007, 46.

58 **a fully domesticated biosphere:** Peter Kareiva et al., "Domesticating Nature: Shaping Landscapes and Ecosystems for Human Welfare," *Science* 316 (2007): 1866–1869.

58 **Kareiva told me:** Interview with Peter Karieva, March 28, 2009.

58 **a baton passing:** This is my way of paraphrasing the multiple-drafts model, explained among other important points on failure of duality in Daniel C. Dennett, *Sweet Dreams: Philosophical Obstacles to a Science of Consciousness* (Cambridge, MA: MIT Press, 2005). The capabilities of intuitive thought are lucidly covered in David G. Myers, *Intuition: Its Powers and Perils* (New Haven, CT: Yale University Press, 2002).

59 **conscious mind concocting reasons:** Mark Buchanan, "Secret Signals," *Nature* 457 (2009): 528–530.

59 **stimulating the parietal cortex:** Michael Desmurget et al., "Movement Intention After Parietal Cortex Stimulation in Humans," *Science* 324 (2009): 811–813.

59 **man is born for justice:** Marcus Tullius Cicero, "On the Laws," *The Treatises of M. T. Cicero*, trans. Charles Duke Yonge (London: Henry G. Bohn, 1853), 411, 406.

60 **One knows virtue:** Ibid., 398–426. I am indebted to Mark Tebbit, *Philosophy of Law: An Introduction* (London: Routledge, 2000).

60 **far more than mental calculation:** Greg Miller, "The Roots of Morality," *Science* 301 (2008): 734–737; H. A. Chapman et al., "In Bad Taste: Evidence for the Oral Origins of Moral Disgust," *Science* 323 (2009): 1222–1226.

60 **experiments with real money:** I'll cover this in depth in Chapter 15.

8. How Captain Cook Saw Alaska

64 **never was seen by Cook:** J. C. Beaglehole, ed., *The Journals of Captain James Cook on His Voyages of Discovery: The Voyage of the* Resolution *and* Discovery, *1776–1780* (1967; Woodbridge, England: Boydell Press, 1999); Cook, pp. 342–344, Samwell, pp. 1106–1107.

64 **Russian fur traders adopted:** Lydia T. Black, *Russians in Alaska 1732-1867* (Fairbanks: University of Alaska Press, 2004), 68.

64 **could have sold in Asia:** The pelts that Cook's men obtained sold for 120 Spanish dollars each in Canton, equivalent to 30 pounds sterling: Robert Kingery Buell and Charlotte Northcote Skaldal, *Sea Otters and the China Trade* (New York: David McKay, 1968), 52.

64 **only two siblings who survived:** J. C. Beaglehole, *The Life of Captain James Cook* (Stanford, CA: Stanford University Press, 1974), 3.

65 **Cook's personality is there:** Ibid., 14–17.

65 **its most successful explorers:** Bernard Smith, "Cook's Posthumous Reputation," in Robin Fisher and Hugh Johnson, eds., *Captain James Cook and His Times* (Seattle: University of Washington Press, 1979), 185.

66 **great and minor artists:** Alan Frost, "New Geographical Perspectives and the Emergence of the Romantic Imagination," in Fisher and Johnson, eds., *Captain James Cook and His Times*, 12–14.

66 **cartographers confidently drew:** Nigel Rigby and Pieter van der Merwe, *Captain Cook in the Pacific* (London: National Maritime Museum, 2002), 10–22; also Beaglehole, *The Life of Captain James Cook*, 107; Dr. van der Merwe kindly hosted me at the Maritime Museum in 2007 and showed me Cook's effects and original art from his trip.

67 **information it did contain was wrong:** George Vancouver, *A Voyage of Discovery to the North Pacific Ocean and Round the World 1791–1795*, W. Kaye Lamb, ed. (1798; London: Hakluyt Society, 1984), 1302–1304. Vancouver politely discounts Cook's negligence, supposing his notes were lost or wrongly edited; however, that wouldn't explain the inaccurate positions Cook did report.

67 **exaggerating the sound's:** Beaglehole, *The Journals of Captain James Cook*, 356–357.

67 **the territory of Tatitlek:** Frederica de Laguna, *Chugach Prehistory: The Archeology of Prince William Sound, Alaska* (Seattle: University of Washington Press, 1956), 24–25.

68 **the Tatitlek Natives retreated:** Beaglehole, *The Journals of Captain James Cook*, 346–348.

68 **"as if nothing had happened":** Ibid., King's journal, 1417.

69 **Makari said husbands:** Birket-Smith, *The Chugach Eskimo*, 96–98.

69 **After a little negotiation:** Susan J. Buck, *The Global Commons: An Introduction* (Washington, DC: Island Press, 1998), 77–80.

69 **"they do not belong to the Church":** Hugo Grotius, *Freedom of the Seas, or, The Right Which Belongs to the Dutch to Take Part in the East Indian Trade (Mare liberum)*, trans. Ralph Van Deman Magoffin (1608; New York: Oxford University Press, 1916; Arno Press, 1972), 16.

69 **the idea of natural law:** Ibid., 1–4.

70 **Possession matters:** Ibid., 25.

70 **many such daydreams:** Derek Hayes, *Historical Atlas of the North Pacific Ocean: Maps of Discovery and Scientific Exploration, 1500–2000* (Seattle: Sasquatch Books, 2001).

70 **Cook now believed:** Glyndwr Williams, "Myth and Reality: James Cook and the Theoretical Geography of Northwest America," in Fisher and Johnson, eds., *Captain James Cook and His Times*, 58–79.

71 **"trifling point of geography":** Beaglehole, *The Journals of Captain James Cook*, 368.

72 **"it may Puzle Antiquarians":** Ibid., King's journal, p. 1421.

73 **"not one would come near us":** Ibid., p. 1422.

73 **land rights as early as 1532:** Buck, *The Global Commons*, 78.

73 **explored by those living on them:** Grotius, *Freedom of the Seas*, 13.

73 **war without cause was unjust:** Ibid., 8–19

73 **"Lords of their own country":** Beaglehole, *The Life of Captain James Cook*, 150.

74 **editor deleted that passage:** Nicholas Thomas, *Cook: The Extraordinary Voyages of Captain James Cook* (New York: Walker, 2003), xxvi–xxvii.

74 **"the two Cardinall virtues":** Thomas Hobbes, *Leviathan*, ed. C. B. Macpherson (1651; London: Penguin Classics, 1982), 187–188.

75 **the essence of legal positivism:** This is, of course, a simplification. For a fuller explanation of the rise of legal positivism over natural law, see Mark Tebbit, *Philosophy of Law: An Introduction* (London: Routledge, 2000).

9. Rethinking the Moon

77 **The moon was a powerful spirit:** Birket-Smith, *The Chugach Eskimo*, 115, 119–120, 175.

77 **as Chaucer mentioned:** Geoffrey Chaucer, *The Canterbury Tales* (1386; New York: Apple-

ton, 1855), 15, lines 390–412. Notably, Chaucer's next profile is of the doctor, an expert in curing ailments using the zodiac and the four humors.

78 **machines that grew:** David Edgar Cartwright, *Tides: A Scientific History* (Cambridge: Cambridge University Press, 1999); my entire discussion of the scientific history of the tides is based on this volume.

79 **established in 1989:** Ibid., 236–237.

79 **"a space-bound ghost":** Walker Percy, *Lost in the Cosmos: The Last Self-Help Book* (New York: Farrar, Straus & Giroux, 1983), 12–13.

80 **literature from before 1800:** Henry Knight Miller, "Augustan Prose Fiction and the Romance Tradition," in R. F. Brissenden and J. C. Eade, eds., *Studies in the Eighteenth Century, III: Papers Presented at the Third David Nichol Smith Memorial Seminar, Canberra 1973* (Canberra: Australian National University Press, 1976), 241–255.

81 **idolized Captain Cook:** Frost, "New Geographical Perspectives and the Emergence of the Romantic Imagination."

81 **values fit for a hippie:** Adam Kirsch, "Avenging Angel: Inside Shelley's Manichaean Mind," *The New Yorker,* August 27, 2007.

81 **Experimental surveys:** Jonathan Haidt, "The New Synthesis in Moral Psychology," *Science* 316 (2007): 998–1001.

82 **"Nobody but a beggar":** Adam Smith, *An Inquiry into the Nature and Causes of the Wealth of Nations* (1776; Washington: Regnery, 1999), 14.

10. Competition in Men and Snails

85 **amazed his professors:** Details on Geerat Vermeij's life story come from my recorded interviews with him on November 9, 2005, and from his vivid autobiography *Privileged Hands: A Scientific Life* (New York: Freeman, 1997).

86 **Vermeij forecast our fate:** Geerat Vermeij, *Nature: An Economic History* (Princeton, NJ: Princeton University Press, 2004).

87 **Mollusks solved advanced problems:** The examples of mollusk adaption in these paragraphs all come from a fascinating book: Geerat Vermeij, *A Natural History of Shells* (Princeton, NJ: Princeton University Press, 1993).

88 **the tooth had evolved:** Vermeij lecture, University of California, Davis, November 9, 2005.

89 **island-dwelling dwarf elephants:** Elizabeth Culotta, "The Fellowship of the Hobbit," *Science* 317 (2007): 740–742; Eleanor M. Weston and Adrian M. Lister, "Insular Dwarfism in Hippos: A Model for Brain Size Reduction in *Homo floresiensis,*" *Nature* 459 (2009): 85–88.

90 **calorie-rich game:** Ann Gibbons, "Food for Thought," *Science* 316 (2007): 1558–1560.

91 **competition drove adaptation:** Vermeij tells the story of this work in *Privileged Hands* and expanded upon it in our conversations in November 2005.

93 **the directionality inherent in history:** Vermeij, *Nature: An Economic History,* xi.

93 **the total human take:** Rodger Doyle, "The Lion's Share: Measuring the Human Impact on Global Resources," *Scientific American,* April 2005.

93 **destroyed its capacity:** Boris Worm et al., "Impacts of Biodiversity Loss on Ocean Ecosystem Services," *Science* 314 (2006): 787–790; Mark Schrope, "The Real Sea Change," *Nature* 443 (2006): 622–624.

93 **the oceans turn acidic:** James C. Orr et al., "Anthropogenic Ocean Acidification over the Twenty-first Century and Its Impact on Calcifying Organisms," *Nature* 437 (2005): 681.

11. The Russian Conquest of Alaska

95   **no otters were seen:** Ivan Veniaminov, *Notes on the Islands of Unalashka District,* trans. Lydia T. Black and R. H. Geoghegan (1840; Fairbanks, AK, and Kingston, ON: University of Alaska Fairbanks/Limestone Press, 1984), 146. As to the harvest on the first expedition, a letter from Shelikhov dated Okhotsk, 1789, is quoted in Hubert Howe Bancroft, *History of Alaska, 1730–1885* (1886; Darien, CT: Hafner, 1970), 192–193. Most sources about Russian America are not contemporaneous, and those that are, are frequently unreliable; all information about the period requires some skepticism.

95   **the sea cow was extinct:** Black, *Russians in Alaska,* 67–68.

95   **frontiers draw psychopaths:** Hector Chevigny, *Russian America: The Great Alaskan Venture, 1741–1867* (Portland, OR: Binford & Mort, 1965), 35.

96   **leaving survivors to starve:** Black, *Russians in Alaska,* 89; Chevigny, *Russian America,* 38.

97   **these businessmen:** Chevigny, *Russian America,* 49–54.

97   **bloody, unprovoked battle:** Lydia T. Black, "The Russian Conquest of Kodiak," *Anthropological Papers of the University of Alaska* 24, no. 1–2 (1992): 165–182.

97   **Baranov pushed:** Black, *Russians in America,* 132–135, 183. As to the comparative size of West Coast towns, Russian America scholar Barbara Sweetland Smith notes that the population of Sitka in 1819 was 759, while San Francisco had 196 people in 1842, and Oregon City had 500 in the 1840s; Smith, personal communication.

98   **when they landed in Canton:** Beaglehole, *The Life of Captain James Cook,* 685.

98   **already married:** Black, *Russians in America,* 123.

98   **Baranov was deeply impressed:** This story is told in Baranov's letters. It seems to have been embellished by Hector Chevigny, *Lord of Alaska: Baranov and the Russian Adventure* (New York: Viking, 1942), 62–74; and Chevigny, *Russian America,* 84–89.

99   **lay readers became chiefs:** Birket-Smith, *The Chugach Eskimo,* 214–215. Birkit-Smith presents strong evidence that Russians introduced the system of chiefs, but is himself unconvinced. Alaskan anthropologist Nancy Yaw Davis said in an interview on October 18, 2007, that the Russians probably installed lay readers who became de facto chiefs, with the lay reader's son-in-law becoming the second chief.

99   **ramshackle trading post:** Jim and Nancy Lethcoe, *A History of Prince William Sound Alaska,* 2nd ed. (Valdez, AK: Prince William Sound Books, 2001), 29–36.

12. The American Conquest of Alaska

102   **Russian American population:** Svetlana G. Fedorova, *The Russian Population in Alaska and California, Late 18th Century–1867,* trans. and ed. Richard A. Pierce and Alton S. Donnelly (Kingston, ON: Limestone Press, 1973).

102   **burnings of entire villages:** Hubert Howe Bancroft, *History of Alaska, 1730–1885* (1886; New York: Antiquarian Press, 1959), 606–619.

102   **another clan in defense:** Donald Craig Mitchell, *Sold American: The Story of Alaska Natives and Their Land: 1867–1959, The Army to Statehood* (Hanover, NH: University Press of New England, 1997), 43–57.

102   **starvation of the Iñupiat:** Ibid., 132–135.

102   **depleted other valuable furbearing animals:** Ernest Gruening, *The State of Alaska,* rev. ed. (New York: Random House, 1968), 72–73. Gruening's book still contains the best concise history of Alaska resource waste.

102 **eliminating nesting geese:** On fox farming, see Lethcoe, *A History of Prince William Sound Alaska*, 93–94. On fox impact on birds, see Union of Concerned Scientists, "Invasive Species Alaska," Cambridge, MA, 2003, http://www.ucsusa.org/assets/documents/invasive_species/Alaska_invasives_1.pdf (accessed January 12, 2008).

103 **seemingly unlimited fish:** Richard A. Cooley, *Politics and Conservation: The Decline of the Alaska Salmon* (New York: Harper & Row, 1963), 23–28, 72–82. This classic tells, with scientific precision, the entire story of the salmon industry before statehood.

104 **more effort to chase fewer fish:** Ibid., 58–61.

104 **business model:** Colin W. Clark, "The Economics of Overexploitation," *Science* 181 (1973): 630. These points are still debated academically, as the practicality of fishermen's decisions depends on various economic inputs. See R. Q. Grafton, T. Kompas, and R. W. Hilborn, "Economics of Overexploitation Revisited," *Science* 318 (2007): 1601.

104 **next lower ecosystem level:** Daniel Pauly et al., "Towards Sustainability in World Fisheries," *Nature* 416 (2002): 689, and Daniel Pauly et al., "The Future of Fisheries," *Science* 302 (2003): 1359.

104 **Sea urchin depletion:** F. Berkes et al., "Globalization, Roving Bandits, and Marine Resources," *Science* 311 (2006): 1557.

105 **reducing predators:** Cooley and Gruening document this history. Bounty hunters killed 100,000 eagles between 1917 and 1953, according to the Alaska Department of Fish and Game's Alaska Wildlife Notebook (Juneau, 1994).

105 **a bounty for each seal nose:** Fishermen and Chugach Natives tell of seal bounty hunts, which sometimes involved dynamite thrown into seal colonies. Killer whale biologist Craig Matkin began his career studying the population impact of killer whale shootings. Alex deMarban reports 45,000 sea lions killed in official state hunts up to 1972, and fishermen slaughtering them into the 1990s; see "Shooting Sea Lions—Hunters, Fishermen Among Suspects in Pinniped Collapse," *Anchorage Daily News,* December 8, 2006.

105 **five hundred pounds of salmon:** Cooley, *Politics and Conservation,* 17–19.

105 **starvation struck Interior:** David M. Dean, *Breaking Trail: Hudson Stuck of Texas and Alaska* (Athens: Ohio University Press, 1988), 276–279.

105 **wiped out the Eyak:** Frederica de Laguna, "Eyak," in Wayne Suttles, ed., *Handbook of North American Indians,* vol. 7: *Northwest Coast* (Washington, DC: Smithsonian Institution, 1990), 195.

105 **Eyak died in 2008:** Debra McKinney, "Last Native Speaker of Eyak Language Dies," *Anchorage Daily News,* January 23, 2008.

106 **fished for the canneries:** Harold Hassen, "The Effect of European and American Contact on the Chugach Eskimo of Prince William Sound, Alaska, 1741–1930," (PhD diss., University of Wisconsin, Milwaukee, 1978), 142–157.

106 **the villages shrank:** W. A. Langille, "Report on the Forest Conditions and Resources of Western Alaska from Dry Bay to Prince William Sound," 1904, National Archives, Anchorage, Alaska, U.S. Forest Service pre-1960, Regional Office Records, Land Case Files 1904–1955, RG 95, Box 23. Also see Lethcoe.

106 **schoolteacher reported:** Andrew Malakoff to Bureau of Education, Department of the Interior, Tatitlek, Alaska, January 31, 1908, National Archives, Anchorage, Alaska, Records of the Bureau of Indian Affairs, Alaska Division, General Correspondence 1908 (calendar year) Tanana-Yukon, RG 75, Box No. 10.

106   **half the sound's Natives had died:** F. M. Boyle, M.D., to William Hamilton, Acting Chief, Commission of Education, Department of the Interior, Valdez, Alaska, November 19, 1908, National Archives, Anchorage, Alaska, Records of the Bureau of Indian Affairs, Alaska Division, General Correspondence 1908, Southwest District, RG 75, Box No. 8.

106   **consolidate the villages:** Malakoff to Bureau, Annual Report, Tatitlek, Alaska, June 1, 1909, RG 75, Box No. 20.

107   **Golden, up Port Wells:** Jim and Nancy Lethcoe, *A History of Prince William Sound Alaska*, 99.

107   **kelp forest destroyed:** James A. Estes and John F. Palmisano, "Sea Otters: Their Role in Structuring Nearshore Communities," *Science* 185 (1974): 1058–1060; Lloyd Lowry and James Bodkin, "Marine Mammals," in Phillip R. Mundy, ed., *The Gulf of Alaska: Biology and Oceanography* (Fairbanks: Alaska Sea Grant College Program, University of Alaska Fairbanks, 2005), 113.

107   **by 1968 few people had even seen one:** Buell and Skladal, *Sea Otters and the China Trade*, ix.

107   **live over muddy bottoms:** Angela Doroff, U.S. Fish and Wildlife Service, interview, August 11, 2007.

108   **acres of herring:** George Rounsefell, "Contribution to the Biology of the Pacific Herring, *Clupea pallasii,* and the Condition of the Fishery in Alaska," *Bulletin of the Bureau of Fisheries* 45 (1929): 227–320. Homer historian Janet Klein called this publication to my attention.

108   **herring is an all-purpose food:** Mundy, *The Gulf of Alaska.*

108   **the runs died out:** Susan Woodward Springer, *Seldovia, Alaska: An Historical Portrait of Life in Zaliv Seldevoe-Herring Bay* (Littleton, CO: Blue Willow, 1997), 121–126; Janet Klein, *A History of Kachemak Bay: The Country, the Communities* (Homer, AK: Homer Historical Society, 1981), 62–67.

109   **energy up the web:** Neil Rooney et al., "Structural Asymmetry and the Stability of Diverse Food Webs," *Nature* 442 (2006): 265–269.

109   **beluga whales:** Janet Klein said that belugas haven't been seen in Kachemak Bay for twenty years or more, based on careful records of sightings that she kept; January 9, 2008.

13. Evolution, Free Will, and Hope

110   **Russian America had voices:** Starting at least as early as 1754, according to Lydia Black, *Russians in America,* 68.

110   **tried to stop the Natives:** Springer, *Seldovia, Alaska,* 121.

110   **salmon were inexhaustible:** George Bird Grinnell, *Alaska 1899: Essays from the Harriman Expedition* (1901; Seattle: University of Washington Press, 1995), 339.

112   **help with Geerat's research:** Details from Vermeij's life are from *Privileged Hands* and our November 2005 interviews.

112   **Wilson predicted:** Edward O. Wilson, *Sociobiology: The New Synthesis* (Cambridge, MA: Belknap Press of Harvard University Press, 1975), 575.

112   **computers taking over:** The 1970 movie is called *Colossus: The Forbin Project.*

112   **A mega-hit scientific paper:** S. J. Gould and R. C. Lewontin, "The Spandrels of San Marco and the Panglossian Paradigm: A Critique of the Adaptationist Programme," *Proceedings of the Royal Society of London, Series B,* 205 (1979): 581–598.

112   **too much scratching:** Rudyard Kipling, *Just So Stories* (Garden City, NY: Doubleday, 1907), 31.

112   **Gould believed adaptation:** The controversy is clearly covered in Michael Ruse, *The Evolution Wars: A Guide to the Debates* (Santa Barbara: ABC-CLIO, 2000), 231–260. Ruse points

out that Gould's rejection of scientific racism also informed his opposition to adaptation as the driver of evolution. As we'll see in Chapter 20, Darwinism has often been used to support the superiority of some people over others.

113 **Creationists grabbed:** Robert Wright, "The Accidental Creationist: Why Stephen Jay Gould Is Bad for Evolution," *The New Yorker,* December 13, 1999.

113 **a book to make peace:** Stephen Jay Gould, *Rocks of Ages: Science and Religion in the Fullness of Life* (New York: Ballantine, 1999), 3–10.

113 **finches in the Galapagos:** Jonathan Weiner, *The Beak of the Finch: A Story of Evolution in Our Time* (New York: Knopf, 1994).

113 **fish called cichlids:** Thomas D. Kocher, "Evolutionary Biology: Ghost of Speciation Past," *Nature* 435 (2005): 29–30.

114 **more closely related:** Annalisa Berta and James L. Sumich, *Marine Mammals: Evolutionary Biology* (San Diego, CA: Academic Press, 2003), 49–79.

114 **the purpose of the markings:** Interview with Bob Pitman of the National Oceanic and Atmospheric Administration, April 11, 2005.

14. Taming the Gold Rush

119 **without losing any:** Langille's letter home describing his trip to Dawson City is reprinted in Lawrence Rakestraw, "A Mazama Heads North: Letters of William A. Langille," *Oregon Historical Quarterly* 76, no. 2 (June 1975). He put down one horse after the trek because of saddle sores and sold the rest for a good price.

120 **the miserable wretches:** Pierre Burton, *The Klondike Fever: The Life and Death of the Last Great Gold Rush* (1958; New York: Knopf, 1984), 155.

120 **he bragged of losing none:** Rakestraw, "A Mazama Heads North."

120 **Jack London followed:** This phase of London's career is detailed in Franklin Walker, *Jack London and the Klondike: Genesis of an American Writer* (San Marino, CA: Huntington Library, 1966), 127–153. For Langille's winter on the Stewart, see: William Langille to his mother, S. Langille, Stewart River, January 22, 1898, W. A. Langille papers, Special Collections, University of Oregon Library. Unless otherwise noted, all the Alaska letters are in the first folder of container AX635. I can find no evidence supporting the statement that London and Langille shared a cabin, found in Lawrence Rakestraw, *A History of the United States Forest Service in Alaska* (1981; Juneau: USDA Forest Service, 2002), 17. However, that book was invaluable in learning about Langille and forest history.

120 **metal sticks to your fingers:** William Langille to his mother, S. Langille, Dawson, Yukon, December 4, 1897, W. A. Langille Papers.

120 **campaigns to protect:** Langille was a founder of the Mazamas, a mountaineering club in Oregon, which sent resolutions to Washington, D.C., on these issues. A March 31, 1896, resolution he signed is contained in the papers of Gifford Pinchot, Library of Congress, container 1007. In the same container is a Mazamas letter to President Grover Cleveland regarding the Olympic Peninsula, which is undated and unsigned.

120: **sometimes intolerant:** A character sketch of Langille and evaluation is included in F. E. Olmsted, Assistant Forester, "Report on Alexander Archipelago Forest Reserve," 1906, 15–20, National Archives, Anchorage, Alaska, RG 95, U.S. Forest Service, Regional Forest Juneau, Historical Files 1903–1911, Box 1, 12/06/13 (4), Folder 4.

121   **every acre of it should:** He made this suggestion more than once, including a letter to Chief
      Forester Gifford Pinchot, Seward, Alaska, January 10, 1905, National Archives, Anchorage,
      Alaska, RG 95, Records of the Forest Service pre-1960, Regional Office, Land Case Files,
      1904–1955, Box 23, Folder 1.

121   **Will roamed the woods:** Joseph Gaston, *Portland Oregon, Its History and Builders,* vol. 3
      (Chicago and Portland, OR: S. J. Clarke, 1911), 816–819. Details also from an unsigned
      memoir by an old Langille family friend, which resides in the W. A. Langille Papers, con-
      tainer AX635, last folder.

121   **He attended school:** Personal Data Memorandum, February 2, 1935, William Langille
      personnel file, National Personnel Records Center (Civilian Personnel Records), St. Louis,
      Missouri.

121   **feminist novel of 1883:** Olive Schreiner, *The Story of an African Farm* (1883; New York:
      Modern Library, 1927).

121   **in 1891 took over management:** Gaston, *Portland Oregon: Its History and Builders.*

122   **150 times between them:** Fred H. McNeil, *Wy'east "The Mountain": A Chronicle of Mount
      Hood* (Portland, OR: Metropolitan Press, 1937), 79–87.

122   **Will installed huge cables:** Anne M. Lang to Mr. A. O. Waha, Mt. Hood Forest, September
      12, 1940, W. A. Langille Papers, container AX635, folder 5.

122   **"awakened better things":** William Langille (signed "Dad") to his daughter, Elizabeth,
      February 12, 1940. W. A. Langille Papers, container AX635, folder 3.

122   **Those two days in the woods:** Harold D. Langille, "Mostly Division 'R' Days: Reminis-
      cences of the Stormy, Pioneering Days of the Forest Reserves," *Oregon Historical Quarterly*
      57 (December 1956): 301–313. The spruce was *Picea engelmannii, var. columbiana.*

123   **dragged Pinchot and Newell:** Details on the hike are in typed extracts of Newell's diary for
      August 19–23, 1901, Pinchot Papers, Library of Congress, container 998. Doug Langille tells
      the story in "Division R. Days." On Pinchot's exhaustion, Gifford Pinchot to H. D. Langille,
      Washington, January 25, 1940; in Pinchot Papers, Library of Congress, container 1003.

123   **of utmost importance to Pinchot:** Gifford Pinchot, *Breaking New Ground* (1947; Washing-
      ton, DC: Island Press, 1998), 289.

123   **with astonishing success:** Langille, "Division 'R' Days"; as to success in turning around
      public opinion, Pinchot's papers include various letters from the west praising Doug, and
      the *Denver Post* of June 15, 1904, pictures Pinchot with a laurel wreath and calls him "well
      known and popular in Colorado."

123   **gold prospects had been exaggerated:** William Langille, "A Mazama Goes North." As to fifty
      summits of Mount Hood, McNeil, *Wy'east "The Mountain,"* 85. As to financial difficulties,
      William Langille to Sarah Langille, Stewart River, January 22, 1898; W. A. Langille Papers. As
      to gold being exaggerated, Will to Sarah Langille, December 4, 1897, W. A. Langille Papers.

123   **never made the money:** William Langille to "Friends," Dawson, September 3, 1899; Will to
      Doug, Dawson, January 16, 1900; both in W. A. Langille Papers.

123   **clear across Alaska:** Travel time to Nome was 329.5 hours according to a résumé of dates
      written by Will; W. A. Langille Papers, container AX635, last folder.

123   **prospecting over the tundra:** Will to Sarah, Nome, August 21, 1900; W. A. Langille Papers.

123   **Will responded:** William Langille to Gifford Pinchot, Hood River, Oregon, March 5, 1903;
      National Archives, Anchorage, Alaska, RG 95, U.S. Forest Service, Regional Forester Ju-
      neau, Historical Files 1903–1911, Box 1, 12/06/13 (4), folder 4.

123 **Pinchot soon hired him:** Federal personnel file.

124 **In less than a week's time:** Langille résumé of dates, W. A. Langille Papers, container AX635, last folder.

124 **summer of 1903:** Rakestraw, *A History,* 15–17.

124 **explore most of Alaska:** Gifford Pinchot to W. A. Langille, April 23, 1904, Washington, National Archives, Anchorage, Alaska, RG 95, U.S. Forest Service, Regional Forester Juneau, Historical Files 1903–1911, Box 1, 12/06/13 (4), folder 3.

124 **wading fully dressed:** Rowing, at Sagamore Hill: Gifford Pinchot's diary, July 3, 1903, Pinchot Papers, Library of Congress. In the canal, E. F. Baldwin to Pinchot (with newspaper clipping), October 21, 1905, New York, and Pinchot to Baldwin, November 2, 1905, Washington, Pinchot Papers, Library of Congress, container 575.

124 **"We seem to win":** Gifford Pinchot, typed summary of letters, June 10, 1905, Pinchot Papers, Library of Congress, container 10.

125 **massive Alaskan land fraud:** Langille to Pinchot, undated (received May 27, 1904); June 14, 1904, Katalla; July 16, 1904, Nome, National Archives, Anchorage, Alaska, RG 95, U.S. Forest Service, Regional Forester Juneau, Historical Files 1903–1911, Box 1, 12/06/13 (4), folder 1.

125 **"bits of emerald":** W. A. Langille, "Report of the Forest Conditions of Western Alaska from Dry Bay to Prince William Sound," 1904; National Archives, Anchorage, Alaska, RG 95, U.S. Forest Service pre-1960, Regional Office Records, Land Case Files 1904–1955, Box 23, 12/06/02 (3–6), folder 3.

125 **"they dismiss the subject":** Ibid., 22.

126 **bound for Unalaska:** W. A. Langille to Pinchot, July 16, 1904 (see above).

126 **a hilarious account:** W. A. Langille to Miss Strause, August 7, 1904, Cement Creek, Koyuk River, W. A. Langille Papers, AX635, folder 3.

126 **Will was in Seward:** W. A. Langille to Pinchot, September 14, 1905, Seattle; October 7, 1904, Seward; National Archives, Anchorage, Alaska, RG 95, U.S. Forest Service, Regional Forester Juneau, Historical Files 1903–1911, Box 1, 12/06/13 (4), folder 1 and folder 3.

127 **pleasure to the world:** W. A. Langille, "The Proposed Forest Reserve on the Kenai Peninsula, Alaska," October–December 1904, 41–42, National Archives, Anchorage, Alaska, RG 95, U.S. Forest Service pre-1960, Regional Office Records, Land Case Files, 1904–1955, Box 23, 12/06/02 (3–6), folder 2.

127 **the word "conservation":** Pinchot, *Breaking New Ground,* 326. Historians contest Pinchot's claim to have "invented" the word, but he clearly brought it to wide usage.

127 **down the Tanana to Fairanks:** W. A. Langille, "A Report on a Forest Reconnaissance from Cook Inlet to Circle City, Alaska," 1905, National Archives, Anchorage, Alaska, RG 95, U.S. Forest Service pre-1960, Regional Office Records, Land Case Files 1904–1955, Box 23, 12/06/02 (3–6), folder 1.

127 **up the Chena:** W. A. Langille to Pinchot, Wrangell, Alaska, June 10, 1905, National Archives, Anchorage, Alaska, RG 95, U.S. Forest Service, Regional Forester Juneau, Historical Files 1903–1911, Box 1, 12/06/13 (4), folder 1.

127 **back to Fairbanks:** W. A. Langille to Chief Section, Reserve Boundaries, March 29, 1905, Copper Center, National Archives, Anchorage, Alaska, RG 95, U.S. Forest Service, Regional Forester Juneau, Historical Files 1903–1911, Box 1, 12/06/13 (4), folder 3.

128 **boat up the Yukon:** W. A. Langille to Pinchot, June 10, 1905.

128 **26.7 million acres:** Size figure is from F. E. Ames inspection report of the Chugach and Tongass national forests, October 1909 report, 5, National Archives, Anchorage, Alaska, Record Group 95, U.S. Forest Service, Regional Forester Juneau, Supervision and Inspection, 1909–1963, Box 1, 12/05/01-02.

128 **many died of exposure:** Lethcoe, *A History of Prince William Sound*, 61–66.

129 **$100 million in shares:** James B. Adams, Assistant Forester, to the Forester, September 4, 1913, supplement to "O, Supervision, Chugach and Tongass," a report of the same date, Pinchot Papers, Library of Congress, container 430.

129 **rails never made it anywhere:** The federal government finally built the line, the Alaska Railroad, which consistently lost money and failed to develop the Alaska economy until it was needed to support World War II construction, as Mitchell documents in *Sold American*, 203–204.

130 **a report written shortly before:** W. A. Langille, Forest Supervisor, "A Statement of Conditions on the Tongass National Forest," February 11, 1911, National Archives, Anchorage, Alaska, RG 95, U.S. Forest Service, Regional Forester Juneau, Correspondence Land Files, 1903–1907, Box 43 12/07/06, folder 4.

131 **Will's supervisor wrote:** George H. Cecil, to the Forester, March 4, 1911 (attached to the Langille report cited in the previous note).

15. The Anatomy of Sharing

132 **complex experiments:** Herbert Gintis et al., eds., *Moral Sentiments and Material Interests: The Foundations of Cooperation in Economic Life,* (Cambridge, MA: MIT Press, 2005); Ultimatum Game, 11–13; other experiments, Ernst Fehr and Urs Fischbacher, "The Economics of Strong Reciprocity," same volume, 151–191; percent of split in Joseph Hernich et al., *Foundations of Human Sociality: Economic Experiments and Ethnographic Evidence from Fifteen Small-Scale Societies* (Oxford: Oxford University Press, 2004), 19.

133 **a new science:** This interdisciplinary movement is summarized in the introductory chapter of N. J. Enfield and Stephen C. Levinson, eds., *Roots of Human Sociality: Culture, Cognition and Interaction* (Oxford: Berg, 2006).

133 **an article in *Nature*:** Ernst Fehr and Urs Fischbacher, "The Nature of Human Altruism," *Nature* 425 (2003): 785–791.

134 **If they could be fined:** Fehr and Fischbacher, "The Economics of Strong Reciprocity," in Gintis et al., eds., *Moral Sentiments and Material Interests*.

134 **explicit incentives:** Elinor Ostrom, "Policies That Crowd Out Reciprocity and Collective Action," in Gintis et al., eds., *Moral Sentiments and Material Interests,* 253–275.

135 **anonymous Prisoner Dilemma:** Fehr and Fischbacher, "The Nature of Human Altruism," in Gintis et al., eds., *Moral Sentiments and Material Interests*.

135 **Altruistic punishers:** Colin F. Camerer and Ernst Fehr, "When Does 'Economic Man' Dominate Social Behavior?" *Science* 311 (2006): 47–52.

136 **I participated in a test:** Marco Janssen, of Arizona State University, and Elinor Ostrom, of Indiana University, hosted me for this test on October 23, 2007; they were testing software for later use in a real experiment.

137 **how selfish evolution could produce:** For a clear and provocative account of these issues, and an alternative solution, see Elliot Sober and David Sloan Wilson, *Unto Others: The Evolution and Psychology of Unselfish Behavior* (Cambridge: Harvard University Press, 1998).

137 **Darwin thought he solved:** Charles Darwin, *The Descent of Man and Selection in Relation to Sex*, 2nd ed. (1874; New York: Appleton, 1909), 134–135.

137 **offered a critical additional element:** This book, explaining the theory, is clear, persuasive, and important: Peter J. Richerson and Robert Boyd, *Not by Genes Alone: How Culture Transformed Human Evolution* (Chicago: University of Chicago Press, 2005).

138 **Brothers might have to decide:** R. Boyd and J. Richerson, "Culture and Evolution of the Human Social Instincts," in Enfield and Levinson, eds., *Roots of Human Sociality*, 469.

138 **where people can tolerate lactose:** Richerson and Boyd, *Not by Genes Alone*, 191–192.

138 **cells' genetic clock:** Hal Whitehead; see note to Chapter 2 ("their rate of genetic change").

138 **getting his work published:** Interview with Joseph Henrich, University of British Columbia, March 26, 2007.

139 **business and economics students:** In a lecture he gave at Indiana University in Bloomington, October 2007, Marco Janssen, of Arizona State University, indicated this is a common finding of behavioral economic experiments.

139 **culture evolved differently:** Joseph Henrich et al., *Foundations of Human Sociality*; and Joseph Henrich et al., "Costly Punishment Across Human Societies," *Science* 312 (2006): 1767–1770.

140 **the number of humans to peak:** Joel E. Cohen, "Human Population: The Next Half Century," *Science* 302 (2003): 1172–1175.

140 **at the end of the fertility study:** Ansley J. Coale and Susan Cotts Watkins, eds., *The Decline of Fertility in Europe: The Revised Proceedings of a Conference on the Princeton European Fertility Project* (Princeton, NJ: Princeton University Press, 1986); Watkins quotation, 447; importance of culture, Barbara A. Anderson, "Regional and Cultural Factors in the Decline of Marital Fertility in Europe," chapter 7; infant mortality correlation, 436; pre-Enlightenment fertility, 2–13. Richerson and Boyd's *Not by Genes Alone* discusses this book and made me aware of it.

16. Catalysts for Conservation

143 **Writers of the era:** One branch of this literature is traced in Roderick Frazier Nash, *Wilderness and the American Mind*, 4th ed. (New Haven, CT: Yale University Press, 2001). A link runs from the Enlightenment through the Romantic poets to the American transcendentalists and to John Muir and his followers. A colder channel of Enlightenment thought leads through the Utilitarian philosophers to Gifford Pinchot and other use-oriented conservationists. A comparison of these two traditions is beyond my scope; I don't identify with either.

143 *The Earth as Modified:* The original edition was titled *Man and Nature*, but Marsh revised and renamed the book, and the later edition is the one that Pinchot read and I used.

143 **"Destructiveness of Man,":** George Perkins Marsh, *The Earth as Modified by Human Action* (New York: Scribner's, 1885), 33.

143 **Pinchot read Marsh's work:** On Pinchot's life, I have relied heavily on a fascinating biography: Char Miller, *Gifford Pinchot and the Making of Modern Environmentalism* (Washington, DC: Island Press/Shearwater Books, 2001); as to Marsh's book, 55–56

144 **"The greatest of all luxuries":** Pinchot, *Breaking New Ground*, 379

145 **his father wrote:** Miller, *Gifford Pinchot*, family history, 15–34; youth and father's letter, 59–64.

145   **in his private diaries:** Pinchot's diaries are available in three forms. The original and barely
      legible microfilm of them are available in the Library of Congress. Pinchot had many ex-
      cerpts typed, which are in his papers at the Library of Congress in container 9. In addition,
      many other excerpts are in Harold K. Steen, ed., *The Conservation Diaries of Gifford Pinchot*
      (Durham, NC: Forest History Society, 2001). On Pinchot's depression, July 3 to 27, 1891; on
      candy, May 5 and May 17, 1894; on reading, January 1, 1904.

145   **Taft wrote to his wife:** This October 1909 letter from William Howard Taft to Helen Taft is
      quoted in Samuel Hays, *Conservation and the Gospel of Efficiency: The Progressive Conser-
      vation Movement, 1890–1920* (1959; Pittsburgh, PA: Pittsburgh University Press, 1999), 170.
      Hays developed the idea of Pinchot's conflation of technical and moral issues in conserva-
      tion.

146   **She instead died:** James G. Brady, "The Mystery of Gifford Pinchot and Laura Houghtel-
      ing," *Pennsylvania History* (Spring 1999): 199–214.

146   **"A clear and beautiful day":** Pinchot diaries, June 8, 1904.

146   **recruiting the Academy:** Harold K. Steen, *The U.S. Forest Service: A History* (Seattle: Uni-
      versity of Washington Press, 1976), 30–32.

147   **Pinchot's minority advocated:** Miller, *Gifford Pinchot,* 136–137.

147   **according to Pinchot's critics:** Hays, *Conservation and the Gospel of Efficiency,* 198.

147   **the Pinchot-Muir dispute:** Pinchot also fought against reserving *any* lands solely for preser-
      vation, most famously advocating the damming of the Hetch Hetchy Valley in Yosemite
      National Park as a reservoir for San Francisco, but also in opposition to the creation of the
      National Park Service itself. Conflicts between the Forest Service and the Park Service, per-
      sonified in the conflict between Muir and Pinchot over Hetch Hetchy, became an insistent
      theme of twentieth-century environmental battles, with Muir ultimately crowned by a halo
      and Pinchot's memory falling into disrepute. Practically, however, Pinchot's style of conser-
      vation meant more, keeping public ownership of enormous forests and other federal lands
      that today offer hope for re-creating the integrity of natural processes and bridging habitats
      affected by climate change, which the national parks do not. After losing about 25 percent of
      forest lands from the beginning of white settlement until the conservation era, the United
      States has kept roughly even over the last century, since Roosevelt's presidency, with forest
      remaining on a third of its total area, about a third of that land in federal ownership. For
      sources and more on these issues, see Steen and Hays. Statistics are from U.S. Forest Service,
      Forest Inventory and Analysis National Program, Resource Planning Act Tables, 2007; and
      Ruben N. Lubowski et al., *Major Uses of Land in the United States, 2002,* USDA Economic
      Research Service, Economic Information Bulletin No. (EIB-14), May, 2006, 25.

147   **Simply opposing waste:** Steen, *The U.S. Forest Service,* 33–37.

147   **Pinchot recalled:** Pinchot, *Breaking New Ground,* 145.

148   **he believed in luck:** Pinchot diary, typescript, July 2–3, 1903.

148   **building his tiny staff:** Gifford Pinchot to the Secretary of Agriculture, February 20, 1899,
      Washington, letter and proposed annual budget, Pinchot Papers, Library of Congress, con-
      tainer 575.

148   **disliked the arrangement:** Hays, *Conservation and the Gospel of Efficiency,* 172.

148   **brushed off Ballinger's concerns:** R. A. Ballinger to Secretary of the Interior, March 7,
      1907, Washington (cover for H. K. Love to Ballinger, February 17, 1907, Juneau); and Gif-
      ford Pinchot to Ballinger, March 13, 1907, National Archives, Anchorage, Alaska, RG 95,

U.S. Forest Service, Regional Forester Juneau, Historical Files 1903–1911, Box 1, 12/06/13 (4), folder 3.

148 **Ballinger finally resigned:** Hays, *Conservation and the Gospel of Efficiency,* 150–152.

149 **Department of Natural Resources:** Pinchot diary, November 21, 1907, typescript.

149 **use his first name:** Pinchot diary, March 6, 1902, typescript.

149 **"T.R. said to me":** Pinchot diary, November 9, 1907, typescript.

149 **damaged his health:** Miller, *Gifford Pinchot,* 174–175.

149 **defended the president to his father:** Gifford Pinchot to James Pinchot, July 26, 1904, Pinchot Papers, Library of Congress, container 575.

150 **Roosevelt needed a Westerner:** Pinchot diary, October 22 and November 5, 1906, typescript.

150 **"energy would expend itself in fighting":** The quotation is Roosevelt to Henry Cabot Lodge, March 1, 1910, found in Miller, *Gifford Pinchot,* 176; as to Roosevelt's autobiography, 35–37.

151 **Taft's private appraisal:** Pinchot, *Breaking New Ground:* Pinchot's notes of the meeting, 375–376; the topic of Taft's election and transition, 373–379; Taft's appraisal of Pinchot, 431.

151 **"Hardly anybody seemed to care":** Pinchot, *Breaking New Ground,* 391.

151 **overturning important conservation:** Hays, *Conservation and the Gospel of Efficiency,* 149–165.

17. Whistle-Blowers to Save Alaska

153 **small percentage became whistle-blowers:** Joyce Rothschild and Terence D. Miethe, "Whistle-Blower Disclosures and Management Retaliation," *Work and Occupations* 26 (1999): 107–128. For more on the social and psychological devastation of whistle-blowing, see C. Fred Alford, *Whistleblowers: Broken Lives and Organizational Power* (Ithaca, NY: Cornell University Press, 2001).

153 **bordering on sainthood:** Colin Grant, "Whistle Blowers: Saints of the Secular Culture," *Journal of Business Ethics* 39 (2002): 391–399.

154 **to assist businesses:** Rothschild and Miethe, "Whistle-Blower Disclosures," 133.

154 **A Forest Service employee association:** Forest Service Employees for Environmental Ethics, Eugene, Oregon, http://www.fseee.org (accessed February 24, 2008).

154 **predictable characteristics:** Rothschild and Miethe, "Whistle-Blower Disclosures."

154 **On psychological tests:** Phillip H. Jos et al., "In Praise of Difficult People: A Portrait of the Committed Whistleblower," *Public Administration Review* (November–December 1989): 552–561.

154 **Perfectionists always fail:** This psychology has been called moral narcissism, the self-identification with unattainable moral perfection; according to this theory, whistle-blowing is an act of narcissistic rage against an impure world: Alford, *Whistleblowers: Broken Lives and Organizational Power,* 63–81.

155 **founder of the Bronx Zoo:** Jonathan Peter Spiro, *Defending the Master Race: Conservation, Eugenics, and the Legacy of Madison Grant* (Burlington: University of Vermont Press, 2008).

155 **food supply became illegal:** Mitchell, *Sold American,* 205–208.

155 **Grant blamed the Natives:** Ken Ross, *Pioneering Conservation in Alaska* (Boulder: University Press of Colorado, 2006), 125.

155 **stopped from reproducing:** Madison Grant, *The Passing of the Great Race: or, The Racial Basis of European History,* 4th rev. ed. (New York: Scribner's, 1921), xxx–xxxiii.

155 **when he reported:** Langille to Pinchot, undated (received May 27, 1904); June 14, 1904, Katalla; July 16, 1904, Nome; National Archives, Anchorage, Alaska, RG 95, U.S. Forest Service, Regional Forester Juneau, Historical Files 1903–1911, Box 1, 12/06/13 (4), folder 1.

155 **he didn't expect anything:** Pinchot to E. T. Allen, August 5, 1904, Denver, National Archives, Anchorage, Alaska, RG 95, U.S. Forest Service, Regional Forester Juneau, Historical Files 1903–1911, Box 1, 12/06/13 (4), folder 1.

156 **limited to 640 acres:** James Penick Jr., *Progressive Politics and Conservation: The Ballinger-Pinchot Affair* (Chicago: University of Chicago Press, 1968), 78; although exclusively taking Ballinger's side, this book contains a clear and well-documented account of events.

156 **even Pinchot recognized:** Pinchot, *Breaking New Ground,* 396.

156 **seams reaching the surface:** Description of the area from Gifford Pinchot's typescript, "Notes of Alaska Trip," 1911, Pinchot Papers, Library of Congress, Box 430; also, from interviews with Rick Steiner.

157 **In short order he modernized:** Ballinger had been a successful mayor and was strongly recommended by Roosevelt's friend and commissioner of corporations, James R. Garfield. Even his opponents noted his success in cleaning up the land office. Penick, *Progressive Politics and Conservation,* 21–25. Louis Glavis agreed in an unpublished memoir of which I have a typescript copy, 6–7 (see note below, "He loved the thrill").

157 **experience as an investigator:** Penick, *Progressive Politics and Conservation,* 21–25; as to experience, Pinchot, *Breaking New Ground,* 426–427.

157 **He loved the thrill:** Glavis's perspective comes from an unpublished memoir he wrote with journalist John Strohmeier, which Strohmeier allowed me to copy. Glavis approached Strohmeier in the mid-1950s when Strohmeier was an investigative reporter in Providence, Rhode Island, and lived for weeks at a time in his attic while working on the manuscript. Glavis lived from 1883 to 1971.

158 **He looked the part:** Glavis description and stutter: James Penick Jr., "Louis Russell Glavis: A Postscript to the Ballinger-Pinchot Controversy," *Pacific Northwest Quarterly* 55 (1964): 67–75. Powerful voice: John Strohmeier, interview May 15, 2007.

158 **he was assigned to the case:** Pinchot, *Breaking New Ground,* 398–403; Glavis, unpublished memoir.

158 **led a team of investigators:** Glavis memoir, 18–30. As to sequence of events, Penick, *Progressive Politics and Conservation,* 84–88.

159 **before Inauguration Day:** Pinchot, *Breaking New Ground,* 377–379.

159 **a proclamation expanding:** Rakestraw, *A History,* 48–54; the proclamation was February 23, 1909; Taft's inauguration was March 4, 1909. Pinchot, *Breaking New Ground,* 426.

160 **trekked for two weeks:** Rakestraw, *A History,* 53–54.

160 **their private communications:** Penick, *Progressive Politics and Conservation,* 102–105.

160 **sneak the coal files out:** Glavis memoir, 52–56.

161 **"It never occurred to me":** Pinchot, *Breaking New Ground,* 434–436.

162 **publish his accusations:** L. R. Glavis, "The Whitewashing of Ballinger," *Collier's,* November 13, 1909.

163 **swallowed his cud:** Glavis memoir, 105–106.

164 **Taft admitted he had exaggerated:** Penick, *Progressive Politics and Conservation,* 137–164; Pinchot, *Breaking New Ground,* 462–497.

164   **they talked all day:** Pinchot, *Breaking New Ground*, 502.

164   **grooming candidates:** James Chace, *1912: Wilson, Roosevelt, Taft & Debs—The Election That Changed the Country* (New York: Simon & Schuster, 2004).

164   **text written by Gifford Pinchot:** Miller, *Gifford Pinchot*, 230–238.

18. Conservation and Eugenics

166   **little altered by Roosevelt:** Miller, *Gifford Pinchot*, 234–236.

166   **the New Nationalism speech:** Louis Auchincloss, ed., *Theodore Roosevelt Letters and Speeches* (New York: Library of America, 2004), 799–814.

167   **flag worship:** Wendy Kaminer, *Free for All: Defending Liberty in America Today* (Boston: Beacon Press, 2002), 62.

167   **centralized monarchial authority:** Pinchot, *Breaking New Ground*, 5.

167   **he felt overwhelmed:** Miller, *Gifford Pinchot*, 78–82.

168   **to mold ideal citizens:** I have relied heavily on a remarkable book: Robert W. Rydell, *All the World's a Fair: Visions of Empire at American International Expositions, 1876–1916* (Chicago: University of Chicago Press, 1984).

168   **grow through racial purification:** Rydell, *All the World's a Fair*: Philadelphia, 25–27; New Orleans, 98–99; Chicago, 63–71.

169   **"our great object lesson":** W. J. McGee, quoted in Rydell, *All the World's a Fair*, 162.

169   **paid him for his autograph:** Mildred Wohlforth, "Geronimo, Princess Alice and a Writer's Career," *New York Times*, Connecticut edition, January 8, 1988.

169   **collected by the Smithsonian:** Rydell, *All the World's a Fair*, 159–165; clothing, 172–174.

170   **The U.S. Department of Agriculture helped:** Edwin Black, *War Against the Weak: Eugenics and America's Campaign to Create a Master Race* (New York: Four Walls Eight Windows, 2003), 32–41, 46.

170   **pictures of its illustrious supporters:** Rydell, *All the World's a Fair*, 224–225.

170   **"fundamentally important documents":** Theodore Roosevelt, *An Autobiography* (New York: Macmillan, 1913), 447. Pinchot wrote this chapter of the autobiography, quoting the president's transmittal letter of the report, which he surely also wrote.

171   **"conservation of the racial stock":** Irving Fisher, *Report on National Vitality, Its Wastes and Conservation,* (Washington: Government Printing Office, 1909), 49–54, 100–101, 124–129.

171   **forced sterilization:** Harry Bruinius, *Better for All the World: The Secret History of Forced Sterilization and America's Quest for Racial Purity* (New York: Knopf, 2006), 219. States were Indiana (1907); Connecticut, Washington, and California (1909); Nevada, Iowa, and New Jersey (1911); New York (1912); North Dakota, Michigan, Kansas, and Wisconsin (1913); Nebraska (1915); and Oregon, South Dakota, and New Hampshire (1917); Black, *War Against the Weak*, 66; figure of 65,000 from Spiro, *Defending the Master Race*, 240.

171   **put antimiscegenation in:** Bruinius, *Better for All the World*, 147.

171   **sheriff had burst into their bedroom:** David Margolick, "A Mixed Marriage," *New York Times*, June 12, 1992.

171   **between Native Americans and whites:** Frederick E. Hoxie, *A Final Promise: The Campaign to Assimilate the Indians, 1880–1920* (Lincoln: University of Nebraska Press, 2001), 235.

171   **both races wanted race purity:** Donald K. Pickens, *Eugenics and the Progressives* (Nashville, TN: Vanderbilt University Press, 1968), 19, 128.

171 **impending "race suicide":** Roosevelt used this phrase, for example, on March 13, 1905, speaking to the National Congress of Mothers. On May 31, 1907, speaking in Lansing, Michigan, he said, "It would be a calamity to have our farms occupied by a lower type of people than the hard-working, self-respecting, independent, and essentially manly and womanly men and women who have hitherto constituted the most typical American, and on the whole the most valuable element of our entire nation. Ambitious native-born young men and women who now tend away from the farm must be brought back to it."

172 **"getting desirable people to breed":** Theodore Roosevelt, *The Outlook,* January 3, 1914.

172 **lesser procreation represented failure:** Pickens, *Eugenics and the Progressives,* 124–127.

172 **Country Life Commission:** Pinchot, *Breaking New Ground,* 340–343. The commission's chair was Cornell eugenicist Liberty Hyde Bailey.

172 **Fitter Family contests:** The contests are described in Bruinius, *Better for All the World,* 235–239; as to Pinchot's membership in the American Eugenics Society, Wayne Evans, *Organized Eugenics* (New Haven, CT: American Eugenics Society, 1931), viii.

172 **a list that included:** These men had large and small roles. To learn more, see the sources cited throughout this chapter.

172 **both were involved:** For Roosevelt, Pickens, *Eugenics and the Progressives,* 124–127. For Sanger, Jonathan Peter Spiro, *Defending the Master Race,* 189–194.

173 **"self-preservation in a racial sense":** Madison Grant, *The Passing of the Great Race,* 90.

173 **American Museum of Natural History:** Donna Haraway, *Primate Visions: Gender, Race, and Nature in the World of Modern Science* (New York: Routledge, 1989), 54–58.

173 **the best to the best:** Fisher, *Report on National Vitality,* 50–51.

174 **without fear of punishment:** Pamela Schmitt et al., "Collective Action with Incomplete Commitment: Experimental Evidence," *Southern Economic Journal* 66 (2000): 829–854.

174 **member of their own group:** Helen Bernhard et al., "Parochial Altruism in Humans," *Nature* 442 (2006): 912–915.

174 **nationalism to transcend race:** Roosevelt spoke on this theme at Oxford University, June 7, 1910. A letter by Roosevelt explaining why he invited Booker T. Washington to dinner at the White House gives some insight. Roosevelt wrote that since the black man "can neither be killed nor driven away," he should be treated according to his merits. Roosevelt to Albion W. Tourgée, November 8, 1901, in *Theodore Roosevelt Letters and Speeches,* 244–245.

174 **"square-headed little man":** Pinchot, *Breaking New Ground,* 395.

174 **calling it "my Bible":** Spiro, *Defending the Master Race,* 356–357

174 **environmental organizations debated opposition:** David M. Reimers, *Unwelcome Strangers: American Identity and the Turn Against Immigration* (New York: Columbia University Press, 1998); Glen Martin and Ramon G. McLeod, "Sierra Club Divided by Vote on Immigration," *San Francisco Chronicle,* February 23, 1998.

175 **impossibility of educating Eskimos:** Henry Fairfield Osborn, *Men of the Old Stone Age: Their Environment, Life and Art* (New York: Scribner's, 1915), 502.

177 **Conservation is a great moral issue:** *Theodore Roosevelt Letters and Speeches,* 808–809.

19. Fantasy Meets Reality at Katalla

179 **Katallans hung the figure:** The only local account of these incidents appears to be in the suspiciously biased *Cordova Daily Alaskan,* May 4, 5, 6, 9, 11, 12, 16, 19, and 25, 1911.

179   **News flashed:** Original newswire reports found in the Pinchot Papers, Library of Congress, container 508.

179   **The *Seattle Times* warned:** *Seattle Times,* June 13, 1911, quoted in Evangeline Atwood, *Frontier Politics: Alaska's James Wickersham* (Portland, OR: Binford & Mort, 1979), 242.

179   **Association issued a statement:** *Cordova Daily Alaskan,* cited above.

180   **the high-class citizenship:** Ibid.

180   **A fight had broken out:** *Washington Post,* February 24, 1911, quoted in Atwood, *Frontier Politics,* 238–240.

180   **Wickersham carried on a campaign:** Clyde Leavitt, Assistant Forester, to District Forester, Portland, Oregon, January 26, 1911, National Archives, Anchorage, Alaska, RG 95, U.S. Forest Service, Regional Forester Juneau, Historical Files 1903–1911, Box 1, 12/06/12 (4), folder 3.

180   **open letter to Taft:** Gifford and Amos Pinchot to Taft, New York, November 7, 1910, Pinchot Papers, Library of Congress, container 429.

180   **with a press release:** Statement given to the press by Mr. Gifford Pinchot at Noon, June 26, 1911, New York City, typescript document, Pinchot Papers, Library of Congress, container 429.

180   **message to Congress:** Taft special message to Congress July 26, 1911, quoted in Atwood, *Frontier Politics,* 207.

180   **original 1909 article:** L. R. Glavis, "The Whitewashing of Ballinger," *Collier's,* November 13, 1909.

181   **West Coast's energy needs:** Gifford Pinchot, "What Shall We Do with the Coal in Alaska?" *Saturday Evening Post,* August 26, 1911.

181   **He stated for the press:** *Cordova Daily Alaskan,* August 24 and 25, 1911.

182   **dipped theirs in pitch:** Interview transcript with W. S. Martin by Gifford Pinchot, aboard the SS *Northwestern,* September 4, 1911, Pinchot Papers, Library of Congress, container 430.

182   **The town blossomed:** *Katalla Herald,* August 10 and November 16, 1907; as to saloons, William C. Hansen, quoted in Lone E. Janson, *The Copper Spike,* 52; as to lack of churches, interview with former Katalla resident Mae Lange, September 15, 2007.

182   **a senator from Washington:** Senator Miles Poindexter.

182   **Pinchot attacked Taft:** Gifford Pinchot, "A Look Ahead in Politics," *Saturday Evening Post,* October 7, 1911.

183   **prevent him from landing:** J. T. Hamilton to Oscar Breedman, Seattle, September 3, 1911, Pinchot Papers, Library of Congress, container 430; John E. Lathrop, "Tried to Keep Pinchot Out," *Newark N.J. News,* September 20, 1911, Pinchot Papers, Library of Congress, container 429.

183   **G. Pinchot, Pinhead:** William Smyth, "What Alaska Thinks of Pinchot," syndicated article, found in *Cleveland Press,* October 25, 1911, Pinchot Papers, Library of Congress, container 429. As to Fisher's comments, E. H. Thomas, *Seattle Post-Intelligencer,* reprinted in *Cordova Daily Alaskan,* September 7, 1911.

183   **ruined the development:** Gifford Pinchot, "Notes of Alaska Trip," September 4–October 12, 1911; typescript document in Pinchot Papers, Library of Congress, Box 430; quotation 23.

184   **in vivid notes:** Pinchot, "Notes of Alaska Trip."

184 **glad to finally leave:** Pinchot, "Notes of Alaska Trip," September 25, 1911, 42; *Newark Evening News*, "Bering River Coal Fields," September 28, 1911; Pinchot diary, September 28, 1911.

184 **won over a great number:** Pinchot to Mrs. James W. Pinchot (mother), Seward, September 16, 1911, Pinchot Papers, Library of Congress, container 576.

184 **hatred for Pinchot:** For example, see the books by Evangeline Atwood and Lone Janson, cited previously.

184 **"a degree of home rule":** Gifford Pinchot, "Who Shall Own Alaska?" *Saturday Evening Post*, December 16, 1911.

184 **Doug Langille was present:** H. D. Langille to Pinchot, Mayfield, Idaho, January 11, 1940, Pinchot Papers, Library of Congress, container 1003.

185 **the three-way race:** The story is well told in James Chace, *1912: Wilson, Roosevelt, Taft & Debs—The Election That Changed the Country* (New York: Simon & Schuster, 2004).

185 **failed a quality test:** Lethcoe, *A History of Prince William Sound*, 90.

185 **Barrett Willoughby:** Willoughby's life story and a good deal of Katalla history is found in Nancy Warren Ferrell, *Barrett Willoughby; Alaska's Forgotten Lady* (Fairbanks: University of Alaska Press, 1994).

187 **against European anti-Semitism:** Char Miller and V. Aleric Sample, "Gifford Pinchot and the Conservation Spirit," in Pinchot, *Breaking New Ground*, xv. Generally, I've relied on Miller, *Gifford Pinchot*; as to Cornelia, 177–181; as to South Pacific, 302–306.

187 **the food floated:** Gray Towers is a National Historic Site administered by the Forest Service, with regular tours in the summertime: http://www.fs.fed.us/na/gt. Curator Rebecca Philpot was a great help. Apparently, Pinchot's papers were purged of most material about Laura Houghteling prior to deposit in the Library of Congress. I suspect he or someone else may have done the same with material on eugenics.

189 **Nothing of value:** Interview with Dave Salmon, March 8, 2008.

189 **a Korean businessman:** Elizabeth Manning, "Battle Vowed on Mine," *Anchorage Daily News*, April 11, 2000.

189 **"Pinchot must be smiling":** Rick Steiner to Bering River Environmental Working Group, December 15, 1997; copy provided by Steiner.

20. Chenega Destroyed

196 **around seventy-five people:** George Plafker et al., *Effects of the Earthquake of March 27, 1964, on Various Communities: Geological Survey Professional Paper 542-G* (Washington, DC: Government Printing Office, 1969). Villagers state the population was closer to 120.

196 **Andy wrote recently:** This and the great majority of material on old Chenega and the earthquake come from a profound book of memories collected by the Chenegans themselves: John E. Smelcer, collector and editor, *The Day That Cries Forever: Stories of the Destruction of Chenega During the 1964 Alaska Earthquake* (Anchorage: Chenega Future, 2006).

199 **into a snowbank:** Frank R. B. Norton and J. Eugene Haas, "The Human Response in Selected Communities," in *The Great Alaska Earthquake of 1964: Human Ecology* (Washington, DC: National Academy of Sciences, 1970), 394.

199 **seventy vertical feet:** Plafker et al., *Various Communities: Paper 542-G*, G16.

200 **a solemn memorial:** Smelcer, *The Day That Cries Forever*.

21. Ownership in a Liquid World

202 **emit and absorb buoyant gas:** Rockfish biology, Alaska Department of Fish Game, http:// www.adfg.state.ak.us/pubs/notebook/fish/rockfish.php, and NOAA Alaska Fisheries Science Center, http://www.afsc.noaa.gov/Rockfish-Game/description/rougheye.htm.

202 **severed two-foot-thick tree trunks:** Plafker et al., *Various Communities: Paper 542-G.*

202 **ten thousand miles of shoreline:** Kirk W. Stanley, *Effects of the Alaska Earthquake of March 17, 1964, on Shore Processes and Beach Morphology: U.S. Geological Survey Professional Paper 543-J* (Washington, DC: Government Printing Office, 1968).

204 **Streams cut deep gullies:** George Plafker, *Tectonics of the March 27, 1964, Alaska Earthquake: U.S. Geological Survey Professional Paper 543-I* (Washington, DC: Government Printing Office: 1969). As to biology: G. Dallas Hanna, "Observations Made in 1964 on the Immediate Biological Effects of the Earthquake in Prince William Sound," and George V. Harry Jr., "General Introduction, Summary, and Conclusions," in National Research Council, *The Great Alaska Earthquake of 1964*, vol. 4: *Biology* (Washington, DC: National Academy of Science, 1971), 1–34.

205 **each grain had been placed:** Stanley, *Shore Processes and Beach Morphology: Paper 543-J.*

206 **Court decided:** *Goins v. Merryman,* 1938 OK 10, 183 Okla. 155, 80 P.2d 268, http://wyom cases.courts.state.wy.us/applications/oscn/deliverdocument.asp?citeid=17382 (accessed March 20, 2008).

206 **ended up owning seafloor:** Stanley, *Shore Processes and Beach Morphology: Paper 543-J,* J19–J20.

206 **the ad hoc group:** Ernest Gruening, *The State of Alaska: A Definitive History of America's Northernmost Frontier,* rev. ed. (New York: Random House, 1968), 73.

207 **stench of rotting clams:** Henry Makarka, "The Long Summer of Clams," in Smelcer, ed., *The Day That Cries Forever,* 38–40.

207 **wasn't suitable for clams:** Rae E. Baxter, "Earthquake Effects on Clams of Prince William Sound," in *The Great Alaska Earthquake of 1964,* vol. 4: *Biology,* 238–245.

208 **too large to repair:** Harry, "General Introduction, Summary, and Conclusions," in *The Great Alaska Earthquake of 1964,* vol. 4: *Biology,* 3–4.

208 **they just worked harder:** Jessie Tiedeman, "A Twist of Fate," in Smelcer, ed., *The Day That Cries Forever.*

208 **fishery disastrously failed:** Thomas A. Morehouse and Jack Hession, "Politics and Management: The Problem of Limited Entry," in Arlon Tussing et al., eds., *Alaska Fisheries Policy: Economics, Resources and Management* (Fairbanks: Institute of Social, Economic and Government Research, 1972), 291–306.

22. Clem Tillion and the Ownership of the Sea

210 **"He's a crazy kid":** Elmer Rasmuson was the banker. I interviewed Tillion at length on July 6, 2007, and March 30, 2008, and have known him for many years; I've also relied on a vivid newspaper profile: Hal Bernton, "Get Out of the Way: Be It for Fisheries or for Family Clem Tillion Does What He Thinks Is Best," *Anchorage Daily News,* August 23, 1992.

212 **fisherman, guide, and pilot:** Jay Hammond, *Tales of Alaska's Bush Rat Governor* (Fairbanks: Epicenter Press, 1994), 41–77.

212 **removed with explosives:** Mark Panitch, "Kachemak Bay: Oil Spill Leads Alaska to Reverse Drilling OK," *Science* 193 (1976): 131.

212   **conservatism with conservation:** Dave Rose as told to Charles Wohlforth, *Saving for the Future: My Life and the Alaska Permanent Fund* (Seattle: Epicenter Press, 2008), 120.

213   **even shot whales:** See notes to Chapter 12 ("reducing predators" and "a bounty for each seal nose").

213   **inevitably spiral into disaster:** Tussing et al., ed., *Alaska Fisheries Policy,* 8–11.

213   **"brings ruin to all":** Garrett Hardin, "The Tragedy of the Commons," *Science* 162 (1968): 1243–1248.

213   **conservation-oriented eugenicists:** This topic is covered in greater depth, with citations, in Chapters 18 and 23.

213   **Hammond initially opposed:** Morehouse and Hession, "Politics and Management," 313.

214   **program finally won approval:** Alaska Commercial Fisheries Entry Commission, "Changes in the Distribution of Alaska's Commercial Fisheries Entry Permits, 1975–2006," CFEC Report Number 07-5N (2007), 21.

214   **from outside the village:** William E. Simeone and Rita A. Miraglia, "An Ethnography of Chenega Bay and Tatitlek, Alaska," Technical Memorandum 5, submitted by Alaska Department of Fish and Game, Division of Subsistence, to U.S. Department of the Interior, Minerals Management Service, Alaska OCS Region, November, 2000, 44–49. Also Karen Oakley, "Permit Drain in the 1980s: A Study of Recent Transfers of Limited Entry Permits for Alaska's Limited Entry Fisheries," House Research Agency, Alaska State Legislature, October 1989.

214   **27 percent of rural Alaska permits:** Alaska Commercial Fisheries Entry Commission, "Changes in the Distribution," 21.

214   **only three in 2006:** Permit numbers from Simeone and Miraglia, "An Ethnology of Chenega Bay," and from Alaska Commercial Fisheries Entry Commission website, http://www.cfec.state.ak.us/cpbycen/2006/261VALDE.htm (accessed March 31, 2008).

214   **the average age:** Wesley Loy, "Waters Not Calm for Young Fishermen; Sea of Gray: New Generation Faces Big Cost to Join Industry," *Anchorage Daily News,* January 27, 2007; statistics, Alaska Commercial Fisheries Entry Commission, "Changes in the Distribution," 42.

215   **high profits for fishermen:** Hal Bernton, "Raising Salmon in the Sound," *Anchorage Daily News,* March 3, 1985; and Hal Bernton, "Koernig Leaves His Mark on Sound Fish Hatcheries," *Anchorage Daily News,* October 17, 1995.

216   **Some biologists believe:** Ray Hilborn and Doug Eggers, "A Review of the Hatchery Program for Pink Salmon in Prince William Sound and Kodiak, Alaska," *Transactions of the American Fisheries Society* 129 (2000): 333–350.

216   **less for other animals:** Veteran NOAA scientist Jack Helle explained his pioneering work showing correlation between larger age classes of salmon and smaller individual fish returning to spawning streams. The phenomenon has since been shown in various settings, confirming that high densities of salmon can eat through the carrying capacity of the ocean ecosystem. John H. Helle and Margaret S. Hoffman, "Size Decline and Older Age at Maturity of Two Chum Salmon *(Oncorhynchus keta)* Stocks in Western North America, 1972–1992," in R. J. Beamish, ed., *Climate Change and Northern Fish Populations* (1995), 245–260; special publication of *Canadian Journal of Fisheries and Aquatic Sciences.*

216   **disposed of whole:** Hal Bernton, "Crews Push to Remove Pinks," *Anchorage Daily News,* August 24, 1991.

217   **kicked off the fisheries council:** Bernton, "Get Out of the Way."

217 **"fencing of the Plains":** Hal Bernton and David Whitney, "Mad Dash for Halibut Is History; Past Fishermen to Split Harvest Under New Rules," *Anchorage Daily News,* January 30, 1993.

218 **"It makes me furious":** Bernton, "Get Out of the Way."

218 **haves and have-nots:** Bonnie J. McCay, "Introduction: Ethnography and Enclosures of Marine Commons," in Marie E. Lowe and Courtney Carothers, eds., *Enclosing the Fisheries: People, Places, and Power* (Bethesda, MD: American Fisheries Society, 2008).

218 **red tape:** Tom Kizzia, "Park Use at Heart of Battle; State, Family Don't Agree on the Rules," *Anchorage Daily News,* March 16, 1990.

220 **other urban junk:** Interview with Bree Murphy, July 12, 2007; she managed the coast walk database for the Center for Alaska Coastal Studies.

221 **from the city dock:** Daisy Lee Bitter, personal communciation.

23. An Alternative to Tragedy

223 **advantage of the defector:** Marco A. Janssen and Elinor Ostrom, "TURFs in the Lab: Institutional Innovation in Real-Time Dynamic Spatial Commons," *Rationality and Society* 20 (2008): 371–397.

223 **share and solve problems:** Elinor Ostrom, *Governing the Commons: The Evolution of Institutions for Collective Action* (Cambridge: Cambridge University Press, 1990).

223 **"'we can do a lot better'":** My interviews and personal impressions of Elinor and Vincent Ostrom are from a visit to Indiana University, October 21–26, 2007, and an extended interview with Lin, April 26, 2007.

223 **nearly equal:** According to google.scholar.com on March 31, 2008: Ostrom, *Managing the Commons,* 5,540 citations; Hardin, "Tragedy of the Commons," 6,344 citations.

223 **published a volume:** Committee on the Human Dimensions of Global Change and National Research Council, Elinor Ostrom et al., eds., *The Drama of the Commons* (Washington, DC: National Academies Press, 2002).

223 *Science* **published:** Elinor Ostrom et al., "Revisiting the Commons: Local Lessons, Global Challenges," *Science* 284 (1999): 278–282.

224 **sustainable institutions:** Ibid.

225 **their own workshop:** Besides interviews, I relied on Pamela Jagger, *Artisans of Political Theory and Empirical Inquiry: Thirty Years of Scholarship at the Workshop in Political Theory and Policy Analysis,* a report issued by the workshop in June 2004.

225 **he was most proud:** Interviews with Vincent Ostrom, September 9, 1999, and October 2008. Observations on the shortcomings of the Alaska Constitution are my own.

228 **other shared realms:** These ideas are eloquently explained in David Boiller, *Silent Theft: The Private Plunder of Our Common Wealth* (New York: Routledge, 2002).

228 **a common pool resource amenable:** Defining types of good is critical to the analysis of how institutions can manage them. I've used the definition of a "common-pool resource" as one from which it is difficult to exclude users, but that can be diminished by overuse. A public good, such as a lighthouse or public radio station, is different because additional use takes nothing away from previous users; likewise, we don't need to be concerned with a good such as a toll road or a potluck dinner, from which users can be excluded if they don't contribute. These points are covered in Margaret A. McKean, "Common Property: What Is It, What Is It

Good For, and What Makes It Work?" in Clark C. Gibson et al., eds., *People and Forests: Communities, Institutions, and Governance* (Cambridge, MA: MIT Press, 2000), 28–34. Vincent and Elinor Ostrom originally advanced these ideas about types of goods in 1977.

228 **a book on the knowledge commons:** Charlotte Hess and Elinor Ostrom, eds., *Understanding Knowledge as a Commons: From Theory to Practice* (Cambridge, MA: MIT Press, 2007).

229 **distributing the profits:** Kathryn Milun, October 22, 2007. The idea originally called Sky Trust has gone through various versions and is now called "cap and dividend" by its sponsor at Tomales Bay Institute; Peter Barnes, personal communication, April 9, 2008.

229 **allowed into the community:** This fishery and its history is explained in Xavier Basurto, "Commercial Diving and the Callo de Hacha Fishery in Seri Territory," *Journal of the Southwest* 48 (Summer 2006): 189–209; I also interviewed Basurto and Lopez on October 24, 2007.

230 **compassion was a weakness:** Garrett Hardin, "Living on a Lifeboat" (1974), and "Population, Environment and Immigration" (1993), reprinted in Hardin, *The Immigration Dilemma: Avoiding the Tragedy of the Commons* (Washington, DC: Federation for American Immigration Reform, 1995).

230 **also supported English-only:** David M. Reimers, *Unwelcome Strangers: American Identity and the Turn Against Immigration* (New York: Columbia University Press, 1998), 43–64.

231 **a national vote:** Martin and McLeod, "Sierra Club Divided."

231 **in the laboratory:** See notes to Chapter 18 ("without fear of punishment" and "member of their own group").

24. Chenega Reborn

233 **"I felt so damn alone":** Larry Evanoff interview, April 8, 2008; Larry's memories are also in Smelcer, ed., *The Day That Cries Forever.*

233 **succumbing to alcohol:** Andy Selanoff, in Smelcer, ed., *The Day That Cries Forever,* 47–48.

234 **absorbed into Chugach National Forest:** Lethcoe, *History of Prince William Sound,* 202.

235 **he became a priest:** Carol Ann Kompkoff informed me about her father, including information from a framed memorial statement she keeps in her house in Chenega.

235 **before limited-entry fishing:** Simeone and Miraglia, "An Ethnography of Chenega Bay," 47.

235 **Natives wielded little power:** Donald Craig Mitchell, *Take My Land Take My Life: The Story of Congress's Historic Settlement of Alaska Native Land Claims, 1960–71* (Fairbanks: University of Alaska Press, 2001), 83–86.

235 **Chugach Native Association:** Lethcoe, *History of Prince William Sound,* 200–203.

236 **freezing disposal:** Mitchell, *Take My Land,* 186–189.

236 **"a society which is essentially his":** Cecil Barnes, "Letter to the Editor," *Cordova Times,* February 16, 1967.

236 **an injunction stopping the project:** Mitchell, *Take My Land,* 318–327.

237 **burying the hot pipeline:** Walter J. Hickel, *Crisis in the Commons: The Alaska Solution* (Oakland, CA: ICS Press, 2002), 109–110.

237 **Native Claims Settlement Act:** As to size of settlement, Mitchell, *Take My Land,* 493.

237 **received 76,000 acres:** Joe Hunt, *Mission Without a Map: The Politics and Policies of Restoration Following the Exxon Valdez Oil Spill,* unpublished manuscript dated October 1, 2003, 132. Hunt wrote this valuable book with funding from the *Exxon Valdez* Oil Spill Trustee Council, which then chose not to publish it and relinquished the copyright to him. My extensive use is with Hunt's permission.

238   **agency employees directed:** Nancy Yaw Davis, "A Village View of Agencies," read at the
      20th Annual Science Conference, University of Alaska, College, August 24–27, 1969. Spon-
      sored by the National Academy of Sciences National Research Council and the Ohio State
      Disaster Research Center. Typescript provided by the author.

25. Native People Become Native Corporations

243   **"He was a preemie":** Interview with Gail Steen and Tina Tapley, August 1992.

245   **corruption at the Land Office:** David Postman, "Inside Deal: The Untold Story of Oil in
      Alaska," *Anchorage Daily News,* February 4, 1990.

245   **mob-infiltrated:** Howard Weaver and Bob Porterfield, "Union Fiefdom Rules Fairbanks
      Warehouse," *Anchorage Daily News,* December 18, 1975.

245   **astronomically overbudget:** John Strohmeyer, *Extreme Conditions: Big Oil and the Trans-
      formation of Alaska* (New York: Simon & Schuster, 1993).

245   **spying and dirty tricks:** David Whitney, "$10,000 Settles Spy Case," *Anchorage Daily News,*
      April 23, 1992.

245   **criminal neglect:** Wesley Loy, "BP Fined $20 Million for Pipeline Corrosion," *Anchorage
      Daily News,* October 26, 2007.

245   **ignoring safety rules:** Kim Fararo, "Alyeska Issues Stop-Work Order," *Anchorage Daily
      News,* November 2, 1995.

245   **punishing whistle-blowers:** Kim Fararo, "BLM Begins Investigation of Alyeska Workers'
      Claims," *Anchorage Daily News,* March 9, 1994; and Richard Mauer, "BP Was Warned of
      Intimidation," *Anchorage Daily News,* September 10, 2006.

245   **bribing politicians:** Tom Kizzia, "Did Oil Producers Know About Veco Illegal Acts?" *An-
      chorage Daily News,* October 2, 2007.

245   **cheating on state taxes:** Ralph Thomas, "Oil-Tax Backlog Swelling," *Anchorage Daily News,*
      February 11, 1994.

245   **impose taxes so low:** Larry Persily, "Democrats Work to Amend Oil Production Tax Laws,"
      Anchorage Daily News, January 12, 2005, and "Oil Producers Profit More in Alaska," *An-
      chorage Daily News,* February 2, 2005.

245   **undermanned supertanker:** National Transportation Safety Board, "Marine Accident
      Report—Grounding of the U.S. Tankship EXXON VALDEZ on Bligh Reef, Prince William
      Sound, near Valdez, Alaska, March 24, 1989," July 31, 1990.

245   **complacency nullified:** Ernest Piper, *The* Exxon Valdez *Oil Spill: Final Report, State of
      Alaska Response* (Anchorage: Alaska Department of Environmental Conservation, 1993),
      13; David Whitney, "NTSB Spreads the Blame for Oil Spill," *Anchorage Daily News,* August
      1, 1990; National Transportation Safety Board, "Marine Accident Report," July 31, 1990.

246   **preempted their authority:** Stan Jones documented the double-hull issue and industry
      capture of the Coast Guard with great clarity in a series of articles in the *Anchorage Daily
      News,* "Blueprint for Disaster: Fighting Flares Anew," October 16, 1989, and "Special Re-
      port: Regulators Serving Public or Industry," October 15, 1989.

246   **knew individual animals:** Interview with Kelly Weaverling, August 15, 1992; like other
      Cordova subjects in this chapter, I've interviewed Weaverling many times over the last
      twenty years.

246   **those cutting the trees:** Hal Bernton, "Major Logging Set for Sound," *Anchorage Daily
      News,* April 30, 1988.

249   **Stevens believed:** Mitchell, *Take My Land,* 236–239. As to Native leaders' support, 155–163. Mitchell admits in these pages that some Native leaders now deny supporting the corporate model, but he documents their support at the time in their own words. In an interview, April 21, 2008, Mitchell pointed out that the Alaska Native Claims Settlement Act does permit village corporations to organize as nonprofits under Alaska law, but also requires them to issue stock; Alaska law does not permit nonprofits to issue stock, so the provision was never operative. In Mitchell's view, mistakes such as this one, and many other poorly considered aspects of ANCSA, came about because of the rush to complete the legislation to permit pipeline authorization.

249   **After their first decade:** Thomas R. Berger, *Village Journey: The Report of the Alaska Native Review Commission* (New York: Farrar, Straus and Giroux, 1985), 26–47.

249   **known as NOLs:** Hal Bernton, "Native Groups Turn Losses into Assets," *Anchorage Daily News,* December 20, 1987.

249   **Congress stopped:** David Whitney, "Congress OKs Stopping Loss Sales," *Anchorage Daily News,* October 22, 1988.

250   **$48 million on loss sales:** Hal Bernton, "Chugach Posts $18.3 Million Profit," *Anchorage Daily News,* March 29, 1988, and "Profits Soar for Native Corporations," *Anchorage Daily News,* March 10, 1989.

250   **a missile testing base:** As to size, Debbie Cutler, "Top 49ers: Alaska's Economic Pipelines to the Future," *Alaska Business Monthly,* October 2007. As to contract work, Chugach Alaska Corporation website, http://www.chugach-ak.com (accessed April 21, 2008).

250   **award no-bid contracts:** Paula Dobbyn, "Some Say No-Bid, No-Limit Government Commissions for Natives Are Unfair," *Anchorage Daily News,* March 19, 2006; and General Accountability Office, "Alaska Native Corportations: Increased Use of Special 8(a) Provisions Calls for Tailored Oversight," Statement of David E. Cooper, Director, Acquisition and Sourcing Management, June 21, 2006 (GAO-06-874T).

250   **litigious maneuvering:** Paula Dobbyn, "Chugach Sees Shift in Power," *Anchorage Daily News,* October 26, 2004.

252   **ill-advised expansion:** Hal Bernton, "Major Logging Set for Sound," *Anchorage Daily News,* April 30, 1988; Tom Kizzia, "Seward Sawmill Gears Up to Cut into Lumber Market," *Anchorage Daily News,* December 3, 1989.

252   **quoted on the corporate website:** Chugach Alaska Corporation website, http://www .chugach-ak.com/landsconserv.html (accessed April 21, 2008).

252   **"raped and pillaged":** I interviewed Bob Sanford aboard his boat on May 10, 2007. I visited again and recorded logging damage in photographs on September 13, 2007.

26. The Broken Covenant

255   *Cato's Letters:* Jackson Turner Main, *The Antifederalists: Critics of the Constitution, 1781–1788* (Chapel Hill: University of North Carolina Press, 1961), 9–13. This fascinating book is full of excellent citations. Here is an excerpt from *Cato's Letters* on the topic: "It proceeds from a consummate Ignorance in Politicks, to think that a Number of Men agreeing together, can make and hold a Commonwealth, before Nature has prepared the Way, for she alone must do it. An Equality of Estate will give an Equality of Power; and an Equality of Power is a Commonwealth, or Democracy. . . . Very great Riches in private Men are always dangerous to States, because they . . . destroy amongst the Commons, that Balance of Prop-

erty and Power, which is necessary to a Democracy, or the democratical Part of any Government, overthrow the Poise of it, and indeed alter its Nature, tho' not its Name." John Trenchard, Thomas Gordon and Thomas Jones, *Cato's Letters*, vol. 3 (London: W. Wilkins, T. Woodward, J. Walthoe, and J. Peele, 1724), 128, 177.

256 **preceded the Constitution:** The difference between the Declaration of Independence and the Constitution marks the shift from a citizen democracy based on universal, natural law to an expansionist national government empowered by an agreement between those concerned about how to allot power among themselves—essentially, a legal positivist contract. American revolutionaries had fought under Jefferson's statement that "We hold these truths to be self-evident, that all men are created equal, that they are endowed by their Creator with certain unalienable Rights, that among these are Life, Liberty and the pursuit of Happiness. That to secure these rights, Governments are instituted among Men, deriving their just powers from the consent of the governed." The Constitution contains no parallel statement about its source of authority. Its preamble merely states the purpose of the document, like the introductory wording of any contract. As such, it implicitly justified exclusion from basic rights those not party to the agreement—those the new nation would exploit—including Native Americans and slaves.

256 **state legislatures:** Isaac Kramnick, introduction to *The Federalist Papers*, by James Madison, Alexander Hamilton, and John Jay (1788; London: Penguin Books, 1987), 22.

256 **already too large:** George Clinton, "Cato, III, in the New York Journal, Thursday, October 25, 1787," in Paul Leicester Ford, ed., *Essays on the Constitution of the United States, Published During Its Discussion by the People, 1787-1788* (Brooklyn, NY: Historical Printing Club, 1892), 255–258.

256 **protect private-property rights:** *The Federalist Papers*, No. 10.

256 **traded more profitably:** Charles A. Beard, *An Economic Interpretation of the Constitution of the United States* (1913; New Brunswick: Transaction Publishers, 1998), 49–50.

256 **lifetime presidents and senators:** Kramnick, introduction to *The Federalist Papers*, 35, 71–73; *Federalist Papers*, No. 11.

257 **as strong as the monarchy:** Kramnick, introduction to *The Federalist Papers*, 29–30.

257 **"If men were angels":** *The Federalist Papers*, No. 51.

258 **The framers settled:** Main, *The Antifederalists*, covers the ratification process well.

258 **only for money or consumption:** Vincent Ostrom, "Report on Clarifying the Meaning of Article VIII: Natural Resources, as Adopted by the Alaska Constitutional Convention, in 1956, with Addenda," February 25, 1994; Workshop on Political Theory and Policy Analysis, Indiana University, 9–11.

258 **constitutional interpretation:** Vincent Ostrom to Professor John M. Gaus, Department of Government, Harvard University, February 28, 1956, in Ostrom, "Report," Appendix III.

259 **"different patterns of order":** Vincent Ostrom, *The Meaning of American Federalism: Constituting a Self-Governing Society* (San Francisco: ICS Press, 1994), 224.

259 **"they form an association":** Alexis de Tocqueville, *Democracy in America and Two Essays on America*, trans. Gerald E. Bevan (1840; London: Penguin Books, 2003), 596.

259 **every two hundred citizens:** National Center for Charitable Statistics, A program of the Center on Nonprofits and Philanthropy at the Urban Institute, http://nccs.urban.org/ (accessed August 30, 2008).

260 **meddle in local affairs:** *Federalist Papers*, No. 17, 156.

260   **down-and-dirty political skill:** Forrest McDonald, *Alexander Hamilton: A Biography* (New York: Norton, 1979), 152–187. Hamilton's immediate priority as the new nation's treasury secretary was to concentrate private capital and create a credit market. He accomplished that goal by persuading a corrupted Congress to pay off its Revolutionary War debt securities and those of the states, certificates which budding financiers as well as congressmen themselves had accumulated for as little as 10 cents on the dollar. Tariffs funded repayment, effectively using federal power to create a new class of rich investors. Government involvement remains among the surest paths to wealth in the United States, but Hamilton's ploy had an ideological purpose, and he didn't exploit the opportunity to make money himself. I find it richly symbolic that this corrupt debate was concluded by a deal that created the capital city of Washington itself.

261   **more than 80 percent:** According to the National Philanthropic Trust, corporations and corporate foundations gave $13.8 billion in 2007, of a total $295 billion in total charitable contributions (a 2006 figure); see http://www.nptrust.org/philanthropy/philanthropy_stats .asp (accessed August 31, 2008).

261   **thwarted their efforts:** Federal preemption on oil spill prevention is partly covered in Chapter 25. As to federal preemption of state climate change efforts, see Zachary Coile et al., "EPA Blocks California Bid to Limit Greenhouse Gases from Cars," *San Francisco Chronicle,* December 20, 2007.

262   **enact his economic program:** Woodrow Wilson, *Congressional Government: A Study in American Politics* (1885; New York: Meridian Books, 1956), 34–46

262   **virtually every aspect:** Vincent Ostrom, *The Meaning of Democracy and the Vulnerability of Democracies: A Response to Tocqueville's Challenge* (Ann Arbor: University of Michigan Press, 1997), 99–101.

262   **make it work efficiently:** Wilson, *Congressional Government,* 205–207.

262   **subscribed to Wilson's ideas, too:** Ostrom, *The Meaning of Democracy,* 22.

262   **Michels wrote:** Robert Michels, *Political Parties: A Sociological Study of the Oligarchical Tendencies of Modern Democracies,* trans. Eden and Cedar Paul (1911; New York: Free Press, 1962), 355.

263   **"what it means to be free":** Ostrom, *The Meaning of American Federalism,* 130–131.

27. Finding Oil

268   **we will be going all-out:** These quotations and most of the material on oil spill events are from my contemporaneous notes.

268   **punished by his superiors:** Stohmeyer, *Extreme Conditions,* 238–240.

269   **buried alive in oil:** Charles Wohlforth, "State Biologists Say Death Toll Extensive," *Anchorage Daily News,* April 2, 1989.

270   **mate struggled on:** National Transportation Safety Board, "Marine Accident Report," July 31, 1990.

271   **self-important fools:** The official Coast Guard history of the oil spill, "T/V *Exxon Valdez* Oil Spill: Federal On-Scene Coordinator's Report," 1993, is an unintentionally funny settling of scores by the admirals who, like Rodney Dangerfield, got no respect. However, I do want to note that Admiral Clyde Robbins proved a thoughtful leader, well aware of the futility of much of what was being done.

273   **became its defenders:** Some of these details are covered in Charles Wohlforth, "Left Out of

Spill's Big Pay," *Anchorage Daily News*, June 25, 1989, and Charles Wohlforth, "Black Gold," *New Republic*, September 18 and 25, 1989.

28. Costs and Values

276 **raises and promotions:** George Frost and Charles Wohlforth, "Exxon Shrugs Off $1 Billion Bill," *Anchorage Daily News*, March 14, 1991; in response to a legal settlement covering spill damages, Exxon's chairman, Lawrence Rawl, said, "The customer always pays everything," and the settlement "will not have a noticeable effect on our financial results." Before the U.S. Supreme Court, February 27, 2008, attorney Jeffrey Fisher said, "Exxon fired one person—Captain Hazelwood. They reassigned the third mate. Everybody else up—further up the chain of command who allowed this to happen received bonuses and raises. They have taken no action inside the company to express in any meaningful way that they've been deterred by what happened in this incident."

277 **nothing would work:** These statements are recorded in my contemporaneous notes.

278 **pies were displayed:** Charles Wohlforth, "Instant Civilization Springs Up in Sound," *Anchorage Daily News*, July 31, 1989.

278 **contaminating seafloor:** Charles Wohlforth, "Hot Water Spill Cleanup Kills Shorelife," *Anchorage Daily News*, February 17, 1990.

278 **died of old age:** These results and much other information I relied on about the biological impact of the oil spill are found in Charles H. Peterson at al., "Long-Term Ecosystem Response to the *Exxon Valdez* Oil Spill," *Science* 302 (2003): 2082–2086.

278 **in 2007 he predicted:** Dennis C. Lees and W. B. Driskell, "Assessment of Bivalve Recovery on Treated Mixed-Soft Beaches in Prince William Sound, Alaska," *Exxon Valdez* Oil Spill Restoration Project Final Report (Restoration Project 040574), National Oceanic and Atmospheric Administration, National Marine Fisheries Service, Office of Oil Spill Damage and Restoration, Auke Bay, Alaska, August 2007.

280 **knocked down a worker:** Charles Wohlforth, "Rescuers Work Hard, But Catch Is Small," *Anchorage Daily News*, April 1, 1989.

280 **if they had been left:** Details about the rescue centers are from my notes at the time and the article cited in the next note. As to the harm of the "rescue" of less-oiled otters, Anthony R. DeGange, "Assessment of the Fate of Sea Otters Oiled and Rehabilitated as a Result of the *Exxon Valdez* Oil Spill: Marine Mammal Study Number 7," U.S. Fish and Wildlife Service, January 12, 1990, 3.

281 **prolonging the agony:** From my notes; additional details and description of the rescue centers found in Charles Wohlforth, "Spill Battle Is Fight Against Frustration," *Anchorage Daily News*, April 9, 1989.

281 **rip through populations:** The state biologists were Lloyd Lowry and Kathy Frost, whom I have interviewed many times, most recently Lowry on June 28, 2008.

281 **the emotional investment:** Interview with Tony DeGange, August 22, 2008; also, Charles Wohlforth, "Otter Rescue Questioned," *Anchorage Daily News*, April 17, 1990.

281 **said biologist Charles Monnett:** Recorded interview with Charles Monnett, June 30, 2008.

282 **died in their first winter:** Robert J. Hofman, "Foreword," in Thomas R. Loughlin, ed., *Marine Mammals and the* Exxon Valdez (San Diego: Academic Press, 1994), xiv.

282 **written up without mention:** Charles Monnett and Lisa Mignon Rotterman, "Mortality and Reproduction of Female Sea Otters in Prince William Sound, Alaska: Technical Report:

Marine Mammal Study Number 6"; Prince William Sound Science Center and U.S. Fish and Wildlife Service, November 1, 1991.

282 **a different supervisor:** The coworker and supervisor were Angela Doroff and Tony De-Gange, respectively, interviewed on August 28, 2008, and August 22, 2008.

282 **the sinking wreckage:** Pamela Doto, "Biologist Swims to Isle After Crash," *Anchorage Daily News,* October 17, 1991.

283 **26 percent didn't know:** Richard T. Carson et al., "A Contingent Valuation Study of Lost Passive Use Values Resulting from the *Exxon Valdez* Oil Spill," A Report to the Attorney General of the State of Alaska, November 10, 1992, 5-82 to 5-83.

284 **yet that price, $2.8 billion:** Carson et al., "A Contingent Valuation Study," chapter 5. My estimate of how much higher the figure might have been comes from my reading of the study itself, which values various choices the researchers made in deciding how to interpret the results.

284 **discarded it as incredible:** Hunt, *Mission Without a Map,* 25–26.

284 **reducing the amount:** *Exxon Shipping Co., et al., v. Baker et al.,* 554 U.S. No. 07-219 (2008).

284 **the most profitable company:** Steve Mufson, "ExxonMobil's Profit in 2007 Tops $40 Billion," *Washington Post,* February 2, 2008.

284 **paid its CEO:** Jad Mouawad, "For Leading Exxon to Its Riches, $144,573 a Day," *New York Times,* April 15, 2006.

29. A Community Collapses

288 **"to keep you whole":** David Postman, "Cordova Fears End of the Line," *Anchorage Daily News,* March 30, 1989. Exxon spokesman Don Cornett may actually have said, "We will consider whatever it takes to keep you whole."

289 **"tearing our little community":** Mike Webber interview, September 15, 2007.

290 **the plaintiff sought nothing more:** Charles Wohlforth, "Suit Over Cordova's Spill Decisions Divides Its People, Drains Its Savings," *Anchorage Daily News,* January 3, 1991; Marilee Enge, "Cordova Mourns Yet Another Loss," *Anchorage Daily News,* May 17, 1993.

291 **conclusions favored by its sponsors:** Lila Guterman, "Slippery Science," *Chronicle of Higher Education,* September 24, 2004, A12–A16.

292 **seine permit had sold:** Alaska Commercial Fisheries Entry Commission, "Estimated Permit Value Report," July 7, 2008, downloaded from http://www.cfec.state.ak.us.

292 **reminders of the painful losses:** J. Steven Picou et al., "Disaster, Litigation, and the Corrosive Community," *Social Forces* 82 (June 2004): 1493–1533.

293 **"litigation is kind of like a disease":** Interview with Steve Picou, April 26, 2007.

30. Rick Steiner and the Fight for the Trees

296 **"what what else can we get":** I interviewed Rick Steiner many times over the years. The majority of the quotations, unless otherwise noted, are from 2007.

296 **bringing a science center:** A detailed history of the Prince William Sound Science Center is found at its website, http://www.pwssc.org.

298 **Rumplestiltskin:** Interview with R. J. Kopchak, March 2, 1990.

299 **job outside the oil industry:** Iarossi's new job was heading the American Bureau of Shipping, a nonprofit that inspects and classifies ships for insurers.

299 **benefited no one but Exxon:** Patti Epler, "Exxon, U.S. Talk About Settling Spill," *Anchorage Daily News,* February 21, 1990.

300 **"thing of beauty":** This incident happened on the tenth anniversary of the spill, but I've described it here for the sake of the story.

300 **approved a restoration plan:** Hunt, *Mission Without a Map,* 106–108.

301 **made environmentalists stronger:** Charles Wohlforth, "Defending the Sound," *Anchorage Daily News,* May 25, 1990.

303 **Her marriage was failing:** Marybeth Holleman, *The Heart of the Sound: An Alaskan Paradise Found and Nearly Lost* (Salt Lake City: University of Utah Press, 2004).

303 **mayor committed suicide:** Bob Van Brocklin's suicide note is quoted in Marilee Enge, "Cordova Mourns Yet Another Loss," *Anchorage Daily News,* May 17, 1993.

305 **"do that ourselves":** Charles Wohlforth, "Broad Ideas Abound to Reclaim Sound," *Anchorage Daily News,* March 28, 1990.

31. Understanding Life as a System

306 **Accounting Office investigated:** U.S. General Accounting Office, "Briefing Report to the Chairman, Committee on Natural Resources, House of Representatives: Natural Resources Restoration: Use of *Exxon Valdez* Oil Spill Settlement Funds," August 1993 (GAO/RCED-93-206BR).

307 **principal investigator:** The cruise was September 2007. Researchers were John Moran and Johanna J. Vollenweider of NOAA's Auke Bay Lab. The contract vessel *Auklet* was operated by Dave Janka. I thank all three for their hospitality and friendship.

307 **half their $1 billion:** Craig Tillery interview, July 10, 2008.

308 **measurement of fish numbers:** Managers of Alaska's wild salmon have it easy because they can count exactly how many fish are escaping upstream to spawn, and even they make mistakes; estimates of marine-spawning species can err by 30 to 50 percent.

308 **rules often change:** This discussion is strongly influenced by James Wilson, "Scientific Uncertainty, Complex Systems, and the Design of Common-Pool Institutions," in Committee on the Human Dimensions of Global Change, Elinor Ostrom et al., eds., *The Drama of the Commons* (Washington: National Academies Press, 2002), 327–360. An excellent discussion is found as well in R. Hilborn et al., "Sustainable Exploitation of Natural Resources," *Annual Review of Ecological Systems* 26 (1995): 45–67.

308 **the weakened stock was fished out:** As to the oil hypothesis, Richard E. Thorne and Gary L. Thomas, "Herring and the *Exxon Valdez* Oil Spill: An Investigation into Historical Data Conflicts," *ICES Journal of Marine Science* 65 (2008): 44–50. As to temperature impacts on juvenile herring, Robert T. Cooney et al., "Ecosystem Controls of Juvenile Pink Salmon."

308 **vetoed long-term studies:** Tom Kizzia, "Measuring Up the Sound," *Anchorage Daily News,* September 5, 1993.

309 **impossible to pool:** R. J. Kopchak interview, September 15, 2007.

309 **To get big numbers:** Hunt, *Mission Without a Map,* 24–26.

309 **not the true casualty count:** The survey script said, "The *only* mammals killed by the spill were sea otters and harbor seals. This card shows information about what happened in Prince William Sound. According to scientific studies, about 580 otters and 100 seals in the Sound were killed by the spill. Scientists expect the population size of these two species will return to

normal within a couple of years" (Richard T. Carson et al., "A Contingent Valuation Study," 3–51). In a court filing at the same time, the government stated that the spill killed 3,500 to 5,500 otters, 200 seals, and unknown numbers of killer whales and river otters (Summary of Effects of the *Exxon Valdez* Oil Spill on Natural Resources and Archological Resources, March 1991, filed April 8, 1991, in U.S. States District Court, District of Alaska, Case A91-082).

309   **dispel improbable theories:** Hunt, *Mission Without a Map,* 181–186.

309   **conclusions correlated mainly:** Guterman, "Slippery Science."

310   **mine useful information:** Remarks by Robert "Ted" Cooney at the Alaska Marine Science Symposium in Anchorage, January 23, 2006.

311   **"repeated it three times":** Ron A. Heintz et al., "Delayed Effects on Growth and Marine Survival of Pink Salmon (*Oncorhynchus gorbuscha*) After Exposure to Crude Oil During Embryonic Development," *Marine Ecology Progress Series* 208 (2000): 205–216. Interview with Stanley "Jeep" Rice and Ron Heintz, February 27, 2007.

311   **remain for thirty years:** Peterson et al., "Long-Term Ecosystem Response."

311   **yet another meeting:** The meeting, in February 2007, was convened in Juneau by the North Pacific Research Board to plan its Integrated Ecosystem Research Program for the Gulf of Alaska.

313   **only 10 percent remaining:** U.S. Fish and Wildlife Service, "Northern Sea Otter (*Enhydra lutris kenyoni*): Southwest Alaska Stock," January 31, 2008 assessment, http://alaska.fws .gov/fisheries/mmm/seaotters/reports.htm#stock.

313   **eaten by killer whales:** A very controversial theory holds this switch was a step in a cascade of changes in the killer whale diet originally caused by depletion of the great whales by commercial hunting fifty years ago; see A. M. Springer et al., "Sequential Megafaunal Collapse in the North Pacific Ocean: An Ongoing Legacy of Industrial Whaling?" *PNAS* 100 (2003): 12223–12228; Mark Schrope, "Killer in the Kelp," *Nature* 445 (2007): 703–705.

314   **"redefined from expert to educator":** Hanna J. Cortner and Margaret Moote, *The Politics of Ecosystem Management* (Washington, DC: Island Press, 1999), 45.

32. Hearts Connect to the Coastal Forest

317   **slipped into legend:** Klein, *A History of Kachemak Bay,* 29.

318   **McBride organized people:** Recorded interview with Michael McBride, July 6, 2007. He also provided his memories in writing and numerous other personal communications.

318   **a million dollars a month:** Hunt, *Mission Without a Map,* 20.

319   **Hickel's volatile chemistry:** I lived through many of these events, including covering the 1990 gubernatorial election as a reporter, and assisting Governor Hickel in writing his book *Crisis in the Commons,* an experience that allowed me to gain both knowledge of and friendship with this extraordinary man.

320   **a snotty letter:** James M. Rockwell to Rick Steiner, July 19, 1990; copy provided by Steiner.

320   **It was Cole who insisted:** Biographical information on Charlie Cole from Sheila Toomey, "The Deal Maker," *Anchorage Daily News,* May 16, 1993. Cole's role in the Trustee Council from Hunt, *Mission Without a Map.*

322   **only the governor himself:** Hunt, *Mission Without a Map,* 107. Hunt cites a personal communication from Cole.

322 **"superficially plausible":** Charles E. Cole, Attorney General, to Honorable Walter J. Hickel, Governor, State of Alaska, July 7, 1992; electronic copy provided by Alaska Department of Law, file 883-92-0141.

323 **the reversal of policy:** Hunt, *Mission Without a Map,* 114–115.

324 **75,423 acres around Cordova:** Again, Hunt's *Mission Without a Map* is the best source concerning the habitat acquisition process. For further details, see Carol Fries, Alaska Department of Natural Resources, "*Exxon Valdez* Oil Spill Restoration Habitat Protection & Acquisition Catalog," *Exxon Valdez* Oil Spill Trustee Council, February 2007, http://dnr .alaska.gov/commis/evos2/hab_cat_intro.pdf (accessed August 21, 2009).

324 **$370 million for habitat:** Carol Fries, "*Exxon Valdez* Oil Spill Restoration Habitat Protection & Acquisition Catalog."

34. How Change Happens

338 **the Iñupiat came to accept:** Charles Wohlforth, *The Whale and the Supercomputer: On the Northern Front of Climate Change* (New York: Farrar, Straus and Giroux, 2004), 134–139, 199–200.

339 **I even published an article:** Charles Wohlforth, "Alaska's Meltdown," *OnEarth* (Summer 2005): 24–25.

339 **an overwhelming majority:** Anthony Leiserowitz and Jean Craciun, "Alaskan Opinions on Global Warming (No. 06–10)" (Eugene: Decision Research, 2006), http://www.decisionre search.org/Projects/Climate_Change/.

339 **In Tony's research:** Interview with Leiserowitz, April 14, 2008.

340 **An academic team:** Amanda H. Lynch and Ronald D. Brunner, "Context and Climate Change: An Integrated Assessment for Barrow, Alaska," *Climatic Change* 82 (2007): 93–111.

340 **meetings without effect:** Ronald D. Brunner, "Science and the Climate Change Regime," *Policy Science* 34 (2001): 1–33.

340 **Senate voted unanimously:** 105th Congress, 1st Session, S. Res. 98.

340 **Hundreds of mayors:** William Yardley, "Mayors, Looking to Cities' Future, Are Told It Must Be Colored Green," *New York Times,* November 3, 2007.

340 **Arnold Schwarzenegger, declared:** Isaac Arnsdorf and Sam Pilku, "Arnold Heads Climate Panel," *Yale Daily News,* April 21, 2008.

340 **Detroit automakers:** Elizabeth Kolbert, "Running on Fumes," *The New Yorker,* November 5, 2007.

340 **testing door-hanger messages:** Robert Cialdini, "Crafting Normative Messages to Protect the Environment," *Current Directions in Psychological Science* 12 (2003): 105–109; Judith Kleinfeld, "Messages Aimed at Followers Work Best," *Anchorage Daily News,* April 25, 2008.

340 **Congress remained incapable:** A weak bill that passed in 2007 was projected to reduce the growth of U.S. carbon emissions by only 4 percent, far in the future, with new car mileage standards weaker than those already adopted by the states, which it preempted, and it lacked provisions to address utility emissions or to promote alternative energy: John M. Broder, "House, 314-100, Passes Broad Energy Bill; Bush Plans to Sign It," *New York Times,* December 19, 2007. As this is being written, President Obama is promising stronger action on climate change.

341 **30 percent of white men:** Anthony Leiserowitz, "Climate Change Risk Perception and Policy Preferences: The Role of Affect, Imagery, and Values," *Climate Change* 77 (2006): 45–72; also April 14, 2008 interview.

35. Chenega's Next Generation

344 **perhaps half a ton:** Statistical information and rich ethnographies on each community affected by the oil spill are found in the reports of the Alaska Department of Fish and Game's Subsistence Division, which I have relied upon throughout. These numbers from James A. Fall, ed., "Update of the Status of Subsistence Uses in *Exxon Valdez* Oil Spill Area Communities; *Exxon Valdez* Oil Spill Restoration Project Final Report (Restoration Project 040471)," Alaska Department of Fish and Game, Division of Subsistence, Anchorage, Alaska (2006), 33.

345 **free groceries:** James A. Fall et al., "Long-Term Consequences of the *Exxon Valdez* Oil Spill for Coastal Communities of Southcentral Alaska; Technical Paper 264," Alaska Department of Fish and Game Subsistence Division, Juneau, Alaska (2001), 182–183.

345 **blew their new income:** Interviews with Davis, and Nancy Yaw Davis, "Preliminary Impacts of the 1989 Oil Spill on Chugach Region Native Residents, Part 3: 'What Do You Think of the Oil Spill?' Responses from a Mailed Survey," Draft report for the North Pacific Rim and Bureau of Indian Affairs, March 1990.

345 **harvests had not returned:** Fall, "Update on the Status."

346 **no fishing permits left:** Permit numbers from Simeone and Miraglia, "An Ethnography of Chenega Bay," and from Alaska Commercial Fisheries Entry Commission website, http://www.cfec.state.ak.us/cpbycen/2006/261VALDE.htm (accessed March 31, 2008).

346 **standing offer of work:** Patricia Liles, "Alaska Native Village Corporations on the Move," *Alaska Business Monthly,* April, 2008.

347 **A Chenegan in the survey:** Fall, "Long-Term Consequences," 270–272.

36. The Hedonic Treadmill

351 **played by Charles Durning:** Children's Television Workshop, *Elmo Saves Christmas,* 1996.

352 **the equity of these costs:** U. Thara Srinivasan et al., "The Debt of Nations and the Distribution of Ecological Impacts from Human Activities," *Proceedings of the National Academies of Sciences* 105 (2008): 1768–1773.

353 **feel angry and tense:** Daniel Kahneman et al., "Would You Be Happier If You Were Richer? A Focusing Illusion," *Science* 312 (2006): 1908–1910.

353 **giving away the windfall:** Elizabeth W. Dunn et al., "Spending Money on Others Promotes Happiness," *Science* 319 (2008): 1687–1688.

353 **behind only dancing:** Several points in this paragraph are covered in Michael Argyle, "Causes and Correlates of Happiness," and David G. Myers, "Close Relationships and Quality of Life," in Daniel Kahneman et al., eds., *Well-Being: The Foundations of Hedonic Psychology* (New York: Russell Sage Foundation, 1999), 354–391.

353 **least happy everyday hours:** Daniel Kahneman et al., "A Survey Method for Characterizing Daily Life Experience: The Day Reconstruction Method," *Science* 306 (2004): 1776–1780.

353 **doubled and redoubled:** R. Kerry Turner and Brendan Fisher, "To the Rich Man the Spoils," *Science* 451 (2008): 1067–1068.

353 **each person occupied:** Alex Wilson and Jessica Boehland, "Small Is Beautiful: U.S. House Size, Resource Use, and the Environment," *Journal of Industrial Ecology* 9, no. 1–2 (2005): 277–287.

353 **millions of storage units:** Self Storage Association, "Self Storage Industry Fact Sheet as of 8/28/08," http://www.selfstorage.org/pdf/FactSheet.pdf (accessed September 11, 2008).

353 **miles driven in vehicles:** Pew Oceans Commission, Leon Panetta, chair, *America's Living Oceans: Charting a Course for Sea Change* (Philadelphia: Pew Charitable Trusts, 2003), 50–52.

353 **we're no happier:** Louise C. Keely, "Why Isn't Growth Making Us Happier? Utility on the Hedonic Treadmill," *Journal of Economic Behavior and Organization* 57 (2005): 333–355.

354 **the United States came in eighteenth:** Ed Diener and Eunkook Mark Suh, "National Differences in Subjective Well-Being," in Kahneman et al., eds., *Well-Being: The Foundations of Hedonic Psychology,* 434–450. Also, Kahneman et al., "Would You Be Happier?"

354 **the want and the need:** This discussion was inspired by Keely, "Why Isn't Growth Making Us Happier?"

355 **paid game players produced:** The interactive game economy is well explored by Julian Dibble, *Play Money: Or, How I Quit My Day Job and Made Millions Trading Virtual Loot* (New York: Basic Books, 2006). The sweatshops are documented in David Barboza, "Ogre to Slay? Oursource It to Chinese," *New York Times,* December 9, 2005.

355 **the worth of each thing:** Keely, "Why Isn't Growth Making Us Happier?"

356 **identifying ecosystem services:** Kareiva et al., "Domesticating Nature."

357 **phobia about germs:** Katherine Ashenburg, *The Dirt on Clean: An Unsanitized History* (New York: North Point Press, 2007).

358 **epidemic of allergies:** Stephen J. Galli et al., "The Development of Allergic Inflammation," *Nature* 454 (2008): 445–454.

37. Making a Difference (Beach Cleaners Part 2)

363 **a corrupt state senator:** Senator John Cowdery later pleaded guilty to bribery: Richard Mauer, "Cowdery Pleads Guilty to VECO Conspiracy," *Anchorage Daily News,* December 20, 2008.

38. Celebrations

369 **John Whitney, a cofounder:** Whitney, a NOAA scientist, quit the group to avoid a conflict of interest in grants from the agency, but continues to volunteer.

374 **he resigned in protest:** Elizabeth Bluemink, "UA Professor in Danger of Losing Federal Funding," *Anchorage Daily News,* March 8, 2009; Associated Press, "Professor to Resign After University Rejects His Grievance," *Anchorage Daily News,* October 24, 2009.

Index